OZONE

A New Medical Drug

OZONE
A New Medical Drug

by

Velio Bocci
Medical Doctor, Specialist in Respiratory Diseases and Haematology and Emeritus Professor of Physiology at the University of Siena, Siena, Italy

Springer

A C.I.P. Catalogue record for this book is available from the Library of Congress.

ISBN 978-90-481-6805-7 (PB)
ISBN 978-1-4020-3140-3 (e-book)

Published by Springer,
P.O. Box 17, 3300 AA Dordrecht, The Netherlands.

Sold and distributed in North, Central and South America
by Springer,
101 Philip Drive, Norwell, MA 02061, U.S.A.

In all other countries, sold and distributed
by Springer,
P.O. Box 322, 3300 AH Dordrecht, The Netherlands.

Printed on acid-free paper

springeronline.com

Printed in the Netherlands.

This book is dedicated to all patients with the wish to regain their health with or without oxygen-ozonetherapy

PREFACE

Is there any reason to write a third book on the topic of oxygen-ozone therapy? The first was written four years ago for italian physicians, who, for some time, had complained about the lack of precise ozone informations. Then it became unavoidable to write a book for english-speaking people and it took a couple of years because i considered worth it while to examine the problem in details with attention to basic findings, so that it could also be useful to scientists and clinicians. The book was briefly reviewed in the new york academy of sciences magazine (1-2, 2003, 14) and was awarded the "Tramezzi" prize in Rome, October 2003. However the book had not been as successful as i had hoped because, particularly american scientists remained sceptical and disinterested. However a letter from prof. L.Packer was encouraging saying: *"there is growing evidence that oxidants like many ros including ozone provide a mild oxidant stress which under certain circumstances actually induce the formation of phase 1 and phase 2 enzymes, strengthening the antioxidant defence system through an adaptation process. Thus it would be rational that a mild treatment with ozone stimulates the antioxidant defence system. So i think some of your ideas have a foundation in current scientific work."* Indeed i have several reasons to believe that one of the most important pedestals of ozonetherapy is based on the induction of oxidative stress proteins.

In Chapter 34 of my previous book (2002), I emphasized the usefulness of biooxidative therapies in poor-countries, where a billion people have no other medical resources but world health authorities, particularly WHO had remained disinterested in this approach. It is depressing to realize how useful proposals remain unheard of and how much time and work are requested before their implementation.

However, the worst disappointment has been on reading a few letters of distinguished American scientists in the field of free-radicals, who were and remained biased towards ozonetherapy. Actually several scientists, who surely received the book did not even bother to answer. Someone, who, by reading his papers, I judged to be very clever, politely wrote me that, although the book presented some interesting aspects, it could not convince him because the ruling dogma in the USA is that "*ozone is toxic and should not be used in medicine*". Although several others were not so naïve to say such a thing, it was clear that the book had failed to remove the prejudice from their heads.

On the other hand, I received a few positive responses from clinicians, who appreciated my effort in creating a rational scheme able to clarify how ozone acts and why toxicity could be avoided. Several practitioners confessed to be disinterested in the biochemical aspects because they hardly had the time to read the practical part.

On the whole these remarks taught me that much more work was needed before reversing the antagonism, for correcting the problem of charlatans and objectively establish the pros and cons of ozonetherapy. During the last three years, personal experience has convinced me that a judicious use of ozone can be very useful in some diseases when orthodox medicine has no further resources and patients are abandoned to their fate. For them ozone is more valuable than gold because, at least for some, it can restore hope and health. I feel I am a lucky man because towards the end of my academic life, in 1988, by mere accident, I stumbled on the ozone problem that although quite controversial, is a real trove. I believe so because ozone, like oxygen, is such a basic molecule able to activate a great number of vital processes. It depresses me to think that ozone, the cheapest drug on earth, is today either badly or minimally used because orthodox medicine refuses to evaluate it and Health Authorities are antagonistic or negligent. Both are responsible for leaving millions of people suffering and dying. I must not get discouraged and continue to work and hope that ozonetherapy will eventually benefit many people. I cannot hide my dream of organizing and gratuitously working in a large clinic with enthusiastic collaborators able to evaluate the full relevance of ozonetherapy. There are many philanthropists helping the development of arts, science and sport and one day we may be lucky to find someone who believes and values this approach.

In conclusion what are the aims of this book?

This has been specifically written for physicians, who want to learn and then perform a correct ozonetherapy. Moreover, by using a plain scientific language, the book should be useful for the layman, who must receive the most objective information avoiding any undue hype. In comparison with the former book, I will strive to be clear without omitting the basic concepts that are essential to understand how ozone works without causing deleterious effects. I will refrain from adding chemical formulae and complex diagrams because I have been told that they are somewhat irritating or useless for inexpert people. I would be enthusiastic to help and collaborate with clinical researchers if they are genuinely interested in evaluating the validity of ozone-therapy.

The greatest effort will be dedicated to clarify in which diseases ozonetherapy has been proved to be really useful and in others where the efficacy remains uncertain. Because mass media and popular books often tend to elicit unrealistic hopes, it appears necessary that the reader is precisely informed on the best available options provided by orthodox medicine and, if necessary, by ozonetherapy that, in any case, remains a complementary approach or the last resort. For the sake of brevity and minimal competence, I will not suggest other important complementary approaches unless scientific data have proven their value.

Velio Bocci

ACKNOWLEDGEMENTS

It has taken some time to have this book ready because I wanted to fully update results of both orthodox and ozone therapies. I would like to thank many collaborators who, throughout the years, have helped me in studying the biological and therapeutic effects of ozone.

Although I tried to do my best in writing the book, I apologise for the poor style and probably some mistakes although the manuscript has undergone the linguistic revision. I am very grateful to Dr. Carlo Aldinucci for the enormous help in typing the bibliography and the skill in preparing the manuscript for printing.

I thank all Authors and Publishers for kindly allowing to publish a few diagrams and to report some unpublished data. I gratefully acknowledge a small support given by Healthzone Clinics Public Ltd. Co., UK.

I am grateful to Mr. Peter Butler, Publishing Manager Springer Science for the enthusiastic support of the project.

Finally I am deeply grateful to my wife Helen and to my children Erica and Roberto for being always very patient with me and my work.

FOREWORD

In 2002, Prof. Bocci published a book entitled "Oxygen-Ozone Therapy. A critical evaluation" with ample scientific data based largely on his experiments indicating the usefulness and atoxicity of ozone therapy in some diseases. However the book contained too many details and it was difficult to read for the busy physician and even more for the layman. He has now written a new version that is shorter and concise. Nonetheless the mechanisms of action of ozone, which are essential for understanding how ozone acts through a number of messengers, are clearly explained in a scientific but plain language. The clinical part has been greatly extended and includes new pathologies as well as diseases previously considered but now amplified with new data.

The book is primarily intended for ozone-therapists, who want to refine techniques, expand treatment modalities and solve problems, but will be equally useful for physicians interested in this approach. However it can also be read by the layman, to provide further insight into this aspect of complementary medicine. The book gives a complete view of this approach including practical aspects of routes of administration and the possibility of simple self-medication.

Against the prejudice that ozone is toxic, the accurate description of side effects reveals that they are of minimal relevance and it is even more surprising that ozone therapy yields a feeling of wellness in the majority of patients. This is a crucial point that has been clearly explained by defining the therapeutic range within which ozone is a very useful drug. It is interesting to learn that ozone can slowly induce an increase of antioxidant defences that is a unique property most important for correcting the chronic oxidative stress plaguing atherosclerosis, diabetes, viral infections and cancer.

I have been impressed by the enthusiasm for this therapy that comes to light on reading the book. Velio told me that he has submitted himself all possible methods of ozone administration and has also performed many experiments. From his results and the often incredible improvements observed in very ill patients, he has the firm conviction that ozone therapy is not a placebo! Jokingly, he also assured me that he is the living proof of ozone therapy atoxicity.

Ozone never finishes surprising us by its versatility recently shown by delivering therapeutic activity in Orthopaedics and Dentistry. I am convinced that Velio's work will dispel misconceptions and scepticism and it will be useful for years to come both to physicians and patients, who like to understand the meaning of the treatment. Indeed this book represents the first comprehensive framework for understanding and recommending ozone therapy in some diseases

The Lord Colwyn. CBE.,BDS,LDS,RCS.
President Natural Medicines Society
President Parliamentary Group for Complementary and Integrated Healthcare.
London, August 2004.

TABLE OF CONTENTS

ACRONYMS

$A^{\cdot -}$	Semidehydroascorbate radical anion
A	Angstrom unit
AA	Arachidonic acid
Aa	Angina abdominis
ABI	Ankle-brachial index
ABTS	2,2'-azinobis-(3ethyl-benzothiazoline-6-sulphonic acid)
AC	Adenylate cyclase
ACD	Citric acid-citrate, dextrose solution
ACE	Angiotensin-converting enzyme
ACR	American College of Rheumatology
ACTH	Adrenocorticotrophic hormone
AD	Atopic dermatitis
ADCC	Antibody-dependent cellular citotoxicity
ADP	Adenosine diphosphate
AGE	Advanced glycation end products
AgII	Angiotensin II
AH^-	Ascorbic acid
AHIT	Autohomologous immunotherapy
AHT	Ozonated Autohaemotherapy, Major and Minor
AIDS	Acquired immune deficiency syndrome
ALS	Amyotrophic lateral sclerosis
AMP	Adenosine monophosphate
AP	Atmospheric pressure
AP-1	Activator protein 1
APC	Antigen presenting cells
APR	Acute phase reactants
ARDS	Acute respiratory distress syndrome
ARMD	Age related macular degeneration
ASA-5	Sulphasalazine
AT	Antioxidant therapy
ATP	Adenosine triphosphate
ATPase	Adenosine triphosphatase
AZT	Azidothymidine (zidovudine)
BALF	Bronchoalveolar lavage fluid
BALT	Bronchial-associated lymphoid tissue
bFGF	Basic fibroblast growth factor
BGBP	Butyl-glycobutyl-phthalate
BLS	Basic life support
BMC	Blood mononuclear cells
BMSC	Bone marrow stem cells
BOEX	Quasi-total body exposure
Ca^{2+}-ATPase	Ca^{2+} adenosine triphosphatase
$CaCl_2$	Calcium chloride

CAF	Cell antiviral factor
cAMP	Cyclic adenosine 3'-5'-monophosphate
CAT	Catalase
CBT	Cognitive behavioural therapy
CCK	Cholecystochinin
$CD4^+$	Helper T lymphocytes
$CD8^+$	Cytotoxic T lymphocytes
cDNA	Complementary DNA
CE	Energetic charge
CFCs	Chlorofluorocarbons
CGMP	Guanosine 3'-5'- cyclic monophosphate
CGRP	Calcitonin gene-related peptide
CH_4	Methane
Cl	Chlorine
CLI	Chronic limb ischaemia
CNS	Central nervous system
CO	Carbon monoxide
CO_2	Carbon dioxide
CoA	Coenzyme A
COPD	Chronic obstructive pulmonary disease
CoQ	Coenzyme Q (Ubiquinone Q-Ubiquinol QH_2)
COS	Chronic Oxidative Stress
CPD	Citrate-phosphate dextrose
CRF	Chronic renal failure
CRH	Corticotrophic releasing hormone
CRP	C-reactive protein
CSF	Cerebrospinal fluid
CTL	Cytotoxic T lymphocytes
Cu	Copper
Cu/Zn-SOD	Copper/Zinc superoxide dismutase
Cys-NO	Cysteine nitrosothiols
Cyt	Cytochrome
DAG	Diacylglycerol
DCHA	Docosahexaenoic acid
DEHP	Di(2ethylesil) phthalate
DHA	Dehydroascorbic acid
DHEA	Dehydroepiandrosterone
DHLA	Dihydrolipoate
DIC	Disseminated intravascular coagulation
DNA	Deoxyribonucleic acid
DNase	Deoxyribonuclease
DNIC	Diffused noxious inhibitory control
2,3-DPG	2,3-Diphosphoglycerate
2,3-DPGM	2,3-Diphosphoglycerate mutase
2,3-DPGP	2,3-Diphosphoglycerate phosphatase
$E^{\cdot}$	Alpha-tocopheryl radical
EBOO	Extracorporeal blood circulation against O_2-O_3

EC	Energy charge
ECs	Endothelial cells
EDCF	Endothelium-derived contracting factor
EDCF-1	Contraction factor I
EDHF	Endothelium-derived hyperpolarizing factor
EDRF	Endothelial-derived relaxing factor
EGF	Epidermal growth factor
EH	α-tocopherol (vitamin E)
ELISA	Enzyme-Linked Immunosorbent Assay
eNOS	Endothelial NO synthase
EP	European Pharmacopea
EPA	Eicosapentanoic acid
EPC	Endothelial progenitor cells
EPO	Erythropoietin
EPR	Electron paramagnetic resonance spin trapping technique
ERG	Electroretinogram
ESR	Erythrocyte sedimentation rate
ET-1	Endothelin-1
EVA	Ethylen vinyl acetate
F_2-IsoPs	F_2-isoprostanes
FAD	Flavin adenine dinucleotide, oxidized form
$FADH_2$	Flavin adenine dinucleotide, reduced form
FCS	Fetal calf serum
FDA	Food and Drug Administration
$Fe^{2+} \leftrightarrow Fe^{3+}$	Iron
FFP	Fresh frozen plasma
FGF	Fibroblast growth factor
FRBM	Free Radicals in Biology and Medicine
GABA	Gamma amino butyric acid
G3-PD	Glyceraldehyde 3-P-dehydrogenase
G-6P	Glucose–6 phosphate
G-6PD	Glucose-6 phosphate dehydrogenase
GALT	Gut-associated lymphoid tissue
GAPDH	Glyceraldehyde 3-phosphate dehydrogenase
G-CSF	Granulocyte- Colony Stimulating Factor
GDP	Guanosine diphosphate
GET	Graded exercise therapy
GGT	Gamma-glutamyl transpeptidase
Glu	Glutamate
Gly	Glycine
GM-CSF	Granulocyte-monocyte Colony Stimulating Factor
GMP	Guanosine monophosphate
GRPs	Glucose-regulated proteins
grp	Gene related peptide
GSH	Glutahione reduced form
GSH-Px	Glutathione peroxidase
GSH-T	Glutathione transferase

GS-NO	Gluthatione nitrothiols
GSSG	Glutathione disulfide
GSSGR	Glutathione reductase
GTP	Guanosine triphosphate
GTPase	Guanosine triphosphatase
GVDH	Graft versus host disease
H.p.	Helicobacter pylori
H_2	Hydrogen
H_2O_2	Hydrogen peroxide
H_2S	Sulphidric acid
HAART	Highly active anti-retroviral therapy
HAV	Hepatitis A virus
Hb	Haemoglobin
HbCO	Carboxyhaemoglobin
HbO_2 or Hb_4O_8	Oxyhaemoglobin
HbS	Haemoglobin sickle cell
HBV	Hepatitis B virus
HCV	Hepatitis C virus
HDL	High-density lipoprotein
HES	Hydroxy ethyl starch
HETE	Hydroxyeicosatetraenoic acid
HG	Hyperglycemia
HGF	Hepatocyte growth factor
HIF-1	Hypoxia inducible factor-1
HIV	Human immunodeficiency virus
HK	Hexokinase
HLA	Human leukocyte antigens
HMG-CoA reductase	Statin, Hydroxy-methyl-glutaryl-CoA reductase
4-HNE	4-hydroxy-2,3-trans-nonenal
HO-1	Haeme-oxygenase I (HSP 32)
$HO_2^{\cdot}$	Hydroperoxy radical
HOCl	Hypoclorous acid
HOT	Hyperbaric oxygen therapy
H-O-U	Heat, ozone and ultraviolet light
HPLC	High pressure liquid chromatography
Hr	Hours
HSPs	Heat shock proteins
HSV-I and II	Herpes virus I and II
5-HT	5-hydroxytryptamine (Serotonine)
HUVECs	Human vascular endothelial cells
HZ	Herpes zoster
IA	Intraarterial
Iat	Intraarticular
IDis	Infectious disease
ID	Intradisc

IF	Intraforaminal
IFN	Interferon
IgE	Immunoglobulin E
IgG	Immunoglobulin G
IKB	Protein blocking NFKB activity
IL	Interleukin
ILs	Intralesional
IL-1Ra	Interleukin 1 Receptor antagonist
IM	Intramuscular
IMOS	International Medical Ozone Society
IOA	International Ozone Association
IP_3	Inositol-1,4,5-trisphosphate
Ipe	Intraperitoneal
IPF	Idiopathic pulmonary fibrosis
IPL	Intrapleuric
IU	International Unit
IV	Intravenous
KDa	KiloDalton
KGF	Keratinocyte growth factor
KI	Potassium iodide
L	Litre/min
LA	α-lipoic acid (Thioctic acid)
LAK	Lymphokine activated killer cells
LD	Leukocyte depletion
LDH	Lactic dehydrogenase
LDL	Low Density Lipoproteins
LE	Lipid emulsion
LMWA	Low Molecular Weight Antioxidants
L-NAME	NG-nitro-L-arginine methyl ester (Nos inhibitor)
LOPs	Lipid oxidation products
LPS	Lipopolysaccharides
LTB_4	Leukotriene B_4
mcg/ml	Micrograms per ml
M	Mean
β_2M	β_2 Microglobulin
MALT	Mucosal-associated lymphoid tissue
MCP-1/JE	Monocyte chemotactic protein 1
MDA	Malonyldialdehyde
MegaU	1 million units
MEM	Minimum essential medium
MHb	Methaemoglobin
MHC	Major histocompatibility complex
Min	Minutes
MIP-1α	Macrophage inflammatory protein 1α
MIP-1β	Macrophage inflammatory protein 1β
MM	Muscularis mucosae
Mn	Manganese

Mn-SOD	Manganese-superoxide dismutase
MPO	Myeloperoxidase
mRNA	Messenger RNA
MS	Multiple sclerosis
MSC	Mesenchymal stem cells
Mx	Mxprotein (IFN marker)
N_2	Nitrogen
N_2O	Nitrous oxide
Na/K-ATPase	Na/K ATPase
$Na_2S_2O_3$	Sodium thiosulphate
NAC	N-acetyl-cysteine
NAD	Nicotinamide adenine dinucleotide, oxidised form
NADH	Nicotinamide adenine dinucleotide, reduced form
NADP	Nicotinamide adenine dinucleotide phosphate, oxidised form
NADPH	Nicotinamide adenine dinucleotide phosphate, reduced form
$NaHCO_3$	Sodium bicarbonate
NaOCl	Sodium hypochloride
NEJM	New England Journal of Medicine
NFKB	Nuclear factor Kappa B
NGF	Nerve Growth Factor
NH_3	Ammonia
Ni	Nickel
NK	Natural Killer
NMR	Nuclear magnetic resonance
$NO^{\bullet}$	Nitric oxide
$NO^{\bullet}_2$	Nitrogen dioxide
NO_x	Nitrogen oxides
NOs	Nitric oxide synthase
NSAID	Nonsteroidal anti-inflammatory drugs
O_2	Oxygen
$O_2^{\cdot -}$	Anion superoxide
O_3	Ozone
$^1O_2$	Singlet oxygen
OFSP	Oedematous-fibro-sclerotic panniculitis
$OH^{\cdot}$	Hydroxyl radical
5-OH-dCyd	5-hydroxy-2'-deoxycytidine
8-OHdG	8-hydroxy-2'deoxyguanosine
8-OHG	8-hydroxyguanine
$ONOO^-$	Peroxynitrite
OSE	Ozone in Science and Engineering
OSP	Oxidative Stress Proteins
OSPs	Oxidative shock proteins
OxLDLs	Oxidized low-density lipoproteins
PAF	Platelet activating factor
PDGF	Platelet-derived growth factor

PEDF	Pigment epithelium-derived growth factor
PEG	Polyethylenglycol, Pegylated
PEG-IFNα	Poliethylenglycol-Interferon α
PEG-IL2	Poliethylenglycol-Interleukin 2
PEG-SOD	Polyethylenglycol-superoxide dismutase
PET	Positron electron tomography
PF_3	Platelet factor 3
PF_4	Platelet factor 4
PFK	Phosphofructokinase
PGI_2	Prostacyclin
PGs	Prostaglandins
6PGD	6-phosphogluconate dehydrogenase
PHA	Phytohaemagglutinin
PHN	Post-herpetic neuralgia
PI	Proliferation index
Pi	Inorganic orthophosphate
PIP_2	Phosphatidyl inositol 4,5-biphosphate
PK	Piruvate kinase
PKC	Protein Kinase C
PLA_2	Phospholipase A_2
PLC	Phospholipase C
pO_2	O_2 partial pressure
POAD	Peripheral occlusive arterial disease
PPase	Phosphatase
ppbv	Parts per billion volume
PPi	Inorganic pyrophosphate
ppmv	Parts per million volume
PRP	Platelet-rich plasma
PS	Physiological saline
PTG	Protein thiol groups
PUFAs	Polyunsaturated fatty acids
PVC	Polyvinyl chloride
PVC-DEHP	PVC-di(2ethylesil)phthalate
pvO_2	O_2 venous partial pressure
Q	Ubiquinone
Q_{10}	Coenzyme Q (Ubiquinone)
QH_2	Ubiquinol
QoL	Quality of life
RA	Rheumatoid arthritis
RANTES	Regulated upon activation, normal T-cell expressed and secreted (Chemokine)
RBP	Retinol binding protein
RCTs	Randomised clinical trials
RDA	Recommended dietary allowances
RES	Reticulo endothelial system
RI	Rectal insufflation
RIA	Radio immuno assay

RNA	Ribonucleic acid
RNAase	Ribonuclease
RNI	Reactive nitrogen intermediates
$RO^•$	Alkoxyl radical
$ROO^•$	Alkoperoxyl radical
ROOH	Hydroperoxide
ROS	Reactive oxygen species
RPE	Retinal pigment epithelium
$RS^•$	Thiyl radical
$RS^•/RSO^•$	Thiyl/sulphenyl radicals
RTLFs	Respiratory tract lining fluids
SAA	Serum amyloid A
SALT	Skin-associated lymphoid tissue
SC	Subcutaneous
SC	Stem cells
SCA	Sickle cell anemia
SCE	Sister chromatid exchange
SD	Standard deviation
Sec	Seconds
SGOT	Aspartate aminotransferase
SGPT	Alanine aminotransferase
SLE	Systemic lupus erythematosus
SODs	Superoxide dismutases
SRA	Scavenger receptor A
T½	Half-life
TAS	Total Antioxidant Status
Tat	Trans-activator of transcription (HIV protein)
TBARS	Thiobarbituric acid-reactive substances (Marker of peroxidation)
TEHT	Tri-(2 etylesil trismellitate)
TGFα	Transforming Growth factor alpha
TGFβ	Transforming Growth factor beta
TIAs	Transient ischemic attacks
TIL	Tumour infiltrating lymphocytes
TM	Thalassaemia
TNFα	Tumor Necrosis Factor alpha
TRALI	Transfusion-related acute lung injury
TRX	Thioredoxin
TRX Px	Thioredoxin peroxidase
TRXR	Thioredoxin reductase
TxA_2	Thromboxane A_2 (active form)
β-TBG	β thromboglobulin
TxB_2	Thromboxane B_2 (stable form)
UDP	Uridine diphosphate
UTP	Uridine triphosphate
UV	Ultraviolet light

VA	Visual acuity
VAS	Visual analogue scale
VEGF	Vascular endothelial Growth Factor
VIP	Vasoactive intestinal peptide
Vitamin A	Retinol
Vitamin E (EH)	α-tocopherol
VLDL	Very low density lipoprotein
WHO	World Health Organisation
WMA	World Medical Association
vWF	von Willebrand factor
WBC	White blood cells
WSC	Work site concentration
XDH	Xantine dehydrogenase
XO	Xantine oxidase
ZIG	Zoster Immune Globulin

INTRODUCTION

Ozone is a natural gaseous molecule made up of three oxygen atoms whereas the oxygen molecule, far more stable, is composed of only two atoms.

Christian Friedrich Schonbein (1799-1868) discovered ozone in 1840, when, working with a voltaic pile in the presence of oxygen, noticed the emergence of a gas with an "electric and pungent smell" that could be a sort of "super-active oxygen". We can smell it during a thunderstorm because the electric discharge of lightning, between the clouds and the earth, catalyses the formation of ozone from atmospheric oxygen. Although Schonbein had probably guessed that ozone could be used as disinfectant, his intuition did not save him when he contracted a Bacillus anthracis infection while exploring a chemical method for preserving meat. The concept that ozone derives from oxygen when an electric discharge was generated by a voltaic arc was practically applied by the chemist Werner von Siemens, who invented the so-called super-induction tube (Siemens's tube), consisting of two interposed electrode plates set at a high voltage which, in the presence of oxygen, could generate some ozone. It became possible to produce ozone at will and clarify that ozone is indeed a very reactive, unstable and unstorable gas that had to be produced "ex tempore" from oxygen and used at once. Industrial ozone generators could then be used for industrial application and disinfection of water, after it was shown the potent and broad bactericidal activity of ozone. Today nobody doubts about its strong disinfectant properties and there are more than 3.000 municipal treatment facilities in the world. As the need of water increases daily and it is indispensable to prevent the spread of infectious diseases, the importance of ozone for practical applications becomes immense. The International Ozone Association (IOA) carefully supervises all the applications and publishes a good scientific journal "Ozone Science and Engineering". So far, one weak point has been not to pay enough attention to the medical applications because this is not IOA's main purpose.

The first medical application seems to have been the use of ozone for treating gaseous, post-traumatic gangrene in German soldiers during the 1st world war. However a big step forward was the invention of a reliable ozoniser for medical use by the physicist Joachim Hansler (1908-1981). The idea to use ozone in medicine developed slowly during the last century and it was stimulated by the lack of antibiotics and the disinfectant properties of ozone. Not surprisingly a Swiss dentist, E.A.Fisch (1899-1966) was the first to use ozone in his practice. By a twist of fate, Dr E Payr (1871-1946), a surgeon had to be treated for a gangrenous pulpite and soon realized the efficacy of the ozone treatment in surgery to become so enthusiastic to report his results at the '59th Congress of the German Surgical Society in Berlin (1935) and write: *"which other disinfectant would be tolerated better than ozone? The positive results in 75% of patients, the simplicity, the hygienic conditions and innocuity of the method are some of the many advantages"*.

In 1936, in France, Dr P. Aubourg proposed to use the insufflation of oxygen-ozone into the rectum to treat chronic colitis and fistulae.

How could ozone be administered for internal use? It seems that Dr. Payr was the first to inject gas with a small glass syringe directly into the vein but he was very careful in slowly delivering a small volume of gas. Unfortunately this route was later on adopted by charlatans and technicians without any medical qualification who, by injecting large volume (up to 500 ml in two hours) have often caused lung embolism mostly due to oxygen and even death. Although this practice has been prohibited since 1984, quacks still uses in third-world countries and certainly it represents one good reason for prohibiting all at once the use of ozone. In most States of USA, the FDA has forbidden the use of ozone and this fact has negatively influenced a correct development of ozonetherapy, that, however, is more or less tolerated in other parts of the world. It is regretful that brilliant pioneers as Fisch, Payr, Aubourg and Dr. H. Wolff (1927-1980), the inventor of ozonated autohemotherapy, have been betrayed by a horde of unscrupoulous and false doctors. If that was not enough, **another serious obstacle has been created in the USA by the ruling dogma that "ozone is always toxic any way you deal with it ".** This was the phrase that one of the best ozone chemists wrote me in 1995. Although I tried to discuss with him showing our data contradicting his dogmatic assertion, he has preferred not to discuss further this issue. When, on June 2002, I sent him my book that critically examined ozone therapy, only his secretary, after a second request, briefly informed me that he had received the book! In the medical field, history has repeatedly shown that not all dogmas are tenable and the one on ozone stands up mostly on the basis of prejudice, medical incompetence and previous bad work. While I fully agree with the experts that ozone is one of the strongest oxidants and an intrinsically toxic molecule, on the basis of our biological and clinical data, **I am sure that ozone, if used in judicious dosages, can be tamed by the potent antioxidant system present in cells and biological fluids.** Obviously, during an inflammatory process, an excessive, continuous and localized release of ozone can be detrimental whereas, depending upon a minimal concentration, short time of exposure and biological location, **the now famous three gaseous molecules: CO, NO and O_3 can act as crucial physiological activators.**

The problem of ozone toxicity is of paramount importance and it will be fully clarified in this book. The interested reader can browse through Chapter 2 of my previous book (Bocci, 2002) where this long controversy was described. I feel confident that slowly in the future, in spite of several drawbacks, the therapeutic value and the lack of adverse effects will become evident to everyone and this complementary approach will be widely used in medicine.

Chapter 1

PHYSICAL-CHEMICAL PROPERTIES OF OZONE. NATURAL PRODUCTION OF OZONE. THE TOXICOLOGY OF OZONE

As I already mentioned, ozone (from the Greek means to give off a smell) is a natural but unstable molecule. The pure gas has a soft sky-blue colour with a pungent, acrid smell. The molecule is composed of three oxygen atoms (O_3) and, the molecular weight, in comparison to the oxygen diatomic molecule (32.00) is of 48.00. Ozone has a cyclical structure with a distance among oxygen atoms of 1.26 Å and exists in several mesomeric states in dynamic equilibrium. For the physician it is useful to know that **the solubility (ml) in 100 ml water (at 0°C) of either ozone or oxygen is either 49.0 ml or 4.89 (ten fold lower), respectively.**

Among oxidant agents, ozone is the third strongest, after fluorine and persulphate, a fact that explains its high reactivity.

Ozone is formed from pure oxygen via an endothermic process allowed by very high voltage gradients set up between the electrodes of the Siemens' tube:

$3\ O_2 \leftrightarrow 2\ O_3 - 68{,}400$ cal.

The reader can note that this reaction is reversible, practically meaning that ozone decomposes spontaneously and therefore it is hardly storable. Moreover the life of the ozone molecule depends on the temperature, so that at 20°C the ozone concentration is halved within 40 min., at 30°C within 25 min., while at -50°C is halved only after three months.

What is known about the natural production of ozone?

In the stratosphere, at about 22 Km from the earth's surface, there is an ozone layer that may reach a maximal concentration of 10 ppmv (parts per million volume, 1:106), equivalent to 0.02 micrograms (mcg/ml). The maintenance of the ozone layer is very important because it absorbs most of the ultraviolet (UV) radiation (<290 nm) emitted by the sun. UV rays include band A (316-400 nm) responsible for suntan and bands B and C (from 100 up to 315 nm), which are far more mutagenic and responsible for enhancing skin ageing and carcinogenesis that has been shown by a progressive increase of carcinomas and melanomas in recent times.

Nature has been provident because, thanks to cyanobacteria, as soon as that oxygen started to increase in the terrestrial atmosphere about 2.3 billion years ago, UV solar emission catalyzed the production of ozone, which then could control the UV irradiation and protect biological systems on earth:

$$O_2 + UV\ (< 242\ nm) \rightarrow O + O$$
$$2O_2 + 2O \rightarrow 2\ O_3$$

The protective ozone layer in the stratosphere was fairly constant as it was the result of a dynamic equilibrium between the ozone-forming reaction and the natural dissociation of ozone. This equilibrium has been partly subverted during the last century owing to a progressive increase of pollutants, namely nitrogen oxides (NOx) and chlorine derived from chlorofluorocarbons (CFCs) used as refrigerant fluids and incautiously dispersed in the environment. One single chlorine atom, through a catalytic chain reaction mechanism discovered by Molina and Rowland (1974), can destroy thousands of ozone molecules before being transported back into the troposphere.

The excessive destruction of ozone has caused the thinning of the protective ozone layer and the famous "Antarctic ozone hole"; only thanks to an international effort to substitute CFC, the ozone layer will be probably restored to normal by 2050 (Schrope, 2000)!

Once again chaotic human activities (industrial processes, vehicular traffic, etc,) have led to a dangerous environmental pollution of the air present in the troposphere, which extends 8-17 Km from the earth's surface. Exaggerated anthropogenic emissions of nitrogen monoxide (NO) and dioxide (NO_2), of carbon monoxide (CO), of methane (CH_4), sulphuric acid and other acid compounds have favoured an almost intolerable increase of ozone concentration up to 0.1 ppmv (0.0002 mcg/ml) or more, while it should be no higher than 0.03 ppmv, i.e. about 300-fold lower than in the stratosphere. In large metropolis, **ozone, mixed with the other compounds, composes the photochemical smog: it has become the main toxicant for the lungs, eyes, nose and, to a lesser extent, the skin because particularly the respiratory mucosa does not contain enough neutralizing substances for this murderous acid mixture.** Indeed the respiratory tract lining fluids (RTLFs) is only an aqueous film layer that is easily overwhelmed by this acidic mixture of strong oxidants. Particularly children, asthmatic and other broncho-pulmonary patients are at risk and the ozone "toxicomania" is well justified (Devlin et al., 1991; Aris et al., 1993; Broeckaert et al., 1999). **Certainly ozone toxicity at the street level has contributed to support the dogma that ozone is always toxic and the layman can well wonder why ozone can be used as a therapeutic agent.** Toxicologists and Health

Authorities are correctly concerned about this problem, which is clearly not only due to ozone and that should not lead to the sweeping conclusion that ozone "is always toxic". The recent surprising findings that human activated leukocytes can produce ozone (Babior et al., 2003; Nieva and Wentworth, 2004) are of critical importance in normal and pathological situations.

We will come back to this point with another two gaseous molecules, namely CO and NO (Moncada, 1992; Verma et al., 1993; Pannen et al., 1998), which also surprisingly can behave at physiological doses as essential effectors and become toxic at high concentrations. In other words, **the concept valid for any molecule is that it is the right dose that differentiates between a therapeutic and a toxic agent.**

Thus, for the safety of patients and personnel, not a trace of ozone should be present and a suitable exchange of air can be insured by an aspirator supplied with an ozone destructor. Moreover a monitor analyzer with warning lights and a loud alarm must be turned on all the time to immediately alert in case of a little contamination. I must say that our odour perception threshold for ozone is about 0.01 ppmv (0.02 mcg/L), ten times lower than the maximum work site concentration (WSC) of 0.1 ppmv (0.2 mcg/L) over a breathing period of one hour. The World Health Organization (WHO) permits to work for 8 hours when the ozone concentration is 0.06 ppmv (0.12 mcg/L) that is well perceived as a fairly strong ozone smell. Needless to say we should never trust our nose because our olfactory receptors become quickly tolerant and, in any case, the air in the clinic must be ozone free.

It is unfortunately confusing that ozone concentrations are reported as either ppmv or as mcg/ml in USA or Europe, respectively. The conversion is as follows:

1 ppmv = 0.002mcg/ml

After prolonged breathing of air contaminated with ozone, the seriousness of symptoms and pathological changes are in relation to the ozone concentration and the exposure time (Table 1)

Table 1. Toxic effects of gaseous ozone in humans

O_3 concentrations in air (ppmv)	*Toxic effects*
0.1	Lachrymation and irritation of upper respiratory airways.
1.0-2.0	Rhinitis, cough, headache, occasionally nausea and retching. Predisposed subjects may develop asthma.
2.0-5.0 (10-20 min)	Progressively increasing dyspnoea, bronchial spasm, retrosternal pain.
5.0 (60 min)	Acute pulmonary oedema and occasionally respiratory paralysis.
10.0	Death within 4 hours.
50.0	Death within minutes.

The toxicological effects are worse if the subject has breathed ozonated air contaminated with NO_2, acidic compounds, CO, etc because the RTLFs of the mucosa have a very weak buffering and antioxidant capacity. It must be emphasized that **the toxicity of ozone for the respiratory tract cannot be extrapolated to blood owing to quite different anatomical, biochemical and metabolic conditions.** An intoxicated patient must lie down and possibly breathing humidified oxygen. A slow intravenous (IV) administration of ascorbic acid and reduced glutathione (GSH) in 5 % glucose solution may limit the damage. Ascorbic acid, vitamin E and N-acetylcysteine (NAC) can also be administered by oral route but this type of treatment is more rational as a preventive than curative therapy. Indeed, the higher is the anti-oxidant capacity of biological fluids; the lower is the possible oxidative damage.

CONCLUSIONS: Ozone is a natural, highly reactive, gaseous molecule produced by an electric discharge or/and UV radiation, alone or with NOx. Remarkably, even activated leukocytes seem to generate ozone in vivo. It can be protective or offensive depending upon its concentration and location. Ozone should never be inhaled because the RTLFs have, in comparison to blood, a negligible protective capacity. Today, the use of ozone for industrial applications and water disinfection has received a wide consensus while its use in medicine remains controversial because medical scientists and clinicians remain sceptical and do not want to learn and understand the usefulness of ozone.

Chapter 2

HOW OZONE IS GENERATED AND ITS CONCENTRATIONS MEASURED?

Owing to ozone instability, it needs to be generated only when needed and used at once. **The ozonetherapist must have an ozone generator that is safe, atoxic, and reproducible**. The instrument must be built with the best ozone-resistant materials, such as Inox 316 L stainless steel, pure titanium grade 2, Pyrex glass, Teflon, Viton and polyurethane avoiding any material that could be released due to ozone oxidation. It is strongly suggested to purchase only a generator that allows to measure in real time the ozone concentration by mean of a reliable photometer.

Unused ozone cannot be dispersed into the environment and it must be decomposed to oxygen by a catalytic reaction inside the indispensable destructor that contains heavy metal oxides maintained at about + 70 °C by an electric thermostat.

The medical ozone generator consists of 2-4 high-voltage tubes connected in series to an electronic programme able to set up voltage differences between 4,000 and 13,000 Volts. In the system defined as the corona discharge ozonator, the ozone is formed when oxygen passes through a gap between high voltage and ground electrodes to create an energy field, denominated corona. The energy from the electric discharge allows the breakdown of oxygen molecules into oxygen atoms which, in the presence of an excess of oxygen molecules, form the three-atom ozone molecule. The generator is fed with pure medical oxygen and, at the supply nozzle, a gas mixture composed of no more than 5 % ozone and 95 % oxygen can be collected at a slightly positive pressure. The synthesis of ozone is allowed by the energy released by the electric discharge while the decomposition of ozone is accompanied by energy release. For medical purposes, air cannot be used because, by containing 78 % nitrogen, the final gas mixture will contain, beside oxygen and ozone, a variable amount of highly toxic NOx

The ozone concentration is determined by three parameters:

1. THE VOLTAGE: the final ozone concentration increases with the voltage, albeit in a non-proportional manner.
2. THE SPACE BETWEEN THE ELECTRODES: this serves to modulate a gradual increase of the ozone concentration.

3. THE OXYGEN FLOW: this is expressed as a volume of litres per minute (L/min) and normally can be regulated from 1 up to about 10 L/min. The final ozone concentration is inversely proportional to the oxygen flow; hence, per time unit, the higher the oxygen flow, the lower the ozone concentration and vice versa.

The criteria for calculating the ozone dose are the following:

A) Total volume of the gas mixture composed of oxygen and ozone.

B) Ozone concentration, expressed as micrograms per ml (mcg/ml).

C) Barometric pressure (mmHg), if different from normal. For safety reasons we must avoid hyperbaric pressure.

THE TOTAL OZONE DOSE IS EQUIVALENT TO THE GAS VOLUME (ml) MULTIPLIED BY THE OZONE CONCENTRATION (mcg/ml).

As an example, if we are using a volume of gas equivalent to 100 ml and the ozone concentration is of 40 mcg/ml, the total ozone dose is: 100 x 40 = 4,000 mcg or 4.0 mg.

A good ozonetherapist with an unreliable ozone generator cannot deliver an efficacious ozonetherapy.

Therefore, it is indispensable that the generator undergoes periodic maintenance including a control by iodometric titration of the photometer, to insure delivery of a precise ozone concentration.

The normal medical generators deliver ozone concentrations from 1 up to 70-100 mcg/ml. As the use of ozonetherapy will soon expand, **I envisage the usefulness of a small, precise and handy generator able to produce ozone concentrations equivalent to 2, 5, 10, 20 and 30 mcg/ml.** This range is suitable for treatments such as rectal insufflation, topical treatments, quasi-total body exposure, preparation of ozonated water and oil for patients to use at home under the supervision of an ozonetherapist. **For several practical reasons, this new device will allow the use of ozone to many chronic patients that, otherwise, find impossible, or time-consuming, or too expensive to be treated, in a clinic.**

Even today there are a number of obsolete or unchecked instruments in use so that clinical results remain questionable and often are not reproducible. If we really want ozonetherapy to progress, we need precise and reliable ozone generators. ***This is so because ozonetherapy is not based on a homeopatic concept that even a trace amount will be active, but on the firm pharmacological basis that ozone is and acts as a real drug and, as such, it must be quantitatively precise.*** Luckily European generators give the ozone concentration directly in terms of mcg/ml and the range 1-100 is sufficient for medical use. Modern generators allow to asses the ozone concentration by photometric determination. This is possible because there is a pronounced absorption of ozone within the Hartley band with a peak at 253.7 nm. At this wavelength, UV radiation (mercury vapour lamp) is linearly absorbed in a concentration-dependent fashion (in agreement with Lambert-Beer's law) on being passed through a tube containing ozone. This system is quite sensitive and precise but the ozonetherapist must know that tends to decay due to lamp ageing. There is also the possibility of measuring ozone concentration at 600 nm (Chappuis band) that, although less sensitive, is more stable. **The great advantage of the photometer is the possibility of checking on a digital display the ozone concentration in the gas mixture flowing into the syringe during withdrawal.** However the photometer must be checked from time to time and possibly adjusted on the basis of ozone concentrations measured by the iodometric method considered the gold standard. The method has been approved by the IOA and the details have been reported by Masschelein in 1996.

When ozone reacts with buffered potassium iodide, iodine is generated and the colourless solution suddenly acquires an amber colour which, upon reduction with a titrated solution of sodium thiosulphate and a starch indicator allows the determination of the ozone concentration in g/L with a reproducibility of about 2% of the measured ozone concentration.

While the need of having a precise instrument is a must, in daily practice a couple of tips can be useful. Firstly, I learnt that what is important is the immediate use of the gas and not so much small changes (±1%) of concentration. Secondly, polypropylene, silicon-coated syringes must be used only once. Finally the generator must be regularly checked because the efficacy of the treatment depends upon the required ozone concentration.

CONCLUSIONS: Ozone must be produced using medical oxygen with a reliable, atoxic generator that allows the measurements of precise ozone concentrations (1-100 mcg/ml) by mean of a photometer often controlled by iodometric titration.

The total ozone dose is equivalent to the gas volume (ml) multiplied by the ozone concentration (mcg/ml). For different medical applications, the ozonetherapist must know the optimal ozone doses and these will be specified in chapter 9.

Chapter 3

PREPARATION OF OZONATED WATER AND OIL FOR THE TOPICAL THERAPY. OZONE AS A DRINKING WATER DISINFECTANT. OZONE DISINFECTION TO PREVENT NOSOCOMIAL INFECTIONS.

In the world, there are millions of people affected by dirty traumatic lesions, infected wounds, chronic torpid ulcers, bed sores, burns, herpetic lesions, fungal infections and insect stings, who suffer for a long time because the conventional topical treatments based on antibiotics and anti-inflammatory drugs are not sufficiently effective. Unfortunately, most physicians and nurses are not aware of the potency and efficacy of both ozonated water and oil. When possible, we can also use the gas mixture: oxygen-ozone, but we must avoid the risk of breathing ozone and not all generators are equipped with a suction pump connected to an ozone destructor. On the other hand, it is easy to apply a gauze compress soaked with ozonated water or oil to any part of the body.

The preparation of ozonated water is carried out by using a glass cylinder about ¾ filled with bidistilled water through which the gas mixture has to be bubbled continuously for at least 5 minutes to achieve saturation. The unused ozone flows out via silicone tubings into a destructor and is converted to oxygen. Some ozone generators have already incorporated the system for ozonating water; if it is not available, it can be simply built with a 500 ml glass bottle that we can fill with 250 ml water and 250 ml of the gas mixture and close with a silicone cork. Also with this rudimentary technique, a vigorous mixing for about 5 min insures a fairly good ozonation of pure water.

Solubilization of ozone in pure water occurs according to Henry's law (1803) that states that the saturation concentration of a gas in water is proportional to its concentration. This is correct only if the water is absolutely pure and the temperature and ozone pressure remain constant. Monodistilled water (or worse, tap water) is unsuitable because, by containing some ions, stimulate the chemical reactivity of ozone with the possible formation of toxic compounds. As an example**, physiological saline**

(0.9 % NaCl) should never be ozonated because of the formation of hypoclorous acid. This compound can cause inflammation and phlebitis upon infusion of ozonated saline as some quacks do. For this reason, **I recommend the use of pure water, which is commercially available. In such a case ozone is simply dissolved in water and its concentration, after 5-6 min of bubbling, is stable and equivalent to ¼ (25%) of the ozone concentration present in the gas mixture.** Thus, if we want a strong preparation of ozonated water, we must use an ozone concentration of 80 mcg/ml of gas that will yield a final ozone concentration of about 20 mcg/ml in the water. This solution is suitable for treating heavily infected wounds in order to eliminate pus, necrotic materials and bacteria. On the other hand, once the wound reaches the proliferation and remodelling stages, we must use a mild solution prepared with an ozone concentration of 20 mcg/ml of gas which will yield an ozone concentration of only about 5 mcg/ml water.

How stable is a preparation of ozonated pure water? Owing to the inherent ozone instability, this is a weak point. The ozonated water must be maintained in a glass bottle tightly closed with a silicone or Teflon cap, possibly in the refrigerator. If it is kept at 5° C, the ozone concentration is halved in some 110 hours, but at 20° C the ozone half-life is only 9 hours! This information has a practical importance because, if maintained properly, it can be used for a couple of days at the patient's home for domiciliary treatment. In contrast, the half-life of ozone solubilised in monodistilled water is less than one hour and therefore it must be used at once.

I cannot lose the opportunity of emphasizing the usefulness of ozonated water for removing thick pus from purulent abscesses and empyemas. After draining the water, the cavity can be insufflated with the oxygen-ozone gas mixture at least twice daily and soon the operator will be surprised to note a rapid healing. I know of several desperate cases where dedicated ozonetherapists were able to eliminate hopeless infections by only using the combination of ozonated water, gas and ozonated oil. Gas insufflation must be performed in a few minutes in a well ventilated room leaving the polyethylene cannula clamped to prevent the exit of gas. Ozone will dissolve into the infected secretions, will sterilize them and will promote the reconstruction of tissues. Obviously one should use at first a high (70-80 mcg/ml) ozone concentration during the septic phase and then progressively lower it as soon as the infection subsides, for enhancing cell proliferation.

It is well known that decubitus and torpid ulcers (in diabetic, venous stasis and chronic limb ischemia patients) require a frustratingly prolonged treatment that often is a failure. Gas can only be used if the ulcer can be contained, with or without slight decompression, by using an ozone-resistant container, such as a polyethylene bag or a Teflon cup. Usually during the day it is more practical to use freshly ozonated water for cleansing, disinfection and stimulation of tissue granulation whereas, during the night, the application of ozonated oil is able to maintain sterile the lesion and enhance healing. In the last decade, there has been a growing interest in the application of ozonated oil: a particular merit goes to Dr. Renate Viebahn and Cuba scientists for developing it and I believe that Renate has patented the production. In Cuba, probably owing to the lack of conventional drugs, ozonated sunflower oil has been employed in torpid ulcers, bacterial, fungal and parasitic infections not only with topical applications but also oral administrations. Results published in Cuban journals seem to be very good.

As a natural preparation, ozonated oil is available in several countries but so far there is not a really standard preparation, which is urgently needed. Recently ozonated sunflower oil (Oleozon) from Cuba was tested by Sechi et al., (2001) and it was found to have valuable antimicrobial activity against all the tested micro-organisms. At our University Hospital, we made our own preparation by bubbling oxygen-ozone in pure olive oil for at least 60 min at room temperature but now we prefer to use a commercial preparation. At the IOA Congress in London (September 14-15, 2001), Miura et al.,(2001) presented an interesting report on the elucidation of the structure of ozonated olive oil: ozonation was carried out for two days until the oil solidified and one gram of oil could absorb up to 160 mg of ozone. A number of analyses led to the conclusion that the prolonged ozonation resulted in exclusive formation of triolein-triozonides, which remained stable in the refrigerator for two years. There is no real need to have a solid oil preparation, except for commercial purposes and long stability. In practice, the pathological situations are so variable to require great flexibility so that the very viscous oil can be either warmed or diluted with pure oil, or better with pharmaceutical *Vaselinum album* (at 50%) when the wound is aseptic.

How ozonated oil acts remains an open question. Probably, when the stable triozonide comes into contact with the warm exudate of the wound, it slowly decomposes to reactive ozone, which readily dissolves in water, generating hydrogen peroxide and LOPs that can explain the prolonged disinfectant and stimulatory activity. If it is correct, this reasoning implies that we should have titrated preparations with high, medium or low triozonide concentrations to be used during the inflammatory septic phase I, regenerating phase II or remodelling phase III, respectively. These phases have been related to the rapidly changing cell types and to the release of cytokines and growth factors that modulate the complex healing process (Chapter 9, Section I).

In the Department of Surgery, Chiba-Tokushkai Hospital in Japan, Matsumoto et al., (2001) tested the efficacy of the oil prepared by Miura et al., in intractable fistulae and wounds after surgical operations (acute appendicitis with peritonitis, intrapelvic, abdominal and perianal abscesses, etc.). In a series of 28 patients, the ozonated oil was fully effective in 27 cases, without adverse side effects. During the last five years, I treated several desperate cases in old people (prevalently bed sores) with a great success so that I can fully confirm Matsumoto's results. I just learnt that patients with radiotherapy skin reactions, treated with ozone, perceived a benefit in terms of pain relief (Jordan et al., 2002). Surprisingly, these results were obtained in Manchester (UK), with an unsuitable method and therefore the use of ozonated oil will likely yield even better results. Ozonated oil has also proved to be very effective in burns and it would be interesting to compare this treatment with the Moist-Exposed Burns therapy and the Moist-Exposed Burns Ointment invented by Xu in China (2004).

Moreover I will mention that there are several pharmaceutical vehicles for the administration of ozonated oil, such as gastro-resistant capsules, pessaries, suppositories and even collyriums, to be used in intestinal, vaginal, anal-rectal and ocular infections. As one can imagine, ozonated oil smells of rancid fat but capsules ingested by mouth have been tolerated by Cuban children. Silvia Menendez et al., (1995) treated 222 children affected by infantile giardiasis, a parasitic disease, obtaining a remarkable cure without toxicity in 76% of children.

Today, we are still using ozonated oil in a very empirical fashion and, when I report these informations, people do not disguise their incredulity and only results obtained after controlled studies will be convincing. However, **once physicians and nurses will realise the therapeutic potential of ozonated water and oil, these products will become a very useful and inexpensive medical treatment.**

In spite of a large use of chlorine, 2.4 billion people or **40 % of the world's population do not have access to adequate sanitation**. Unfortunately chlorine has unsatisfactory organoleptic characteristics and it is being widely substituted by ozone all over the world. Ozone is possibly an even more potent drinking water disinfectant able to inactivate several human pathogens, e.g. as many as 63 different bacteria (*Salmonella, Shigella, Vibrio, Campylobacter jejuni, Yersinia enterocolitica, Legionella,* etc.), some 15 viruses (polio-, echo-, Coxsackie viruses, etc.), some 25 fungi and mould spores (*Aspergillus, Penicillium, Trichoderma,* etc.), several yeast varieties, and up to 13 fungal pathogens (*Alternaria, Monilinia, Rhizopus,* etc.). More recently, due to contamination of groundwater with faecal material, the problem of disinfection has become more complex, since encysted protozoa, such as *Giardia lamblia, Cryptosporidium parvum* oocysts and helminth eggs (*Ascaris suum and Ascaris lumbricoides*), require a much longer time of contact with ozone than bacteria and viruses. Every year Cryptosporidium causes outbreaks of sickness, which can be fatal for elderly and very ill patients (AIDS).

Water is rapidly becoming a precious commodity and wastewater from cities, animal breeding (particularly cattle, sheep, and swine) and industrial plants must be reused for irrigation in order to increase agricultural production. This happens most frequently in underdeveloped countries, but also in the USA and Italy, and poses a health risk by causing serious gastro-intestinal diseases (Stein and Schwartzbrod, 1990; Ayres et al., 1992;

Johnson et al., 1998; Orta de Velasquez et al., 2001; Liou et al., 2002). Toze (1999) has reported that, **in countries with poor sanitation systems, about 250 million people are infected each year by waterborne pathogens, with about 10 million deaths.**

The oxidation of organic and inorganic materials during ozonation (gas to water phase) occurs via a combination of molecular ozone and hydroxyl radicals. Water companies throughout the world are evaluating several methods to optimize the various steps of the water-treatment process, which varies in different countries depending on the quality of the water, concentration of organic matter, turbidity and salt content (Kadokawa et al., 2001; Evans et al., 2001; Courbat et al., 2001; Hijnen et al., 2001). Ozone appears very effective in inactivating most bacteria and viruses, while protozoan cysts and helminth eggs are far more resistant; only by using realistic ozonation conditions can one achieve a moderate degree of inactivation (Graham and Paraskeva, 2001; Lewin et al., 2001). This is an important problem that requires more intensive sanitation of wastewater, particularly from animal breeding.

Another aspect for prevention of outbreaks of intestinal infections is the possibility of using ozone as an antimicrobial agent in direct contact with food and fruits. On June 26, 2001, the U.S. Food and Drug Administration (FDA) formally approved the use of ozone, in the gaseous and aqueous phase, as an antimicrobial agent for the treatment, storage and processing of foods (Rice, 2001). It must be mentioned that, in addition to the disinfection of drinking water, the use of ozone can also improve its organoleptic properties. In fact, it enhances the coagulation and flocculation process, oxidizes bad taste and odour compounds (as well as iron and manganese), and improves particle removal in filters or through bioactive granular activated carbon. **The efficacy of ozone has now been validated by more than 3,000 municipal water treatment plants around the world.**

During the last decade nosocomial infections have become common because the resistance of pathogens to antibiotics has increased to a point where we no longer have an effective drug for some strains. This is a complex story, partly due to the extensive use of antibiotics in animal food and the improper use in patients. The result is dramatic because almost every month, we hear of a series of deaths due to incontrollable infections breaking out in hospitals after more or less complex operations and in intensive therapy units. With some approximation, it seems that several thousand deaths could be avoided each year if we could eliminate the resistant bacteria. The problem is so important that some 1000 papers per year report relevant data (Aitken and Jeffries, 2001; Guerrero et al., 2001; Kollef and Fraser, 2001; Olsen et al., 2001; Shiomori et al., 2001; Slonim and Singh, 2001; Stephan et al., 2001; Stover et al., 2001; Wenzel and Edmond, 2001).

Applications for ozone can be divided into two phase:

1) The gas to gas phase,
2) The gas to water phase (liquid phase-ozone).

The first phase is widely used to remove as many as 272 organic odours and pollutants: these range from acrolein to bathroom smells, body odours, cigarette smoke, decaying substances, ether, exhaust fumes, faecal and female odours, hospital odours, medicinal odours, mould, putrefying substances, sewer odours, toilet odours, waste products, etc. Ozone is proficiently used in hospital wards and nursing homes to get rid of the smells caused by incontinent patients. In air conditioning systems (cooling towers, etc.), a small amount of ozone rids the recirculating air of odours, bacteria (Legionella pneumophila, etc.) and viruses. Moreover, ozone is providential for fumigation of bedding, bedclothes and treatment of air in operating rooms. Ozone is effective but it is necessary to take precautions:

a) To allow enough time, even days if necessary, for the ozone gas (which is less active and slower than aqueous solubilized ozone) to be in contact with the contaminants to be oxidized and destroyed.

b) When confined spaces are treated with gaseous ozone, people must not be present. The ozone generator must be regulated by a timer, which can be operated by every user. Ozone release must stop well before people re-enter the facility.

c) Prior to returning the air mixed with ozone into the atmosphere, the gas mixture must pass through an ozone destructor. Personnel can usually re-enter an area treated with ozone, after appropriate de-aeration, after a short while.

d) To prevent lung toxicity, an ozone monitor must be installed to check for any residual ozone concentration.

Ozone fumigation of bedding, bedclothes and any other object can be carried out according to the instructions given by Inui and Ichiyanaghi (2001). Ozone is used in conjunction with a negative ion generator and, if necessary, a heater to control mites and ticks.

Several pharmaceutical firms in the USA have recently started to package pharmaceutical products in an ozone-containing atmosphere to maintain a sterile packaged product line.

The gas to water phase has been adopted in the USA by a number of laundries to effectively launder and sterilize various linens used in health care facilities. It seems that, although this process is not energy efficient, it does extend linen life by 25-50%. Moreover, ozone washing provides a good alternative to conventional linen processing, since it is more effective in preserving the environment from contaminated water. All these innovative technologies increase health care costs, but the quality of service is improved and, more importantly, nosocomial infections can be minimized.

A full report informing about how to improve safety in hospitals can be found online at http:// www.ahrq.gov/making health care safer: a critical analysis of patient safety.

CONCLUSIONS: Ozonation of either bidistilled water or olive oil is performed by bubbling the gas mixture (O_2-O_3) for either five min or up to two days, respectively. The ozone concentration in pure water, due to solubilised ozone, corresponds to 25% of the used ozone concentration, which is more than enough for an optimal disinfection. One gram of oil can bind up to 160 mg of ozone. While ozonated water remains efficacious for one-two days, the oil remains stable for two years in the refrigerator. Both acts as potent disinfectants and enhance healing by stimulating cell proliferation. As soon as the medical community will appreciate their efficacy, both ozonated water and oil will become indispensable tools in chronic wound healing units. *I would like to predict that the application of ozonated oil, a simple and inexpensive remedy, will become far more useful than expensive pharmaceutical creams and will herald a medical revolution for the topical treatment of torpid ulcers and wounds. Under these terms,* ***it is not exaggerated to proclaim ozone as "the wonder drug of the XXI century".***

The problems of the disinfection of drinking water and the prevention of nosocomial infections have become of primary importance because their solution means life ore death for many people. In comparison to chlorine, the versatility and efficacy of ozone is widely acknowledged.

Chapter 4

HOW DOES OZONE ACT? HOW AND WHY CAN WE AVOID OZONE TOXICITY?

This is one of the most important chapters because I believe that, if the ozonetherapist understands how ozone reacts with body fluids and cells, he can achieve useful therapeutic results. The patient represents the substrate yielding a number of biochemical, pharmacological and psycho-neuro-immunological reactions and as such, she/he is an essential part of the process.

Although oxygen represents the bulk (95-98 %) of the gas mixture, by considering the enormous dilution of the small reinfused oxygenated-ozonated blood with venous blood, it has a negligible role. While, only thanks to oxygen we can live, this gas has a negative effect on the long run because cell respiration allows the formation of reactive oxygen species (ROS), among which, hydroxyl radical ($OH^{\cdot}$) is one of the most destructive radical compounds for precious enzymes and DNA. Almost every one knows that ageing, the metabolic disorders (atherosclerosis, diabetes, cell degeneration) can be worsened by ROS and, only in part, we can prevent their damageable effects. Ironically, even the partial lack of oxygen (hypoxia), observable in ischemic vascular diseases, represents the cause of death due to limb ischaemia, heart infarction and stroke. Moreover, hypoxia enhances neoplastic metastatisation and ultimately leads to death.

Ozone, the triatomic oxygen, synthesized in the stratosphere to protect us from excessive UV radiation, can be precisely produced with a medical generator but it is up to us to use it proficiently as a real drug. As ozone is one of the most potent oxidants, we must learn how to tame it and **the scope of this chapter is to define its therapeutic coefficient, or**, in simple words, **to distinguish the therapeutic from the toxic dose.**

When I ask physicians how ozone acts, I receive odd answers: a favoured one is the esoteric idea that ozone, during its decomposition to oxygen, will transfer some energy to the body thus invigorating it, and another is that ozone will be absorbed and, after entering into the cells, will turn them on. In comparison to other complementary approaches based on philosophical postulations, a positive characteristic of **ozonetherapy** is that it **can undergo the most objective scientific investigation carried out with normal**

biological and clinical methods. It has been unfortunate that for several decades, empiricism and the lack of basic studies have delayed an understanding of the mechanisms of action. Moreover, dangerous, even deadly infusion of ozone by quacks, a good dose of prejudice and **the inconsistent dogma that "ozone is always toxic"** are responsible for the strong and dull opposition of conventional medicine to the use of ozonetherapy. However I will persevere in my endeavour and I feel confident that this wrong belief will change in the near future.

At the moment my duty is to schematically try to demonstrate that ozone obeys perfectly well to common physical, chemical, physiological and pharmacological notions and that its activities modulating several cellular functions are already known.

First of all, ozone, as any other gas, dissolves in the water either of the plasma (the liquid part of blood), or into the extracellular fluids, or into the thin layer of water covering the skin and particularly the mucosae of the respiratory tract, gut, vagina, etc. At normal temperature and atmospheric pressure, owing to its high solubility and depending upon its relative pressure, some ozone dissolves into the water but, unlike oxygen, DOES NOT EQUILIBRATE with the ozone remaining in the gas phase. This happens because **ozone, being a potent oxidant, REACTS IMMEDIATELY with a number of molecules present in biological fluids, namely antioxidants, proteins, carbohydrates and, preferentially, polyunsaturated fatty acids (PUFAs).**

The reaction of ozone with so many molecules implies two fundamental processes:

A) I call **the first** "THE OZONE INITIAL REACTION" because some of the ozone dose is unavoidably consumed during oxidation of ascorbic and uric acids, sulphydryl (SH)-groups of proteins and glycoproteins. Although albumin, ascorbic and uric acids tame the harsh reactivity of ozone (Halliwell, 1996), they allow this first reaction that is important because it generates reactive oxygen species (ROS), which triggers several biochemical pathways in blood ex vivo (ie, in the glass bottle). ROS are neutralized within 0.5-1 minute by the antioxidant system.

B) **The second**, well characterized reaction is known as "LIPID PEROXIDATION" (Pryor et al., 1995). In the hydrophilic plasma environment, one mole of an olefin (particularly arachidonic acid present in plasma triglycerides and chylomicrons) and one mole of ozone give rise to two moles of aldehydes and one mole of hydrogen peroxide (H_2O_2). These two reactions, completed within seconds, use up the total dose of ozone that generates hydrogen peroxide, an oxidant but not a radical molecule (usually included in the ROS family) and a variety of aldehydes known as LIPID OXIDATION PRODUCTS (LOPs).

FROM NOW ON, NOT OZONE, BUT ONLY ROS (MOSTLY HYDROGEN PEROXIDE) AND LOPs ARE RESPONSIBLE FOR

THE SUCCESSIVE AND MULTIPLE BIOCHEMICAL REACTIONS HAPPENING IN DIFFERENT CELLS ALL OVER THE BODY.

Therefore it should be clear that **a good deal of ozone is consumed by the antioxidants present in plasma and only the second reaction is responsible for the late biological and therapeutic effects**. This should clarify why a very low ozone dose can be ineffective or equivalent to a placebo.

ROS include several radicals as anion superoxide ($O_2^{\cdot -}$), nitrogen monoxide ($NO^{\cdot}$), peroxynitrite ($O{=}NOO^{-}$), the already mentioned hydroxyl radical and other oxidant compounds such as hydrogen peroxide and hypoclorous acid (HClO). ***All of these compounds are potentially cytotoxic*** (Fridovich, 1995; Pullar et al, 2000; Hooper et al., 2000)***, luckily have a very short half-life (normally a fraction of a second) and both the plasma and cells have antioxidants able to neutralize them, if their concentrations do not overwhelm the antioxidant capacity.***

LOPs generated after peroxidation of a great variety of PUFAs are heterogenous and briefly are represented by peroxyl radicals ($ROO^{\cdot}$), a variety of hydroperoxides (R-OOH) and a complex mixture of low molecular weight aldehydic end products, namely malonyldialdeyde (MDA), and alkenals, among which 4-hydroxy-2,3 transnonenal (4-HNE), is one of the most cytotoxic. The chemistry and biochemistry of these compounds has been masterfully described by Esterbauer's group (1991). If one thinks about the wealth and chemical heterogeneity of lipids, glycolipids and phospholipids present in plasma, it becomes difficult to imagine how many potent, potentially noxious, compounds can be generated by the lipids reacting with ozone. During one of my several disputes with American referees, a distinguished scientist wrote: "*It is grotesque to think that any Western World Drug Regulating Agency would condone infusing the hodgepodge of ozonized products to treat diseases, although it is probable that the products would initiate and/or modulate a wide spectrum of inflammatory-immune processes to varying degrees*".

In my opinion, **this referee missed what I believe is the formidable strength of ozonetherapy: provided that we can control (by using precise ozone concentrations exactly related to the blood volume and antioxidant capacity) the amount of LOPs, we can achieve a multitude of biological effects unthinkable with a single drug** (Figure 1).

The scheme ought to fix in the reader's mind this crucial point and the sequence of events eventually leading to the therapeutic results: ROS are produced only during the short time that ozone is present in the glass bottle, *ex vivo*, and they yield EARLY biological effects on blood, whereas LOPs, which are simultaneously produced, have a far longer half-life and, during the reinfusion of ozonated blood in the donor, they reach the vascular system and practically all the organs where they trigger LATE effects (Figure 2).

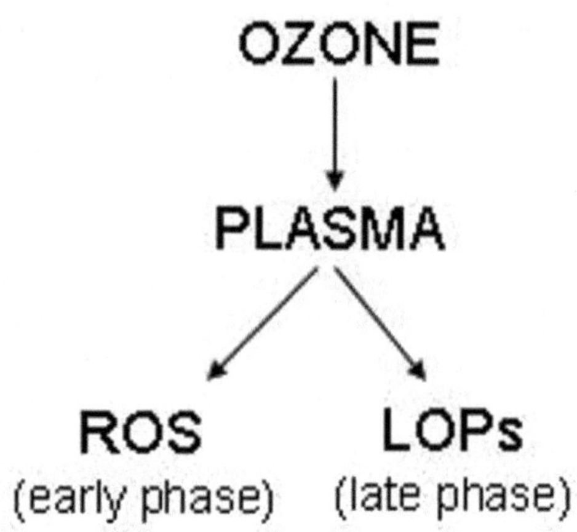

Figure 1. The scheme intends to show that ozone dissolved in the plasmatic water reacts immediately with a number of biomolecules and disappears. The compounds generated during the reactions (ROS and LOPs) represent the "ozone messengers" and are responsible for the biological and therapeutic effects.

We have come to a critical point: how can we reconcile the production of toxic compounds with the idea that these compounds exert important biological and therapeutic effects?

Let us first examine the behaviour and pharmacodynamic of hydrogen peroxide, which in practical terms is the most important ROS. As soon as ozone dissolves in the plasmatic water and reacts with PUFAs, **the concentration of hydrogen peroxide starts to increase but, just as rapidly, decreases because this unionized molecule diffuses quickly into erythrocytes, leukocytes and platelets, where it triggers several biochemical pathways.**

Does the increased intracellular concentration of hydrogen peroxide become toxic for the cell? **Absolutely no! Because, at the same time, it undergoes reduction to water** in both plasma and intracellular water, thanks to the presence of powerful antioxidant enzymes such as catalase, glutathione-peroxidase (GSH-Px) and free reduced glutathione (GSH). Perhaps for one second, the plasma-intracellular concentration has been estimated to range from 1 to 10 micromolars, which avoids any toxicity (Stone and Collins, 2002). The transitory presence of hydrogen peroxide in the cytoplasm means that it acts as one of the ozone chemical messengers and that **its level is critical: it must be above a certain threshold to be effective but not too high to become noxious. In our studies,** performed with human blood exposed to ozone concentrations ranging from 20 to 80 mcg/ml per ml of blood, **the process of hydrogen peroxide generation, diffusion and reduction was found always extremely transitory** (Bocci et al.,1993a; b; 1998a;b).

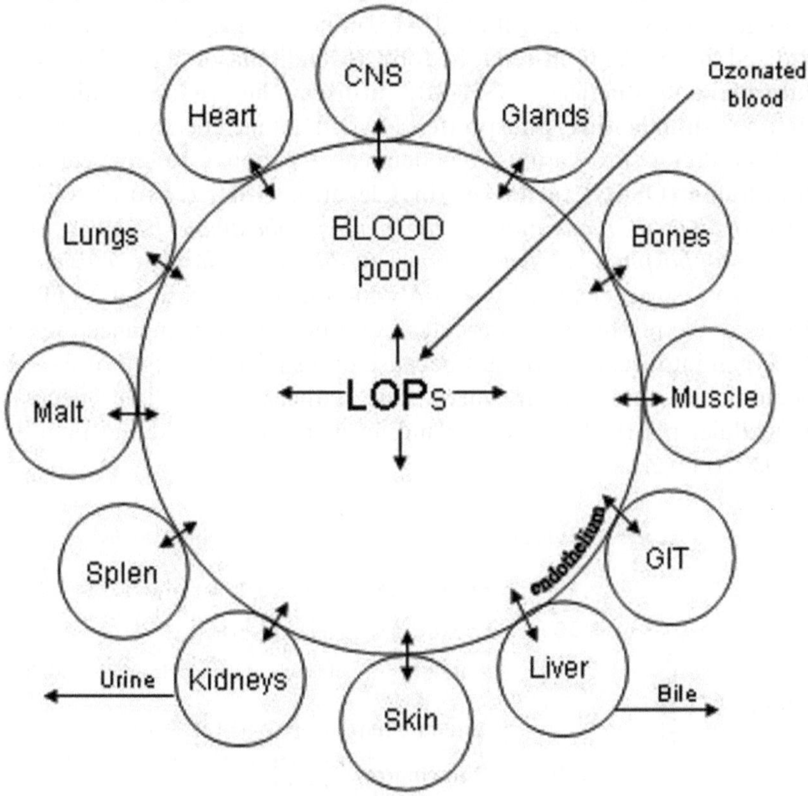

Figure 2. The multivaried biological response of the organism to ozonized blood can be envisaged by considering that ozonized blood cells and the generated LOPs interact with a number of organs. Some of these represent real targets (liver in chronic hepatitis, vascular system for vasculopathies), while other organs are probably involved in restoring normal homeostasis.

ER: erythrocytes, PLAT: platelets, BMC: blood mononuclear cells, GRAN: granulocytes, CNS: central nervous system, GIT: gastrointestinal tract, MALT: mucosal associated lymphoid tissue.

Hydrogen peroxide is now widely recognised as an intracellular signaling molecule able to activate a tyrosine kinase, which phosphorylates a transcription factor (Nuclear Factor KB, NFKB), which allows the synthesis of a number of different proteins (Baeuerle and Henkel, 1994; Barnes and Karin., 1997). Basically hydrogen peroxide functions by oxidizing cysteines (Rhee et al., 2000), and we and Others have found that it acts on blood mononuclear cells (Bocci and Paulesu, 1990; Bocci et al., 1993b; 1998a; Reth, 2002), on platelets (Bocci et al., 1999a), on endothelial cells (Valacchi and Bocci, 2000) and on erythrocytes (Bocci, 2002).

ROS entering into the erythrocytes are almost immediately reduced (hydrogen peroxide to water and lipoperoxides to hydroperoxides) at the expense of GSH. The enormous mass of erythrocytes can easily mop up hydrogen peroxide and, within 10-15 minutes, marvellously recycle back oxidized antioxidants in reduced form (Mendiratta et al., 1998a, b). While **glutathione reductase** (GSH-Rd) **utilises the reduced nicotinamide adenine dinucleotide phosphate** (NADPH, this coenzyme serves as an electron donor for various biochemical reactions) **to recycle oxidized glutathione (GSSG) to the original level of GSH,** *the oxidized NADP is reduced after the activation of the pentose phosphate pathway, of which glucose-6-phosphate dehydrogenase (G-6PD) is the key enzyme.* Thus, glycolysis is accelerated with a consequent increase of ATP levels. Moreover the reinfused erythrocytes, for a brief period, enhance the delivery of oxygen into ischemic tissues because of a shift to the right of the oxygen-haemoglobin dissociation curve due either to a slight decrease of intracellular pH (Bohr effect) or/and an increase of 2,3-diphosphoglycerate (2,3-DPG) levels.

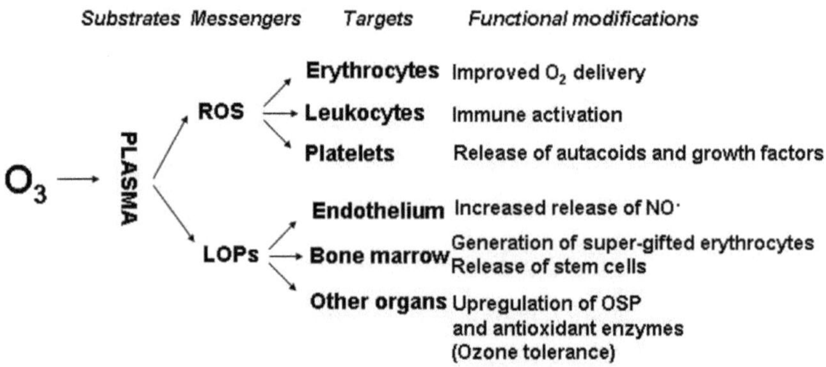

Figure 3. A summary of the biological effects elicited during exposure of human blood to oxygen-ozone, ex vivo and during its reinfusion in the donor.

There is an ample literature regarding the cytotoxicity of LOPs. These compounds, when tested either in tissue culture, or examined in the context of the delicate respiratory system, are toxic even at a concentration of 1 micromolar. Surprisingly, submicromolar concentrations (0.01-0.5 microM) tested in several cell types can stimulate proliferation and useful biochemical activities. These findings lead to believe that toxicity of ozonated lipid products depends upon their final concentrations and tissue-localization, so that they can act either as injurious or useful signals (Dianzani, 1998; Parola et al., 1999; Bosch-Morell et al., 1999; Larini et al., 2004). **Blood, in comparison to the lungs, is a much more ozone-resistant "tissue" and we have never observed any damage.** However, **when we reinfuse ozonated**

blood, what is the fate of LOPs? We have often measured the kinetic of their disappearance from blood and their half-life in six patients with age-related macular degeneration (ARMD) was equivalent to 4.2±1.7 min. On the other hand, if the same ozonated blood samples were incubated in vitro, levels of LOPs hardly declined during the next two hours, a result clarifying their toxicity in static cell cultures. As far as cholesteryl ester hydroperoxide is concerned, Yamamoto (2000) has emphasized the role of the enzymatic degradation and hepatic uptake. **Thus LOPs toxicity in vivo is most likely irrelevant for the following reasons:**

1) DILUTION (up to 150-200 folds) of these compounds in blood and body fluids rapidly lowers their initial concentration to pharmacological, but not toxic levels. Obviously the ozone dose must be within the therapeutic range.

2) NEUTRALISATION of LOPs due to the antioxidant capacity in body fluids and cells.

3) DETOXIFICATION of LOPs (scarcely observable in vitro) due to the interaction with billions of cells endowed with detoxifying enzymes such as aldehyde- and alcohol-dehydrogenases, aldose reductase and GSH-transferases (GSH-T).

4) EXCRETION of LOPs into the urine and bile after hepatic detoxification and renal excretion.

5) BIOACTIVITY without toxicity. As already mentioned, **submicromolar concentrations of LOPs can act as physiological messengers able to reactivate a biological system gone awry.**

From a pharmacokinetic point of view, trace amounts of LOPs, can reach all organs and particularly the bone marrow and the Central Nervous System (Figure 2). **LOPs are extremely important in informing specific cell receptors of a minimal and calculated oxidative stress eliciting the adaptive response.** In regard to erythrocytes, LOPs can influence the erythroblastic lineage, allowing the generation of cells with improved biochemical characteristics. These "supergifted erythrocytes" as I called them, due to a higher content of 2,3-DPG and antioxidant enzymes, during the following four months, are able to deliver more oxygen into ischemic tissues. The consequence of repeated treatments, obviously depending upon the volume of ozonated blood, the ozone concentration and the schedule is that, after a few initial treatments, a cohort (about 0.8 % of the pool) of "supergifted erythrocytes" will enter daily into the circulation and, relentlessly, will substitute old erythrocytes generated before the therapy. This means that, during prolonged ozonetherapy, the erythrocyte population will include not only cells with different ages but, most importantly, erythrocytes with different biochemical and functional capabilities. In the course of ozone therapy, we have already measured a marked increase of G-6PD and other antioxidant enzymes in young erythrocytes. (Bocci, 2004). The process of cell activation is very dynamic and don't last for ever because blood cells have a definite life-time and a limited biochemical memory; therefore, the therapeutic advantage MUST BE MAINTAINED WITH LESS FREQUENT TREATMENTS.

Ozone toxicity to blood, biological fluids and internal organs can be totally avoided when the ozone dose reduces only in part and transitorily the multiform and potent antioxidant capacity. The antioxidant system has evolved during the last two billions years as an essential defence against oxygen: it is made up of scavengers components, namely albumin, vitamins C and E, uric acid, bilirubin, cysteine, ubiquinol, alpha-lipoic acid and of intracellular antioxidants, such as GSH, thioredoxin and enzymes (superoxide dismutase, SOD; GSH-Px, GSH-Rd, GSH-T, catalase, etc.,) and proteins such as transferrin and caeruloplasmin, able to chelate free iron and copper that, otherwise, can favour the formation of hydroxyl radicals. **The wealth and the variety of extracellular and intracellular antioxidants,** thoroughly described by Chow and Kaneko (1979), Halliwell (1994; 1999a, b; 2001), Frei (1999), Holmgren, (1989), Di Mascio et al., (1989), Jang et al., (1997), Packer et al., (1997), Bustamante et al., (1998) and Chae et al., (1999), **are able to explain how bland amounts of ozone can be tamed with the results of stimulating several biological systems without deleterious effects. *Until this key point is understood, the dogma of ozone toxicity will continue to linger.***

The reader can appreciate the complexity of this system in Table 2

Table 2. The antioxidants system

NON ENZYMATIC			*ENZYMATIC*
Hydrosoluble	*Liposoluble*	*Chelating proteins*	
Uric acid Ascorbic acid Glucose, Cysteine, Cysteamine, taurine, Tryptophane, Hystidine, Methionine GSH Plasma proteins	Vitamin E Vitamin A, Carotenoids Coenzyme Q α-lipoic acid Bilirubin Thioredoxin Bioflavonoids Melatonin Lycopene	Transferrin Ferritin Caeruloplasmin Lactoferrin Haemopessin Albumin	Superoxide dismutases (SOD) Catalase Glutathione peroxidases Gluthatione redox system Reducing equivalents via NADPH and NADH

The interaction among antioxidants, enzymes and the metabolic system is very important as it allows their rapid regeneration and the maintenance of a normal antioxidant status. The following scheme, drawn by Prof. L. Packer, beautifully illustrates the cooperation among various antioxidant system in order to neutralize a lipoperoxide radical $ROO^{\cdot}$ (shown on the left hand side) to a less reactive hydroperoxide, ROOH. The reducing activity is continuously generated by cellular metabolism via the continuous reduction of $NAD(P)^+$ to NAD(P)H.

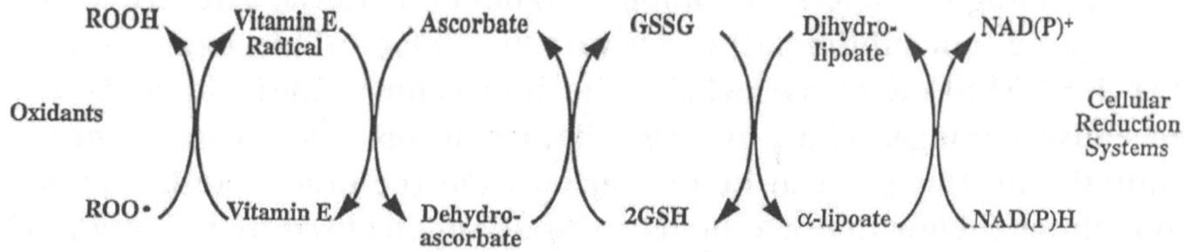

It suffices here to say that, during the transient exposure of blood to appropriate concentrations of ozone, the antioxidant reservoir decreases between 2-25 % in relation to ozone doses between 10-80 mcg/ml of gas per ml of blood. **It is important to add that this partial depletion is corrected in less than 20 min thanks to the recycling of dehydroascorbic acid, GSSG, alpha-tocopheryl radical to the reduced compounds.**

CONCLUSIONS: What happens when human blood is exposed to a therapeutic dose of oxygen-ozone?

Both gases dissolve in the water of plasma depending upon their solubility, partial pressure and temperature. While oxygen readily equilibrates between the gas and the blood phases, the ten-fold more soluble ozone cannot equilibrate because IT REACTS with biomolecules (PUFA, antioxidants) present in the plasma. The reaction yields hydrogen peroxide (among other possible ROS) and lipid oxidation products (LOPs). The sudden rise in plasma of the concentration of hydrogen peroxide generates a gradient, which causes its rapid transfer into blood cells where, in a few seconds, it activates several biochemical processes and simultaneously undergoes reduction to water by the efficient intracellular antioxidant system (GSH, catalase, GSH-Px).This critical step corresponds to a controlled, acute and transient oxidative stress necessary for biological activation, without concomitant toxicity, provided that the ozone dose is compatible with the blood antioxidant capacity.

While ROS are responsible for *immediate* biological effects (Figure 1), **LOPs are important as *late* effectors, when the blood, ozonated ex vivo, returns into the circulation upon reinfusion** (Figures 2 and 3).

LOPs can reach any organ, particularly the bone marrow where, after binding to receptors in submicromolar concentrations, elicit the *adaptation to the repeated acute oxidative stress,* which is the hallmark of ozonated autohemotherapy. Upon prolonged therapy, LOPs activity will culminate in the upregulation of antioxidant enzymes, appearance of oxidative stress proteins (haeme-oxygenase I as a typical marker) and probable release of stem cells, which represent crucial factors explaining some of the extraordinary effects of ozonetherapy (Chapter 8).

It must be emphasized that BLOOD EXPOSED TO OZONE UNDERGOES A TRANSITORY OXIDATIVE STRESS necessary to activate biological functions without detrimental effects. The stress must be adequate (not subliminal) to activate physiological mechanisms, BUT NOT EXCESSIVE to overwhelm the intracellular antioxidant system and cause damage. Thus, an excessive ozone dose or incompetence in manipulating this gas can be deleterious. On the other hand, very low ozone doses (below the threshold), are fully neutralised by the wealth of plasma antioxidants and can produce only a placebo effect.

The concept that ozonetherapy is endowed with an acute oxidative stress bothers the opponents of this approach because they consider it as a damage inflicted to the patients, possibly already under a chronic oxidative stress. THEY DO NOT BELIEVE THAT OZONETHERAPY INDUCES A MULTIVARIED THERAPEUTIC RESPONSE ALREADY WELL DOCUMENTED IN SOME DISEASES. Moreover THEY DO NOT DISTINGUISH *THE CHRONIC OXIDATIVE STRESS (COS)* DUE TO AN ENDOGENOUS AND UNCONTROLLED HYPEROXIDATION *WITH THE SMALL AND TRANSIENT OXIDATIVE STRESSES that we can precisely perform EX VIVO with the ozone dose.*

The THERAPEUTIC RESPONSE achieved after these repeated oxidative stresses can be envisaged as a PRECONDITIONING EFFECT eventually able to reequilibrate the redox system altered by pathogenetic stimuli.

Chapter 5

HOW IS OZONE ADMINISTERED?

Except for the inhalation route (prohibited by tracheo-bronchial-pulmonary toxicity), many parenteral and topical routes are used to administer ozone without toxic effects and minimal discomfort (Table 3).

Table 3. Routes of ozone administration

Parenteral	*Topical or locoregional*
Intravenous (IV)	Nasal†
Intra-arterial (IA)*	Tubal†
Intramuscular (IM)	Auricular
Subcutaneous (SC)	Oral†
Intraperitoneal (IPE)	Vaginal
Intrapleural (IPL)	Urethral and intrabladder
Intra-articular (IAT)	
A) Periarticular	Rectal
B) Myofascial.	
Intradisc (ID)	Cutaneous
Intraforaminal (IF)	Dental
Intralesional (IL)**	

* It is no longer used for limb ischaemia. Hepatic metastasis could be embolized via the hepatic artery.
**intratumoural or via an intra-abscess fistula
† to be performed during 30-60 sec apnoea

The gas mixture, composed of no less than 95% oxygen and less than 5% ozone has a slight positive pressure and can be collected either with a calibrated syringe (glass is ideal but impractical and has been substituted with disposable, polypropylene, silicone-coated syringes), or, if a continuous flow of gas is needed, by inserting a stainless steel connection to the exit valve of the ozone generator. **RUBBER TUBINGS CANNOT BE USED** because ozone disintegrates the rubber; but a silicon tubing is ideal.

Although ozone is a potent disinfectant, medical oxygen, O-rings and taps are not sterile and, except for rectal insufflation, ozone, as a dry gas, should be filtered for any other application to avoid any possible infection. We are currently using an antibacterial, ozone-resistant, hydrophobic filter with a porosity of 0.2 micron.

Owing to several deaths due to lung embolism, **THE DIRECT INTRAVENOUS AND INTRA-ARTERIAL ADMINISTRATION OF THE GAS MIXTURE**, containing variable amounts of ozone, **IS PROHIBITED** since 1984. Although the gas is injected very slowly, it favours the formation of a train of bubble gas, where ozone (more soluble than oxygen) dissolves and reacts with blood, while oxygen reaches the right ventricle and then the pulmonary circulation. It ought to be kept in mind that oxygen solubility at 37°C is only about 0.23 ml per 100 ml of plasmatic water and therefore venous plasma cannot dissolve oxygen quickly enough, leading to the formation of a gas embolus. It is a disaster that naturalist, practitioners and quacks without any medical qualification continue to perform this practice in Kenya, Canada, Jamaica, India, etc. and propagate this technique in other underdeveloped countries where there is no medical control.

The crazy idea of the direct IV gas administration is that ozone, once dissolved in the plasma, inactivates HIV present as free virus particles, just as ozone is used to purify water flowing in an aqueduct.

This idea is totally wrong on the basis of the following calculation: about 500 ml of gas mixture are injected in about four hours (2 ml per min) with a total ozone dose of 35 mg (70 mcg/ml X 500 ml = 35,000 mcg). A normal 70 kg man has about 5 L of blood which, at rest, circulate entirely in one minute. This means that a total blood volume of about 12 00 L circulates in 4 hours. The plasma volume is about 3 L but it continuously exchanges components (and antioxidants) with some 12 L of extravascular fluids. This implies that the total ozone dose of 35 mg will finally dissolve and react with an actual water volume near 15 L. Therefore, the final ozone concentration may range between 0.3-3.0 mcg/ml, which is an absolutely ineffective virucidal concentration also because most of the ozone is quenched by the antioxidants. Even more important is that THE BULK OF INFECTIVE VIRUSES AND PROVIRUSES IS INTRACELLULAR **and, ironically, remains well protected by the intracellular antioxidant system.** The other argument of charlatans is that the direct IV gas administration procures a good blood oxygenation. A simple calculation demonstrates that this is wrong too and, in any case, oxygen therapy can be performed efficaciously and safely by breathing humidified oxygen for a couple of hours at home or under pressure in a hyperbaric chamber according to a standard procedure.

Because of the small gas volume and the gas fragmentation into the limb capillary bed, IA administration does not involve a risk of embolism, but it has been proven to be of no advantage in comparison to the classical ozonated autohemotherapy or even the rectal insufflation of gas. Therefore it is no longer used, also because repeated arterial punctures should be avoided. On the other hand the practice of therapeutic (with alcohol and cytotoxic compounds) embolism for hepatic metastasis is now in current use and appears to be useful. It is then possible to consider the slow intra-arterial (via hepatic artery) administration of 20-40 ml of gas with an ozone concentration up to 70 mcg/ml. The risk of producing oxygen embolism is minimal because the gas will be dispersed into the sinusoidal and tumor capillaries, **possibly with direct ozone cytotoxicity on neoplastic cells and without side effects, as may occur with chemotherapeutic compounds.** So far I am testing this procedure in a patient with diffused hepatic metastasis without any adverse effects.

What is today's state of the art in terms of other parenteral routes?

In the past, particularly quacks have used both the IM and SC ROUTES for the treatment of chronic viral hepatitis and vasculopathies. Volumes of up to 300 ml of gas were subdivided in different sites using ozone concentrations of 10-15 mcg/ml. In Italy, during the last two decades, many ozonetherapists have earned their livings by performing SC administration of gas (ozone concentration: 2-4 mcg/ml) for the treatment of lipodistrophy, vulgarly known as cellulite. Unfortunately this treatment has become so popular to be carried out in beauty centres by inexpert beauticians, who dared to inject up to 500 ml of gas in a single session. **Even with the utmost care, these large volumes of gas can easily cause pulmonary embolism.** At least **two young women, treated for this trivial pathology, have been killed in Italy** in March 1998 and December 2002, so that the Ministry of Health has correctly prohibited the use of ozone therapy in all cosmetic and beauty centres and, incorrectly, the use of ozonated AHT in all public hospitals.

Obviously, even if this enormous gas volume is fractionated into various areas of SC tissue of the lower part of the body, it can converge into a deadly embolus. Both the direct IV gas administration and the just described practice have deeply compromised the future of ozonetherapy in Italy and it is disgraceful that so much importance has been given to ozone applications in cosmetic treatments. The future of ozonetherapy, if any, will not certainly come by treating cellulite!

To my knowledge the INTRAPERITONEAL and INTRAPLEURAL ROUTES have been used by Russian physicians by using first ozonated water to wash out purulent material and then insufflating into the cavities 100-300 ml of gas with ozone concentrations from 50 down to 5 mcg/ml depending on the gravity of the infection. Ozone dissolves quickly and, by reacting with exudates, can eliminate the infection. Furthermore, by stimulating vasodilation and cell proliferation, can lead to a rapid healing. This treatment does not damage the peritoneum, as we have observed after insufflating up to 300 ml of gas (!) into the rabbit peritoneal cavity testing an ozone concentration of 20 mcg/ml. Neither animal discomfort, nor any damage to the peritoneal lining was noted at autopsy after 24 and 48 hours. In my opinion, **these routes deserve to be evaluated in peritoneal and pleural carcinomatosis and it is regretful that oncologists disregard this approach:** daily insufflations of 200-300 ml of gas could be easily performed upgrading ozone concentrations from 5 up to 40 mcg/ml on the basis of the patient's reactivity. Ozone can be directly cytotoxic to neoplastic cells as chemotherapeutic compounds, with the advantages of avoiding chemoresistance, not causing toxic effects, bone marrow depression, mucositis and costing almost nothing. The risk of embolism is practically nil and the advantage of a local, albeit transitory, hyperoxia cannot be neglected. A permanent silicone cannula can easily be inserted permanently in the cavities for daily administration.

Another interesting possibility, applicable to patients frequently affected with chronic viral hepatitis, while undergoing peritoneal dialysis, is to insufflate oxygen-ozone every day intraperitoneally via the catheter already implanted. With a suitable ozone generator at home, autotherapy could be easily performed by the patient between peritoneal dialysis sessions, perhaps reducing the incidence of occasional peritonitis and the loss of permeability. Insufflated volumes could be of 200-300 ml starting with an ozone concentration of 5 mcg/ml and slowly upgrading it to 8-10 mcg/ml, thus allowing the induction of ozone tolerance. Obviously antiviral therapy can be complemented by using interferon or/and ozonated autohemotherapy and rectal insufflation of gas. Clinical experience has taught me that the combined use of several approaches is more proficient than a single one.

Intra-articular, intradisc and intra-foraminal administration will be discussed in the context of orthopaedic diseases.

TOPICAL APPLICATIONS can be performed with the gas mixture and ozonated water and oil.

Nasal, tubal and oral (gingival, mucosal and tonsillary) affections can be treated with suitable metal or silicone cathethers. If the gas is used, a volume of about 20 ml (ozone concentrations from 5 up to 20 mcg/ml) can be sufficient but the patient, after taking a deep breath, must remain in apnoea for 40-60 sec and then expire deeply. Aphthous ulcers in the oral cavity can be treated with intralesional mini-injections of ozone (concentration: 5-10 mcg/ml) followed by daily application of ozonated oil. At the moment it has become fashionable to use a silicon cup tightly enclosing and exposing a herpetic lesion to ozone for 20-30 seconds (Chapter 9, Section XVI). The application of ozonated oil in the lesion is far more practical and inexpensive.

Ozone treatment of chronic rectal and vaginal infections (bacterial, viral, fungal and protozoan), resistant to conventional treatments, respond very well to ozonetherapy. After inserting about 10-20 cm of a polyethylene catheter (lubricated with oil), we can start to wash the cavities with ozonated water for removing purulent secretion. Then, we can insufflate either 50 or 300 ml of gas (for vaginal or rectal cavities, respectively) for a few minutes and then, with a syringe, apply 5-20 ml of ozonated oil, being careful to lower the ozone concentration as the infection recedes. Vaginal and rectal pessaries and suppositories of ozonated oil can be applied before the night rest. A similar strategy can be used to treat either urethral or bladder infections keeping in mind to lower ozone concentrations between 3 and 10-15 mcg/ml, respectively.

Cutaneous applications regard all kinds of infections (from soreness to diabetic ulcers, insect and jellyfish stings and burns), accidental and war trauma. The gas can be used but the lesion must be sealed hermetically with various ozone-resistant devices to prevent ozone breathing. With a rigid Teflon cup, a slight vacuum can be achieved which, according to Werkmeister (1995), favours local vasodilation. In such a case, the ozonetherapist need an ozone generator equipped with a suction pump. If a dynamic exposure is not feasible, the static system can be achieved with a large polyethylene bag sealed with a wide adhesive tape, not too tight to cause venous stasis. All of these systems must contain water because ozone hardly acts in dry form and must dissolve in water. Recently, various types of chambers, with or without vacuum and with temperature control have been proposed but, all of these complex gadgets do not really improve the basic treatment of cleaning the wound, by applying twice daily a compress wet with ozonated water for about 20 min and then applying the ozonated oil throughout the night.

No one doubts the potent disinfectant activity of ozone (probably slightly inferior to iodine, which is actually too harsh) in regard to Gram negative and Gram positive bacteria, mycetes, viruses and protozoa. The simple and inexpensive treatment with ozonated water and oil is well tolerated, does not have noxious effects and the healing time is far shorter than with any conventional treatment. The latter advantage is due to a number of concomitant factors such as the disinfection, vasodilation, and oxygenation with normalisation of tissue acidosis and reabsorption of oedema.

The theoretical sequence of **wound healing** has been schematically represented to happen in three successive stages (Martin, 1997). The scheme presented in Figure 4 shows three phases: **Phase I indicates the inflammation stage,** normally lasting 2-3 days. The bacterial infection successive to a trauma, diabetes, local ischaemia and possibly antibiotic resistance, can become chronic unless we intervene with ozonetherapy. **Phase II corresponds to the intermediate stage** and normally lasts two weeks. The synthesis of extracellular matrix (fibronectin, collagen III/I, hyaluronic acid and chondroitin sulphate) is accompanied by an active proliferation of fibroblasts and keratinocytes. The use of ozonated oil, not only prevents a superinfection, but stimulates the initial tissue reconstruction. **The restitutio ad integrum, i.e. phase III, includes the final healing and scar tissue remodelling** and may take a long time in elderly and/or diabetic patients. In some cases, excessive release of Transforming Growth Factor (TGF beta 1) may stimulate excessive fibrogenesis with cheloid formation.

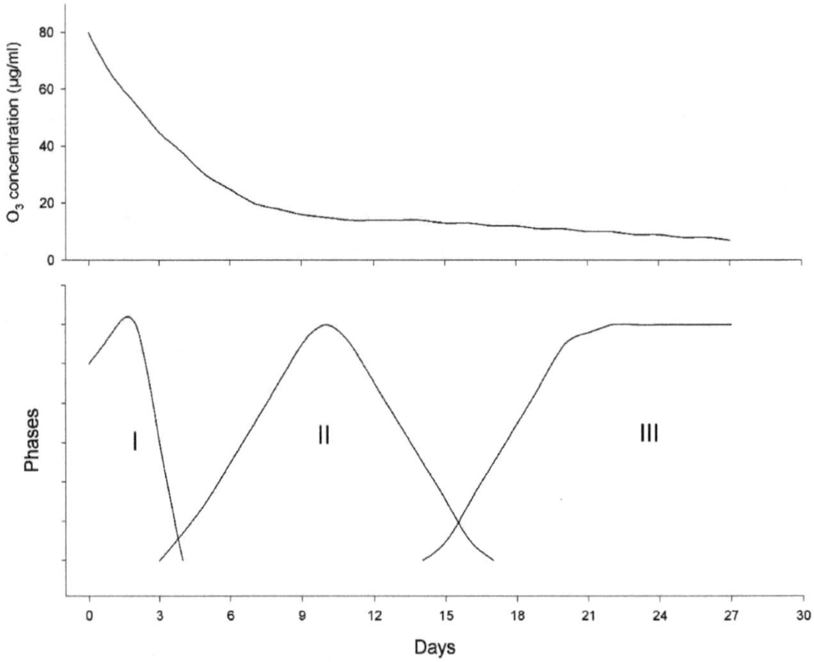

Figure 4. The three phases of wound healing. In the first (I) phase, inflammation prevails, with the presence of neutrophils, macrophages, mastocytes, platelets, bacteria and toxins. Application of ozone inhibits the infection and promotes the second (II) phase, lasting about two weeks. During this phase, the constant application of ozone at progressively lower concentrations not only prevents a superinfection but stimulates cell proliferation, the synthesis of fibronectin, collagen III/I, hyaluronic acid and chondroitin sulphate. Macrophages are still present but there is an active proliferation of fibroblasts and keratinocytes. The restitutio ad integrum, i.e. Complete reconstruction of the wound, occurs during the last (III) phase. However, excessive release of TGFβ1 may stimulate excessive fibrogenesis with cheloid formation. The above diagram shows the approximate ozone concentrations that must be progressively lowered to avoid inhibition of healing

In my experience, the successful and fairly rapid healing of a necrotic ulcer in arteriopathic, diabetic and immunosuppressed patients can be achieved by combining the parenteral treatment (ozonated autohemotherapy) with the appropriate application of progressively lower ozone concentration of ozonated water and oil. A tight control of glycemia and the combination of these therapies appear to act synergistically.

CONCLUSIONS: The reader will be amazed by the variety of routes of ozone administration. In spite of its intrinsic toxicity, if it is used at judicious doses, ozone is a versatile drug, which can be surprisingly useful in several diseases. Even local infections or neoplasms at the oral-nasal-pharyngeal site can be treated, provided the patient can remain in apnea for about 40 seconds or has been intubated. Owing to charlatans' false claim that direct IV gas administration could cure HIV infection,

this route, in spite of having caused many accidents and deaths, is still used in third-world countries. Even though death is due to oxygen embolism and not to ozone toxicity, it must be proscribed because there are other safe methods for ozone administration.

Regarding the SC administration, ozonetherapists treating lipodistrophies must be warned to inject small volumes (2-4 ml) of gas in multiple sites for a total of only 80-100 ml. This is already somewhat dangerous but it has never caused death as it has occurred after injecting 300-500 ml. Intraperitoneal and intrapleural administrations have been hardly used by practitioners but they are of great interest for treating life-threatening peritonitis, empyema, peritoneal and pleural carcinomatosis and chronic viral hepatitis in patients undergoing peritoneal dialysis.

Accidental and war trauma, burns and all sorts of acute and chronic cutaneous infections can be proficiently treated with ozonated water and oil that, in comparison to conventional creams, deserve great attention. The topical use of ozone in chronic and torpid ulcers and wounds present in diabetic patients and elderly people allows such a rapid improvement and healing to promote ozone to the rank of "WONDER" drug.

Chapter 6

THE ACTUAL SIX THERAPEUTIC MODALITIES

Parenteral administration of ozone may represent the key to solve some medical problems when orthodox medicine has failed to do so. To the old procedures: major and minor ozonated autohaemotherapy (AHT) and rectal insufflation, our work has permitted the addition of three new options, all of them will be critically examined in this chapter.

1. MAJOR OZONE AUTOHAEMOTHERAPY (AHT)

This term indicates the classical and unsurpassed procedure by which a volume of blood is drawn from an arm vein, exposed to oxygen-ozone for at least five min with gentle mixing and reinfused either IV (major AHT) or IM (minor AHT) into the donor. **"Major"** and **"minor"** are only meant to indicate a different volume of blood: 50-270 ml for the former and 5-10 ml for the latter. The original idea to expose blood ex vivo to a gas mixture was proposed by Wehrli and Steinbarth (1954), who published the method of irradiating blood with UV light in the presence of pure oxygen. This procedure, called HOT (Hamatogene oxidations therapie), is no longer used because it was uncertain with regard to the real concentration of ozone during irradiation of oxygen and was cumbersome and risky because the quartz ampulla had to be cleaned and sterilized after each treatment. Indeed a few cases of cross-infection with HCV, due to imperfect sterilization, were widely publicised to denigrate modern ozone therapy (Gabriel et al., 1996), that has nothing to do with HOT. In the 1960s, reliable medical generators became available and HANS WOLFF PROPOSED THAT BLOOD BE EXPOSED DIRECTLY TO OZONE, with the advantage of knowing its exact concentration. As early as 1974, he reported that he had used this method in many patients without any problem.

Unfortunately, **modifications were subsequently introduced that worsened the procedure;** for example, the use of only one tube to collect and reinfuse the blood, (involving the risk of a clot formation and the disadvantage of an imperfect mixing of blood with gas) and even worse, since 1991 in Italy, the substitution of neutral glass bottles, perfectly ozone-resistant, with plastic bags because they are cheaper and easier to stow away. These bags are made of about 55% polyvinyl chloride (PVC) mixed with a number of additives, among which about 43% of phthalates (Valeri et al., 1973; Lewis et al., 1977; Lawrence, 1978; Thomas et al., 1978; Callahan et al., 1982; Labow et al., 1986; Whysner et al., 1996). These compounds make the PVC elastic but a minimal amount of phthalates is released into blood. This little contamination is permissible and bags are commonly used for storage of blood but **the problem arises after the addition of ozone into the bags because ozone causes a huge release of plastic microparticles and phthalates into the blood with worrisome consequences for the patient after reinfusion.** After my notification to Health Authorities, the Italian Ministry of Health established very clearly that plastic bags should never be used for ozonetherapy. In spite of this precise regulation, **some Italian ozonetherapists, shamefully unconcerned about the patient's safety, continue to use them!** Fortunately, this does not seem to happen in other European countries but, once again, this reprehensible behaviour discredits this approach. Phthalates may not be toxic but plastic microparticles, taken up by the reticulo-endothelial system in the spleen liver and bone marrow, may represent a cancerogenic stimulus.

After several years of laboratory experimentation and clinical work, we have now optimised an autohemotherapeutic method that is fairly simple, ozone-resistant, absolutely atoxic and flexible in the sense that one can use a blood volume from 100 to 270 ml (depending on the patient), a suitable volume of sodium citrate (3.8 %) solution and the necessary gas volume without increasing the atmospheric pressure in the glass bottle. Our device consists of **1)** a neutral 500 ml glass bottle (sterile and under vacuum) where we inject, as a first thing, the chosen anticoagulant, **2)** a new atoxic tubing with an Y form where one tubing (Segment A, when connected with the Butterfly G19) collects blood and the other (Segment B) is used for insufflating sterile-filtered O_2-O_3 via an antibacterial (0.2 micron), hydrophobic ozone-resistant filter. As one can see in the Figure 5, both Segment A and B are connected to Segment C, which carries firstly blood and then gas inside the glass bottle, **3)** a standard tubing (Blood filter) that is used, firstly for infusing saline and, secondly, for returning the ozonated blood to the donor. In this way, we perform only one venous puncture because, while we carry out the ozonation of blood, the patient receives a slow infusion of saline. It is important that the exposure of blood to the gas mixture lasts at least 5 min because mixing of blood MUST be gentle to avoid foaming. Because blood is very viscous, it takes 5-10 min to achieve a complete and homogenous equilibrium. It can be noted that the pO_2 slowly reaches supraphysiological values (up to 400 mmHg) and then it remains constant: on the other hand, ozone progressively dissolves in the water of the plasma but then reacts instantaneously with biomolecules so that the entire ozone dose is practically exhausted within 10 min. **The visible clamps in Segments A, B and C are open or shut throughout the procedure for allowing the passage of blood and gas without losing the vacuum.**

The ozonetherapist must follow this procedure for avoiding either negative effects on the patients, or being found guilty of medical malpractice. I can assure the ozonetherapist that, after a preliminary experience, this procedure apparently complicate is indeed easy, rapid and clean.

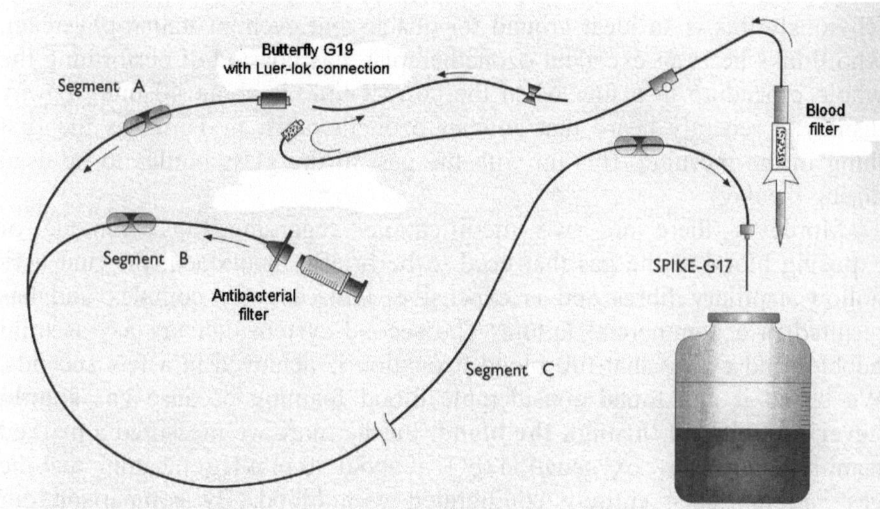

Figure 5. A schematic drawing of the components necessary to perform the ozonated autohaemotherapy with a glass bottle.

A brief digression is necessary in regards to blood anticoagulants. Is sodium citrate the best anticoagulant? It is, provided is not in excess (Chapter 7) to avoid a transitory hypocalcemia and it is safe in practically all patients, including those already under either anticoagulants (Warfarin, heparin, hirudin), or antiplatelet drugs (aspirin, dipyridamole, ticlopidine, clopidogrel), or thrombolytic agents (streptokinase, tissue plasminogen activator), or patients with hepatic diseases and a low prothrombin level. In these cases, use of heparin may aggravate the dyscoagulation and cause severe haemorrhages. I will remind that heparin can induce thrombocytopenia (Warkentin, 2003) and platelet aggregation (Bocci et al., 1999a) using high ozone concentrations (near 80 mcg/ml per ml of blood). Nevertheless heparin is regularly used during EBOO and dialysis and, bearing in mind the above indicated restrictions, may be useful in vascular diseases and cancer because of the increased release of a number of growth factors from platelets (Valacchi and Bocci, 1999) and cytokines from leukocytes (Bocci et al., 1993a, b). Thus, only after a careful analysis of the patient, the ozonetherapist can select the most idoneous anticoagulant.

ONE BIG PROBLEM THAT WEAKENS THE VALIDITY OF OZONETHERAPY IS THAT OUR METHOD IS NOT USED BY ALL OZONETHERAPISTS. How can we compare anecdotal results (already questionable), if ozonetherapists disagree about the blood and gas volumes, ozone concentrations and exposure times? What is most disheartening about this chaotic situation is that behind it there are commercial interests (plastic or glass, small or large bottles, lack of appropriate transfusion tubing with a filter, etc.), mental reservations, lack of basic knowledge and plain stupidity. Obviously this is an ideal ground for quacks but even an Italian physician, who thinks he is an excellent ozonetherapist, has boasted of performing the whole procedure in 6 min when the correct time is about 40 min! To my dismay, I recently heard that another ozonetherapist in Turin, as the first thing in the morning, fills up with the gas all the glass bottles to be used during the day!

Moreover there are two modifications regarding the technique of exposing blood to the gas that need to be briefly mentioned: the first uses hollow capillary fibres and is expensive, unnecessarily complex and has resulted in a commercial failure. The second system delivers gas as mini bubbles and claims that full blood ozonation is achieved in a few seconds. We tested it and found considerable blood foaming because **gas should never be bubbled through the blood**. Furthermore we measured a marked hemolysis and a low oxygenation (pO_2 at about 90 mm Hg) meaning that the gas had not been entirely equilibrated with blood. By comparison our method requires at least 5 min of gentle mixing (to avoid foaming), but allows complete ozonation and oxygenation as it is well demonstrated by a very high pO_2. Haemolysis remains negligible.

Another critical issue that remains to be scientifically settled is the volume of blood to be collected for each treatment. Needless to say the volume of blood should not be imposed by any commercial purpose or by a trivial timing aspect. The volume of blood must be flexible and must be in relation to the patient's body weight, sex, stage and type of disease. To avoid any risk of lipothymia, no more than 270 ml blood should be withdrawn and a 500 ml glass bottle appears suitable in all cases. In Germany, some practitioners believe that 50 ml, or at most 100 ml, is optimal. There is neither experimental nor clinical support for this contention and this belief disagrees with the classical biochemical and pharmacological concepts expressed in the previous chapter. *If we accept the evidence that ozone generates crucial messengers, such as ROS, LOPs, metabolic intermediates and autacoids that undergo dilution, degradation and excretion but that, after binding to cell receptors, can express pharmacological effects, we have to consider that a minimal stimulation or a small blood volume, may correspond only to either a placebo or to a homeopathic effect. Our contention is supported by the experimental finding that, in critical stages of hind limb ischemia, a dramatic improvement was observed immediately after the first treatment performed with large volumes (810-4800 ml of blood).*

Our standard approach has been to perform 2 or 3 treatments weekly, usually using 225 ml of blood each time, for 13-15 sessions. This schedule is practical, appears effective in most patients but can be modified to satisfy individual requirements.

Has the classical AHT any other disadvantage? The limitation of blood volume can be easily overcome by performing successively up to three AHTs, within two hours, on the whole ozonating about 750 ml of blood without any side effects, as I have tested on myself and in several patients.

Unless the ozonetherapist owns a reliable portable generator, domiciliary treatment, that could be very useful in some emergencies, cannot be performed. Nevertheless, superficiality and malpractice are endless and one German ozonetherapist boasted of performing several AHTs every morning by first loading with ozone small glass bottles at his clinics and then going around town to the patients' homes to give treatments, disregarding the fact that ozone concentration halves every 30-40 min.

A correct reinfusion of 250 ml blood plus Na citrate takes about 20 min and then we must carefully check the haemostasis and avoid haematic extravasation which may compromise the continuation of the therapy. Great care must be exercised to maintain the venous access in the best condition, particularly in women. Risk of infections (HIV, HCV, etc.) among patients and ozonetherapist must be prevented and we fully agree with Webster et al. (2000) that some mistakes, e.g. repeatedly using a contaminated needle, or a syringe, or a solution, are inadmissible.

If, SEVERAL AHTs ARE PERFORMED SIMULTANEOUSLY, ALL GLASS BOTTLES MUST HAVE THE PATIENT'S NAME to prevent mistakes during reinfusion, with possible dramatic consequences. In any case, we write the name even for a single AHT.

One difficult question to answer is if we can perform an ozonated well-characterised, **allogeneic** blood transfusion in cachectic, anemic or in AIDS patients. While, after having performed 8000 autologous transfusions, there has not been a single case of transfusion-related acute lung injury or other noxious effects, **we need to be very careful regarding a blood allogeneic transfusion deleterious effect**. If it is absolutely necessary, blood must be subjected to a leukocytes and platelets depletion step (Williamson, 2000) and then, after ozonation, must be infused very slowly. Provided it is done with great caution and very SLOW infusion, the ozonated allogeneic blood transfusion may help critical patients.

Finally **AHT has a few potential drawbacks**: the first is that AHT is not a simple little pill to swallow at home because the patient must go to a public or private clinic to receive the treatment. As a consequence, AHT can be advised only when absolutely indispensable and not replaceable by an equally effective conventional medication. However I have learnt that patients, once realize the clear-cut efficacy of AHT, do not hesitate to continue the maintenance treatments for years. Obviously the ozonetherapist must have a perfect competence for performing AHT in the smoothest possible way. Indeed some ozonetherapists do not feel skilled enough and prefer to perform other more rewarding tasks. The second problem is that medical personnel working in infectious disease wards are somewhat reluctant to deal continuously with infected blood and needles and the third is the occasional lack of venous access. These are not trivial problems: one can be frequently solved by the use of an idoneous blood substitute, that can be slowly injected into small veins; in the case of a difficult venous access, we can propose three options: a) cannulation of a central vein, keeping in mind some risks (Renaud and Brun-Buisson, 2001; Castagnola et al., 2003), b) quasi-total body exposure to oxygen-ozone in a cabin, c) rectal insufflation of gas.

2. MINOR OZONE AUTOHAEMOTHERAPY

In the 1950, when I was a medical student, we used to do IM injections of either autologous freshly drawn blood or sterile milk as unspecific immunomodulators. This practice is then very old and continues to be used also without ozone (Olwin et al., 1997). Wolff may have had the idea of ozonating blood in the hope of activating its components.

The technical procedure is empirical and simple: firstly, I collect the blood (5 ml) in a 10 ml syringe, and secondly, via a two-way stopcock, I add an equal volume of filtered oxygen-ozone at ozone concentrations between 40-80 mcg/ml depending upon the scope of the treatment and the disease. One can, more simply, first collect the 5 ml of gas and then withdraw, less precisely, about 5 ml of blood. In both cases, the blood, vigorously mixed with the gas, develops abundant foaming and certainly in this case the whole ozone dose reacts in less than one minute. After disinfecting the buttock skin and checking to not have penetrated a vessel, I inject, either in the subcutis or in the muscle, blood and foam in one site, usually without causing pain. We can do multiple injections or repeat them 2-3 times weekly. We do not know whether the IM or SC administration in multiple sites is more effective. Under an ozonetherapist guidance, a nurse's help and the availability of an ozone generator, the patient can easily do her/his own therapy at home.

What is the rationale of this sort of unspecific proteintherapy coupled to ozone remains conjectural and a scientific investigation will be useful. At the moment I can only speculate that blood, without anticoagulant, will infiltrate into the muscle tissue or the subcutis and will undergo coagulation due to platelet and prothrombin activation. If we delay IM injection, this can happen already in the syringe!

Several processes, such as fibrinolysis, serum reabsorption via lymphatic vessels and a mild sterile inflammatory reaction, are likely to take place as occasionally suggested by a slight swelling at the injection site reported by some patients during the next few days. Chemotactic compounds released at the site may stimulate the local infiltration of monocytes and neutrophils, which take up haemolysed erythrocytes and denatured proteins. Activated monocytes and lymphocytes may release interferons and interleukins either in loco or along the lymphatic system, upregulating the physiological cytokine response (Bocci, 1981c; 1988). Thus it would be quite interesting to evaluate some immunological parameters and ascertain if there is a simultaneous induction of HO-1 and some other heat shock proteins (Tamura et al., 1997) that may enhance immune reactivity and explain the beneficial effects.

The minor AHT is easy to perform, atoxic, inexpensive and, if we could perform a controlled clinical trial, it could become a very useful tool in some affection. So far we have only anecdotal data in patients with herpes I and II, acute herpes zoster and post-herpetic neuralgia (Konrad, 2001). A similar approach has been publicized by Cooke et al., (1997), who claim great advantages in Raynaud's disease, by using a particular formulation in which blood is treated with ozone, heat and UV light; a similar methodology was proposed by Garber et al., (1991) and uselessly tested in AIDS patients. I feel that we should test seriously only ozone before complicating the problem with the seemingly superfluous addition of heat and UV irradiation. During the last year, in almost all patients, I started to perform both major and minor AHT at the same time and I have noted a marked improvement of the therapeutic response suggesting a synergistic effect and ABSOLUTELY NO ADVERSE EFFECTS.

The problem of new vaccines is becoming urgent and I would like to propose the use of ozone as an agent able to eliminate the infectivity, while enhancing the immunogenicity of a pathogen.

Once we have demonstrated the ozone capacity to inactivate a virus, the idea of a possible autovaccination, by heavily ozonating small volumes (3-5 ml) of infected plasma with ozone at high concentration (100 or more mcg/ml per ml of plasma) does not seem farfetched. The oxidation of viral components may represent an effective immune stimulant in several chronic viral diseases, from herpes to cytomegalovirus, HIV, HCV, just to cite a few because there are many pathogenic agents. Infected blood may even be better because it may well contain intracellular pathogens as well and displays an adjuvant activity. The autovaccine can be either injected via IM or SC or intra-epidermal injections for facilitating the uptake by Langerhans cells. For some pathogens we could also use the oral route. I HAVE

APPLIED THE SAME REASONING FOR CANCER PATIENTS and I use the minor AHT as a sort of autovaccine.

The minor AHT has no record of side effects. This corresponds very well with my experience. However I cannot omit to report the excellent paper by Webster et al. (2000), who described the careless and unforgivable performance of some incompetent operators in a naturopathic clinic in London (!!). They were treating patients by using the old minor AHT, WITHOUT OZONE, and they were diluting blood (WHY?, WAS IT NECESSARY?) WITH SALINE COLLECTED ALWAYS FROM A CONTAMINATED BOTTLE. In this way they infected more than 70 patients with HCV!!!

It is most unfortunate that incompetent mass media, when they heard about this misdeed, which occurred with autohaemotherapy, attributed the fault to ozonetherapy when clearly **ozone was NOT GUILTY and actually, if present, might have blocked the infection!**

3. THE BIOOXIDATIVE THERAPY WITH HYDROGEN PEROXIDE DISSOLVED IN THE ISOTONIC GLUCOSE SOLUTION. THE CONTINUOUS SEARCH OF AN EFFICACIOUS BLOOD'S SUBSTITUTE.

Particularly in infectious diseases wards, I often found that both physicians and nurses are reluctant to perform AHT because they are afraid of accidentally pricking themselves with an infected needle. That is the main reason why I have often been asked to find an efficacious solution to be directly infused in the place of blood.

A decade ago I spent considerable time searching a suitable solution and eventually I found that the simple saline (NaCl: 0.9%) could trap more ozone than any other electrolyte solution. At first it appeared to be a possible solution because it was easily administered with a very thin needle (G27) in patients with a poor venous access. However, after ozonating saline with an ozone concentration of 80 mcg/ml, I tested it and, in spite of a considerable blood dilution during the slow infusion, the next day I felt a painful irritation along the venous path up to the axilla. I realized that the ozonated saline was somewhat caustic and could cause a chemical phlebitis. Then I went to my lab and I measured the formation of hydrogen peroxide, that was a good thing, but also of hypochlorous acid (HOCl), that was a bad thing. Traces of Fe^{++} could also catalyze the formation of hydroxyl radicals. Years later, I was surprised to learn that ozonated saline is widely used in Russia and also in Italy by a few charlatans. However, at least in Russia, it does not seem to procure significant damage because the ozonation is performed at extremely low level (about 2 mcg/ml), so that it works as a placebo.

Now I must strongly recommend to avoid the use of ozonated saline owing to inherent toxicity or/and minimal activity.

There is now a far better solution: Pryor et al. (1995) and our study (Bocci et al., 1998 a) have made sure that hydrogen peroxide is one of the most important ROS (generated by ozone), that can physiologically activates several targets although, in excessive amounts, can also be a damaging oxidant. Interestingly, Dr. I.N. Love (1888!) working in St. Louis published a note entitled "Hydrogen peroxide as a remedial agent" after he had obtained beneficial effects after topically using a diluted solution of hydrogen peroxide. We must admire Love's insight into a problem that, only not long ago, has clarified that our white cells can win their daily battle against pathogens only if they can deliver and kill bacteria with ROS (anion superoxide, hydrogen peroxide, single oxygen, etc,).

Today, for the disinfection of wounds, everyone uses the 3.6% solution of hydrogen peroxide, which, among disinfectants, is one of the most efficatious. Subsequently, Dr. C.H. Farr (1993) promoted the use of IV administration of a dilute solution of hydrogen peroxide in several illnesses, very similar to those treated with ozonetherapy. Almost needless to say, hydrogen peroxide must be considerably diluted before its contact with blood in order to avoid dangerous oxygen embolism and endothelial damage. Dr. Farr is acknowledged as one of the founders of bio-oxidative therapy, included among the complementary medical approaches by the National Institutes of Health.

The precise formulation of the solution (that I briefly call the GLUCO-PEROXIDE solution) for IV administration, first elaborated by Dr. Farr, consists of a few steps that I have simplified and improved:

1) A 15% solution is prepared by diluting 30% reagent grade H_2O_2 with an equal volume of apyrogenic, sterile bidistilled water. I never store this solution and I use it immediately.

2) In order to prepare the final solution when needed, it is necessary to dilute 0.5 ml of the sterile 15% H_2O_2 solution with 250 ml of 5% sterile glucose solution. I would like to recommend: **a**) to withdraw the 0.5 ml without the use of a metal needle because iron (from the needle) will contaminate the solution and enhance formation of hydroxyl radicals; **b**) there is no need to filter the 15% solution, that is directly injected (0.5 ml), via a sterile plastic spiked cannula, into the isotonic 5% glucose solution flask; **c**) to never dilute hydrogen peroxide into saline, to avoid the risk of HOCl formation; **d**) to administer intravenously as a common drip infusion. In the rare case of an intrahepatic arterial catheter, it must be slowly administered with a syringe to counteract the arterial pressure; **e**) to use it during the day. I would like to point out that **the hydrogen peroxide titre remains quite stable, even at 20° C, for at least three days.** (Bocci et al,

manuscript in preparation). My procedure eliminates one step and avoids iron contamination. The final hydrogen peroxide concentration is equivalent to 0.03%, is isotonic and suitable for direct slow (15-20 min) IV infusion. It may be worthwhile reminding physicians, who like to make strange solutions, to **avoid mixing the 0.03% H_2O_2 solution with antioxidants (vitamin C, GSH), amino acids, minerals**, etc., to avoid negative interference. Depending on the stage and type of disease, treatments can be carried out daily, every other day or twice weekly.

I have been told that, for serious illnesses, Dr. Farr has slowly infused a five-fold greater concentration (0.15%, i.e. 2.5 ml of the 15% H_2O_2 solution diluted into 250 ml of 5% glucose solution), with "excellent results". In order to avoid toxicity and to allow adaptation to COS, I would suggest a gradual increase of the total volume (from 125 to 250 ml) and an increase of the concentration to no more than 0.15%. I have tried on myself several 250 ml preparations at 0.15 % infused in 30-40 min, without any adverse effects, in contrast to ozonated saline. To be very cautious, one can prolong the infusion for one hour. Dr. Farr has performed IV infusions of a 0.03% solution in many patients affected by several diseases. The IV administration of gluco-peroxide solution in arterial and heart ischaemia and in cancer has been reported by Urschel Jr. (1967). Interesting studies on the antitumoural effects of H_2O_2 have been reported by Sasaki et al. (1967), Nathan and Cohn (1981) and Symons et al. (2001). Los et al., (1995) and several other Authors have shown that hydrogen peroxide is a potent activator of lymphocyte functions *in vitro*.

While this approach has been widely used in the USA, Canada and Mexico, it has not been used in Russia and Germany. To my knowledge, I am the only one to use it in Italy and, after thousands of infusions, I am sure of its effectiveness and atoxicity. **I have found that the"gluco-peroxide" solution, in the range 0.03-0.12% is practically as effective as the classical AHT in ARMD women with small venous accesses incompatible with HAT. I have also already carried out almost a thousand infusion of the solution (range: 0.03-0.15% or 8.8-44.0 mM) in cancer patients without any adverse effects**

Work in progress aims to compare laboratory and clinical results by testing the classical AHT and the "gluco-peroxide"solution in chronic limb ischaemia, age-related macular degeneration and chronic C hepatitis. Such a study appears difficult because to achieve clear statistical significance, it may be necessary to evaluate thousands of patients. Indeed the crucial question is: can the "gluco-peroxide"solution satisfactorily substitute AHT or other approaches using ozone? I must add that in seriously ill patients, I am already using routinely the gluco-peroxide solution to supplement the AHT, particularly when patients can only come twice weekly for the treatment.

The proposal of this solution is not senseless, particularly since we know that this compound is one of the early ozone messengers. However, the question whether it is as effective as AHT remains open because late products, like LOPs, may be scarcely generated in vivo owing to rapid reduction of H_2O_2. On the other hand, during blood ozonation, hydrogen peroxide is generated *ex vivo,* in the bottle but is rapidly reduced, whereas the direct infusion of the "gluco-peroxide" solution implies an immediate interaction with the blood pool. It must be added that the infusion, although it does not present the risk of the direct gas administration, must be performed slowly to avoid any risk of oxygen embolism. **A disadvantage to bear in mind is that the "gluco-peroxide" solution cannot be used in diabetics.** Nevertheless, I believe that this approach deserves to be pursued because it has potential advantages:

- **Ozone generators, with all their problems and cost, would become superfluous.** Electric energy is unnecessary.
- **The cost of the "gluco-peroxide" solution is almost negligible. Preparation of the solution is simple, well standardized and reliable, and the solution is far more stable than ozone. Moreover, it can be transported and can be administered everywhere.**
- **One needs reagent grade H_2O_2 (30%), sterile bidistilled water, a 5% glucose solution, and a few plastic disposable tools. The advantage is that the therapy can be performed in poor countries in the most remote corners of the Earth, particularly to alleviate some diseases. I will do the best I can to promote its application by the WHO, which has overlooked this possibility.**

In 1993, Dr. Farr reported that injection of a 0.03% H_2O_2 solution into joints and muscles relieved pain quickly. This paradoxical result is similar to the one I will discuss with ozone injection (Orthopaedic diseases, Section XIII). Three years ago, the Ethical Committee of Siena University approved my protocol for the IM administration of a 0.12-0.24 % solution (35.2-70.4 mM H_2O_2). Preliminary results have shown that these concentrations are suitable (depending on the patient's reactivity) for IM injection (5 ml per site) into trigger points present in paravertebral muscles, as a substitute for gas injection (O_3 at 20 mcg/ml), in patients with backache. The effect of so-called "chemical acupuncture" with O_2-O_3 is attributed to the local release of hydrogen peroxide acting on nociceptors and eliciting the analgesic response. This study has the scope to clarify the role of H_2O_2 as an "antinociceptive" drug.

Besides the gluco-peroxide solution, there are two other potential possibilities of blood substitutes:

a) Fresh frozen plasma (FFP);

b) A lipid emulsion made of medium- and long-chain fatty acids and phospholipids, currently used for total parenteral nutrition.

After blood, FFP seems a reasonable solution because it contains all the basic reactants preferred by the solubilized ozone. However, as blood cells are absent, the formed H_2O_2 will not diffuse into them and will not activate metabolic pathways ex vivo. As it is described in Chapter 4, H_2O_2 will be reduced in 1- 2 minutes after ozonation and the infused plasma will contain LOPs and will have a reduced antioxidant capacity. It is unlikely that it will be as effective as ozonated blood. Yet perhaps, if alternated with the gluco-peroxide solution, it may represent a good compromise. However, while the gluco-peroxide solution is sterile, FFP can still transmit infections, in spite of a highly reduced risk. To enhance its validity, FFP should be obtained after strict screening and appropriate controls only from leukocytes-depleted blood. Moreover, it should be subjected to one of the currently used and expensive methods to ensure viral inactivation, such as solvent-detergent or methylene blue treatment, unless the ozonation process has an equivalent potency, an interesting possibility not yet fully ascertained. If this can be proved, it would be useful and reduce the cost. Even so, there remains the problem of the availability of FFP, as it is widely employed to obtain precious plasma components.

The final option is a lipid emulsion. There are several already employed for parenteral nutrition. Indeed we have spent some time evaluating one, which I will simply indicate as LE, rich in phospholipids, partly unsaturated medium and long-chain triglycerides, glycerol and water. It is isotonic, practically ion-free and obviously sterile. When exposed to O_2-O_3, ozone dissolves as usual, reacts immediately with PUFAs and forms ROS and LOPs, which by mixing with blood during reinfusion may at least partly activate blood cells. Thus, it shows advantages and is a promising solution. After obtaining permission from the Ethical Committee and the Ministry of Health in April 1998, we conducted a preclinical study to assess the toxicity in rabbits (manuscript in preparation). Initially, we investigated which ozone dose (20, 40, 60, 80 mcg/ml) would be most suitable for the ozonation of LE. More recently, we examined the effect of 5, 11 and 21 treatments (within 56 days) (slow infusion via the ear marginal vein) of LE exposed to oxygen-ozone or only oxygen. Results showed that a medium ozonation (40 mcg/ml of LE) significantly enhanced (in comparison to control) the animal's body weight (mean increase of 550 g). Haematological parameters, TBARS, PTG and TAS plasma levels did not show abnormal variations. Histological examinations performed at the end of the experimental period on many organs from each rabbit group failed to show any pathological variations.

We are now characterizing the chemical change in composition of LE after ozonation. This line of research is interesting and we will take a step forward if we can use ozonated LE in patients, thus avoiding the problem of blood handling. Moreover, I envisage the possibility of dissolving a precise volume of filtered 15% H_2O_2 solution directly in the LE, thus excluding the use of ozone and extending its therapeutic use to poor countries. This study is in progress in our laboratory because we feel it is important to develop a useful possibility for patients who are not treated today. **I would like to remind that only about 20% of the world population receives proper medical attention and we ought to make an effort to help the remaining majority.**

4. RECTAL INSUFFLATION OF OXYGEN-OZONE (RI)

Payr and Aubourg, in 1936, were the first to suggest the insufflation of this gas mixture into the colon-rectum and today this approach has been adopted world-wide because it is easy to perform, is inexpensive, practically risk-free, often beneficial and because most people, recognising the advantage, do not object to rectal medication. Even in several states of the USA, where ozonetherapy, owing to misuse by quacks, has been prohibited, many HIV patients used to do their own auto-insufflation using an often imprecise portable ozonator. In California, Carpendale et al (1993) were allowed to perform a study in AIDS patients with profuse diarrhoea due to opportunistic *Cryptosporidium* infection; as it was expected, they reported only a temporary improvement in some of the patients.

The main field of application is represented by rhagases, anal and rectal abscesses with fistulae, proctitis, bacterial and ulcerative cholitis, Crohn's disease and chronic B and C viral hepatitis. Even ischaemic diseases and dementias have been treated with RI, which was postulated to have a systemic effect. Indeed a surprisingly rapid systemic effect seems supported by recent studies in the rat (Leon et al., 1998; Barber et al., 1999; Peralta et al., 1999, 2000; Borrego et al., 2004; Gonzalez et al., 2004), in which it was shown that RI for two weeks induced adaptation to chronic oxidative stress.

In spite of the fact that hundreds of thousands of treatments are performed every year, it was unclear whether and how these gases could affect some physiological, biochemical and immunological parameters. Although mainstream medicine, as usual, scorns this simple treatment, I feel it is important to address the following questions:

1) Are oxygen and ozone absorbed by the intestinal mucosa?

2) Does RI have only local effects or systemic ones as well?

3) Bearing in mind the toxic effects of ozone on the respiratory tract, we must ascertain if ozone negatively affects the intestinal mucosa.

Knoch et al., (1987) examined the PvO_2 modifications after rectal insufflation in the rabbit. They found increased oxygen content of 230, 121 and 127% in a mesocolonic vein, portal vein and liver parenchyma, respectively, 8-20 min after rectal insufflation of 150 ml of gas. The values returned to baseline after 50 min. This result is not new because we know that several gases, such as carbon dioxide, methane, hydrogen, oxygen, nitrogen and hydrogen sulphide, either ingested or produced by the bacterial flora are partly absorbed or excreted or even exhaled with expired air. Obviously we are interested in the fate of ozone introduced in the gut lumen. In Chapter 4, it has been clarified that ozone, firstly dissolves in water but, unlike oxygen that freely diffuses into other compartments, reacts immediately with any biomolecule, particularly PUFA producing ROS and LOPs. Thus we can determine the fate of ozone by measuring LOPs in the intestinal-portal and peripheral circulation. While the respiratory mucosa is overlaid by a very thin and hardly protective film of fluids, the gut mucosa is abundantly covered by the glycocalyx and a thick coating of water containing mucoproteins and other secretion products with potent antioxidant capacity (Halliwell et al., 2000). Besides this gel-mucous layer, a variable faecal content is present and can quench the oxidant activity of ozone. It becomes clear that **this unpredictable parameter represents the weak point of RI because we cannot ever be sure of the ozone dosage really available.** However we felt worth while investigating in the rabbit whether ozone has, through the LOPs either a local, or/and a systemic activity. Results have been enlightening and have reported in extenso by Bocci et al., 2000 and Bocci, 2002.

It suffices here to sum up the following data:

1) After rectal insufflation, **we measured increased oxygen content both in the portal vein (20-35 min later) and in the jugular vein (35-40 min later).** There were no significant variations of $PvCO_2$ and pH.

2) Concomitantly, there was **a constant increase of LOPs' values up to 60 min after gas insufflation,** when they started to decline. **Values were markedly higher in the portal than in the jugular blood due to dilution in the general circulation.** Conversely, values obtained by measuring oxidation of protein thiol groups showed an opposite trend, i.e. reached a minimum after 90 min. Both parameters returned to baseline 24 hours thereafter.

Therefore, it appears that RI can exert a local and a rapid systemic effect due to absorption of ROS and LOPs generated by the interaction of ozone with biomolecules present in the luminal content. ***The quantity of absorbed ROS and LOPs are however unpredictable due to the variable content of fecal material.***

Figure 6 attempts to suggest that ozone dissolves rapidly in the luminal water, but, in comparison to oxygen, it is not absorbed because it partly reacts with mucoproteins lining the mucosa, partly may react with fecal material and partly can be reduced by antioxidants. LOPs, like oxygen, pass through the muscularis mucosa (MM) and enter the circulation via lymphatic and venous capillaries. **This conclusion** is relevant and **would support the contention that the beneficial effect of RI in chronic limb ischaemia may**

be similar or equivalent to major AHT. If this result can be confirmed in a controlled, randomised clinical trial, it will be helpful for patients because they will be able to do automedication and avoid repeated venous punctures. Moreover it does explain why prolonged (up to 13 weeks) RI in aged subjects cause an increase of both ATP and 2,3-DPG in erythrocytes (Viebahn, 1999a,b). These results are the more surprising because, in comparison to the precise volumes and ozone concentrations in major AHT, we know very well how imprecise the application of ozone can be and particularly the volume of gas retained and effectively acting in the gut lumen.

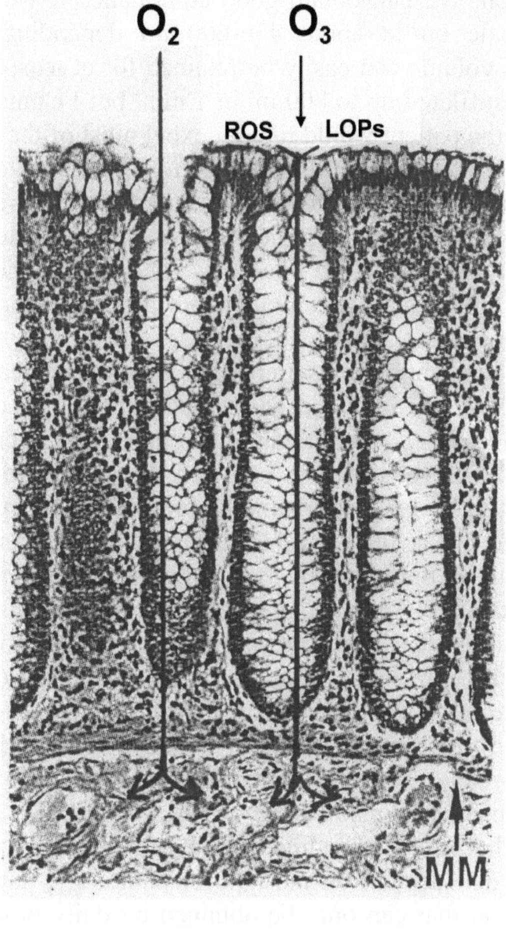

Figure 6. A schematic view of the transfer of the O_2-O_3 gas mixture from the colonic lumen into the submucosa. Both gases dissolve in the luminal mucous layer, but ozone reacts immediately and decomposes into a number of ROS and LOPs. These are absorbed with water via venous and lymphatic capillaries in the submucosa below the muscularis mucosae (MM).

This leads to the discussion of some technical details in terms of gas volume, ozone concentration and schedule of administration. **RI should be done after defecation or after an enema, when the rectal ampulla is empty.** The patient must lie on one side and try to relax; often he/she prefers to personally insert the disposable, oil-lubricated polyethylene (rubber must never be used) catheter (30-40 cm long). The insertion is easy and it should not stimulate peristalsis. To this end, the gas has to be introduced slowly and in steps of 50-100 ml every 1-2 min. If it is done quickly, the gas will be expelled at once. The gas can be introduced via: **a**) a manual two-way silicone pump connected to the gas just collected in a polyethylene bag, or with **b**) a 50 ml silicone-coated syringe, clamping the catheter each time after insufflation. We can obtain good compliance if we start with 150 ml and slowly scale up to about 450-600 ml depending on the patient's tolerance. This volume can easily be retained for at least 20-30 min. Knock et al. (1987) insufflated up to 800 ml in 1 min, but I cannot confirm this and it is likely that the patient would rapidly expel most of the gas. Carpendale et al. insufflated from 700 to 1300 ml of gas (up to 30 mg ozone daily) in AIDS patients, hoping the gas would diffuse into the whole colon. This was a desperate, almost useless enterprise because Cryptosporidium contaminates the whole gastro-intestinal and bile ducts. The patient should be left to rest for at least 15 min after RI to avoid rapid gas expulsion and to allow the reaction of ozone with the luminal contents.

The ozone concentration is important to induce local and generalized effects but there is a general consensus that it should not exceed 40 mcg/ml. In my experience, this concentration often elicits painful cramps, particularly in patients with ulcerous cholitis or when the application is done after an enema, suggesting a dangerous stimulation of the local gut reflexes. If the overlaying mucus has been washed away, this high concentration might cause direct damage to the enterocytes and we should not forget that ozone is potentially mutagenic. Thus I suggest beginning treatments with 3-5 mcg/ml and slowly scale up to 30 mcg/ml if the patient tolerates it well. It has been written (D'Ambrosio, 2002a) that, in the case of haemorrhagic ulcerative cholitis, an ozone concentration of 70-80 mcg/ml should be used for haemostatic purposes, but this could induce cytotoxic damage and is not advisable. Moreover, on the basis of the concept of inducing ozone tolerance, it appears reasonable to reach the concentration of 30 mcg/ml in 2-3 weeks. Whether it is worthwhile reaching the highest ozone concentration of 40 mcg/ml will depend on the type of pathology, patient tolerance and other information that can only be obtained by daily observations during a well controlled clinical study. Treatment can be done daily or every other day. Table 4 provides an example of a flexible schedule.

Table 4. A possible schedule of ozone administration by RI.

Weeks	*Days*	*Concentration O_3 (mcg/ml)*	*Gas volume (ml)*	*Total Ozone dose (mg)*	*Range*
1	1	3	100	0.3	Low-Medium
	3	5	150	0.75	
	5	8	200	1.6	
2	1	10	200	2.0	
	3	10	250	2.5	
	5	15	250	3.75	
3	1	20	300	6.0	Medium-high
	3	25	350	8.75	
	5	30	400	12.0	
4	1	35	400	14.0	
	3	35	450	15.7	
	5	35	500	17.5	

If the patient responds positively to the therapy, it could be continued 2-3 times per week, maintaining a high or medium ozone concentration. Although I am not enthusiastic of the IR approach because the effective ozone dose is never known due to the fecal contents and other variables, I admit that it is the simplest and most practical option to be adopted in poor countries. **In order to prevent cross contaminations, the catheter and syringe must be disposed of after each treatment.**

If, by an appropriate randomized clinical trial (RCT), we can prove that IR also has therapeutic activity in vascular disease, chronic hepatitis and intestinal diseases, we will have to promote RI, as the Cinderella of approaches, to the rank of AHT. Moreover the possibility of an easy and safe automedication by the patient at home for prolonged periods cannot be underestimated. Sixty-six years after the introduction of RI and after millions of applications with no cause for complaint, we can say that this approach, if properly performed, does not seem to induce adverse local effects. It appears reasonable to think that a judicious ozone dosage, the mucous layer, the antioxidant system and the adaptive response of enterocytes are all responsible for the lack of toxicity. However we must keep in mind that Eliakim et al., (2001), after repeated enema in rats with ozonated water (20mcg/ml), have reported the appearance of a microscopic colitis. Although gas insufflation is probably less irritating than the enema, this result reinforces my suggestion to use low doses of ozone at least initially for inducing the tolerance phenomemon.

In Chapter 9, we will briefly examine the pathogenesis of the diseases where RI is best employed, but here it may be useful to speculate about the local effects of ozone. These may be as follows:

a) **Biochemical effects.** In the studies already cited (Leon et al., 1998; Barber et al., 1999; Peralta et al., 1999, 2000; Borrego et al., 2004; Gonzalez et al., 2004), RI in rats upgraded the enzymatic antioxidant response in liver and kidney but the viability of enterocytes was not examined.

b) **Bactericidal effects.** The human colon-rectum contains up to 600 g of about 400 species of mostly anaerobic bacteria, and ozone may partly change the environment for a short while. Except in particular conditions, like clindamycin-associated enterocolitis (Schulz, 1986), bactericidal activity per se is probably unimportant but may cause the release of LPSs and muramyl peptides. These compounds are among the most potent cytokine inducers and in large amounts are responsible for the toxic shock syndrome and likely death. However, in physiological conditions, the daily absorption of traces of LPSs bound to specific proteins and to lipoproteins is considered essential for maintenance of the basic cytokine response and an alert immune system (Bocci, 1981b, 1988b, 1992c). Particularly in the last paper, it was postulated that the somewhat neglected gut flora has a crucial immunostimulatory role. This idea remains valid today and it is possible that RI favours a slight increase of LPS absorption with the consequence of enhanced activation of intrahepatic lymphocytes, Ito's and Kupffer's cells (O'Farrelly and Crispe, 1999), which may change the evolution of chronic hepatitis.

c) **Modification of the bacterial flora equilibrium.** Owing to the multiplicity of bacterial species, this remains a complex area. However, the normal flora contains *Lactobacillus (Lb) acidophilus, Lb. bifidus, Lb. fermentum, Lb. casei, Streptococcus faecalis, S. thermophilus, S. bulgaricus, Escherichia coli, Proteus and a variety of enterocci.* The bacteria and their products interact with each other and with the enterocytes, goblet and enteroendocrine cells (producing a myriad of hormones) and the gut-associated lymphoid tissue, GALT (Hooper and Gordon, 2001). On the other hand, it is well known that contaminated food, water and antibiotics can subvert this dynamic symbiosis by allowing the establishment of pathological bacteria and fungi like *Candida albicans, C. tropicalis, Torulopsis glabrata,* etc. The successive dysmicrobism usually has far-reaching deleterious consequences, ranging from transient to chronic enterocolitis and to autoimmune reactions and therefore we must try to correct it in order to restore normal homeostasis. Whether RI with a daily input of oxygen-ozone can re-equilibrate the bacterial flora and lead to normal immunoreactivity remains to be demonstrated (and explained), although anecdotal results suggest a beneficial effect.

d) **Effects on the GALT.** The gastrointestinal compartment represents almost 40% of the whole immune system. Besides the famous plaques described by Johann Konrad Peyer (1653-1712), over a total intestinal surface of some 300 m^2, there are about 10^{11} immunocytes per m^2 or about one per 6-7 enterocytes.

Intra-epithelial immunocytes are mainly T lymphocytes, either α-β of thymic origin or γ-δ of local origin. The latter induce a Th-2 type response that is anti-inflammatory and immunosuppressive, quite important to prevent excessive stimulation due to alimentary, bacterial, viral and toxic antigens. Perdue (1999) has emphasized that **a continuous cross-talk between immunocytes and enterocytes may maintain a healthy homeostasis and prevent breakdown of the mucosal barrier and inflammation.** In spite of interesting hypotheses (Fiocchi, 1998, 1999; van Parijs and Abbas, 1998; Okabe, 2001; Shanahan, 2002; Ardizzone and Bianchi Porro, 2002), the etiology and pathogenesis of both ulcerative colitis and Crohn's disease remain uncertain and it is difficult to identify the culprits that, step by step, cause the disease. Using the current paradigm of T-cell homeostasis, ulcerative colitis seems compatible with a poorly polarized Th-2 response while Crohn's disease is characterized by an excessive Th-1 response. In other words, any alteration of the balance between pro-inflammatory (IL-1, IL-2, IFNγ, TNFα) and anti-inflammatory cytokines (IL-10, TGF-β) appears critical (Schreiber et al., 1995), and an excessive release of IL-4, which affects the enterocytes, also appears important in ulcerative colitis (Perdue, 1999).

Another piece of the puzzle is represented by a more or less adequate synthesis of Hereman's "protective vernix" i.e. A-type immunoglobulins (Ig) produced by plasma cells (B lymphocytes). IgAs have a critical role in neutralizing foreign antigens and this may limit the onset of an autoimmune process. Once this starts, the vicious circle is complicated by other cells, namely cytotoxic lymphocytes, monocytes, macrophages and granulocytes, and by the release of other inflammatory compounds such as ROS, proteinases, eicosanoids and platelet-activating factor (PAF).

During the last twenty years, official medicine has made a great effort to sort out this intricate problem. Yet still today Crohn's disease remains a serious affliction. D'Ambrosio (2000 a and b), in a open study, has shown thar RI can lead to a marked improvement of these affections. If his results could be confirmed, no patient should miss this opportunity and we ought to present a rational basis for using ozonetherapy. Table 4 shows a possible treatment scheme that could be adopted for a randomized clinical trial. Intuitively I feel that the local treatment should be combined with two-three AHTs weekly, plus a supporting therapy with antioxidants, probiotics and omega-3 PUFA. It will be important to perform at least a pilot trial and investigate whether AHT coupled to RI will be able to re-equilibrate the immune response and lead to normal mucosal metabolism. Official medicine is really struggling to find an effective treatment as critically examined by

Hanauer and Dassopoulos (2001), who have reviewed pros and cons of as many as twenty possibilities. In Chapter 9, Section V, there is an ample discussion regarding the novel therapy with antibodies to TNF alpha.

5. QUASI-TOTAL BODY EXPOSURE TO OXYGEN-OZONE (BOEX)

Some eight years ago, we raised the possibility of exposing the body (excluding the head and neck to avoid pulmonary toxicity) in an ozone-resistant container (a large polyethylene bag would be a poor solution) for patients who refused rectal insufflation and for those who had no previous venous access for AHT (Bocci 1996c, d). The problems inherent in this approach are discussed here.

1) Is ozone as toxic for the skin as it is for the respiratory mucosa? (Lippman, 1989; Kelly et al., 1995) In common with ozone, chronic UV irradiation of the skin generates ROS, which after life-long exposure can result in skin changes such as wrinkles, pigmented spots and possibly cancer. Further studies have shown that both ozone treatment and UV-irradiation of epidermal layers of murine and human skin cause peroxidation and depletion of vitamins C and E (Thiele et al., 1997a, b; Podda et al., 1998; Fuchs and Kern, 1998; Valacchi et al., 2000, 2002, 2003). It has also been shown that these oxidizing agents, hence ROS and LOPs, activate NFKB and activator protein-1 (AP-1), but that alpha-lipoic acid (LA), n-acetyl-cysteine (NAC), Thioredoxin (Trx) and Selenium can inhibit the activation to a large extent and induce adaptive protection, such as over-expression of MnSOD and GSHPx as a response to oxidative damage (Haas et al., 1998; Saliou et al., 1999; Meewes et al., 2001; Didier et al., 2001). It is clear that the skin has a multiform antioxidant defence system, far more potent than that present in RTLF, and that it cannot be overwhelmed provided the attack by ozone or UV irradiation is not too harsh. These findings lend support to the empirical observation that **during topical ozonetherapy of necrotic ulcers, we have never noticed any damage to normal skin.** Moreover, during balneotherapy with slightly ozonated water, no local or generalized untoward effects have been reported.

2) Are there anatomical-physiological reasons for the relative tolerance of skin to ozone? Yes, if one examines the scheme of Figure 7 owing the structure of skin, with the epidermis, the derma and the disposition of the vascular system. The most external layer is the stratum corneum, i.e. the end product of keratinocyte function, which is a compressed and tough layer. **This "dead layer" is more or less covered by a very dynamic film, containing some proteins and water, due to the secretion of the eccrine glands.** It is partly responsible for thermoregulation, since it allows cooling of the skin surface (-580 cal/g) as the water changes from liquid to vapour. Moreover, the layer of lipids, produced by sebaceous glands, consists of unusual oily material, partly modified by the resident microflora (Nicolaides, 1974); in our opinion, this represents the first line of defence against ozone and UV rays. Progressing towards the dermis, there are the stratum granulosum, the stratum Malpighi and the proliferating basal cell layer. The dermis and the subcutaneous tissue contain a very flexible vascular system with a heat-exchanger, represented by capillaries and mainly by the venous plexus associated with the opening of arteriovenous shunts. It is able to accommodate up to 30% of the cardiac output so that heat transfer through the skin can increase up to 8 fold from a state of total vasoconstriction to extreme vasodilatation.

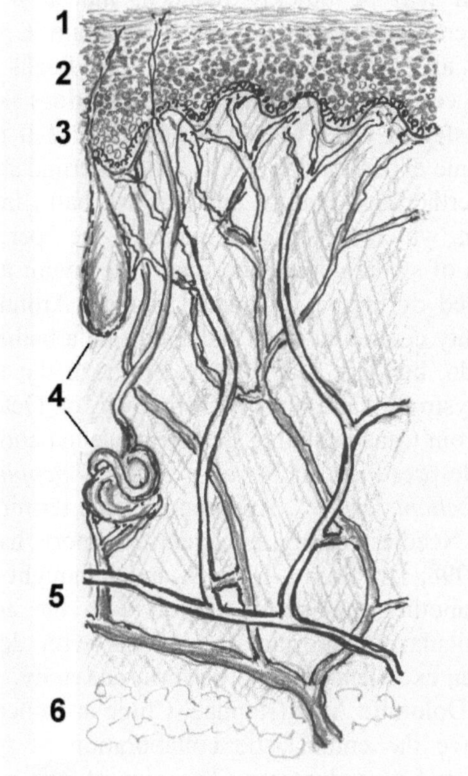

Figure 7. A schematic view of the skin and cutaneous circulation. Numbers indicate: 1) The stratum corneum overlaid by a superficial hydrolipid film, in which ozone dissolves and generates ROS and LOPs. 2) Malpighi's layer. 3) The basal cell layer and basement membrane. 4) The dermis, with a sebaceous gland and a sweat gland. 5) The arterial and venous vasculature with arteriovenous anastomosis. 6) The subcutaneous fatty tissue.

3) A crucial question is: when the skin is exposed to oxygen-ozone, do these gases penetrate all the cell layers to reach the dermis and enter the capillaries? It has been said that ozone reaches the blood circulation and has a cleansing effect, with the elimination of viruses and toxins. Yet this claim is not correct and has only commercial purposes. Only oxygen and carbon dioxide can move easily through cell membranes. However, **owing to its dipolar moment and high solubility, ozone dissolves in the superficial water film and reacts immediately with PUFAs of the sebum, generating ROS, hence H_2O_2 and an array of LOPs.** Therefore, it is more than likely that ozone does not even reach the phospholipids of the outer corneocytes, a conclusion already advanced by Pryor in 1992 for the pulmonary air-tissue boundary. However, the generated ROS and LOPs can be partly absorbed and pass through the epidermis, derma and capillary wall to enter into the blood stream. Obviously hydrogen peroxide and other ROS have a very short half-life and will be quickly reduced; indeed it has been clearly reported that several antioxidants (vitamins E and C, etc.) are readily oxidized (Thiele et al., 1997a, b; Podda et al., 1998; Fuchs and Kern, 1998).

4) The obvious corollary that comes to mind is: **does skin vasodilatation enhance the transfer of O_2, CO_2, ROS and LOPs?** It probably does and we will discuss some experimental results. The "thermal stress" that is easily induced with hyperthermia (Finnish and Turkish bath) increases cutaneous capillary perfusion, which may greatly increase the "perspiratio sensibilis" through activation of sweat glands and may also favour absorption of ROS and LOPs produced during an "ozonated sauna". Around 1995, we were informed that beauty centres in Italy had used sauna bathing with a trace of ozone for a decade, but this had remained only in the realm of cosmetic treatment of lipodystrophy and obesity. Moreover, on October 10, 1997, we received a letter from Canada stating that steam sauna combined with ozone had come into widespread use and *"well over 2,000 people had been treated with uniformly excellent results"*. Apparently some terminal cancer patients had been cured!!! Needless to say, no scientific reports had been published. Nonetheless, in 1998, Dr. Emma Borrelli and I thought that the ozonated sauna might be another therapeutic option with the advantage of non-invasiveness, particularly important in patients with deteriorated venous access.We found an excellent place to perform our study: a thermal resort in the middle of the Dolomite Alps (Raphael Clinic at Roncegno, Trento). We were lucky to have the enthusiastic collaboration of seven middle-aged physicians who acted as volunteers. The aim of our programme was to evaluate the following aspects:

a) **possible variations of arterial and venous pO_2, pCO_2, pH, examined before (pre), immediately after (end) and then 0.5, 1.0 and 24 hours after a period in a sauna cabin in the presence of either oxygen-**

ozone (May 1998) or only oxygen (control, September 1998). Unfortunately, only venous pO_2 values were obtained because our colleagues objected to the arterial blood collection.

b) **Modifications of body mass, oral temperature, diastolic and systolic blood pressure and ECG pattern.**

c) One important question was **to examine any possible variations of peroxidative markers in plasma during and after treatment.** In other words, we wanted to ascertain whether a 20 min exposure to ozone of almost the entire cutaneous surface could induce an oxidative stress and, if so, if this would be tolerable and lead to a therapeutic benefit. All details can be read in the original paper (Bocci et al., 1999).

The cabin was made of laminated plastic and, after subtraction of the body volume, had an internal residual volume of about 440 L. The flow of gas through the cabin (either a mixture of about 97% O_2 and 3% O_3 or pure medical O_2) was 1 L/min. The volume of gas must be limited for avoiding any risk of explosion. The ozone concentration was assessed in real time with a portable photometer. Any internal increase of barometric pressure in the cabin was prevented by an external silicone tubing connected to an ozone destructor. The maximum ozone concentration was reached at the end of the session and was estimated to be no higher than 0.90 mcg/ml, i.e. many times lower than the minimal ozone concentration used during local treatment of torpid ulcers for the same period (Werkmeister, 1995). Steam was generated in the cabin by a thermostatically controlled heater set at 90° C and turned on 10 min before the subject entered the cabin. Two towels and one polyethylene sheet were wrapped around the subject's neck to prevent breathing ozone. Although the doors were tightly closed by means of ozone-resistant gaskets, they were further insulated with the polyethylene sheet and towels to avoid any leakage of gas into the room. An improved, better insulated cabin is now being tested. The session lasted 20 min, during which the maximum temperature inside the cabin reached 46-50 C° with a humidity of 100% comparable to a Turkish bath. Just before the doors were opened, the gas flow was interrupted and the internal gas was rapidly aspirated via the outlet to prevent any breathing of ozone by the subject and the assistant. Determination of several variables was performed before, immediately after, and then 0.5, 1.0, 24 hours after the session. Body (oral) temperature was also measured in the middle of session. Standard 12-lead electrocardiograms were recorded before and after the session. Body mass was assessed with an electronic balance with an error of ± 50 g. Blood gas analysis was performed with a standard blood gas analyser. Systolic and diastolic arterial blood pressures were measured with a standard cuff sphygmomanometer.

Each volunteer was subjected to one 20-min exposure in the water vapour-saturated cabin, in the presence of either oxygen-ozone or oxygen only, i.e. he served as his own control.

There was a significant increase in body temperature, which reached a peak at the end of the treatment and declined rapidly thereafter. The maximum oral temperature ranged between 37.5 °C and 39.3 °C. There was a concomitant reduction in body mass (200-600 g). Similarly, blood pressure decreased slightly, but recovered within the next 30-60 min.

There was **a significant increase of PvO_2 and decrease of $PvCO_2$ at the end of the session and for 1 h after exposure to either oxygen-ozone or oxygen alone**; the increase in PvO_2 after exposure to oxygen alone was not significantly higher than that after exposure to both gases. Values for both erythrocytes and haematocrit increased immediately after the 20-min exposure. They decreased thereafter, probably due to rehydration, and were almost normal after 24 hours

We noted **an initial significant increase in leukocytes**, followed by a decrease 1 h after oxygen-ozone exposure.

The experimental data regarding the plasma levels modifications of total antioxidant (TAS), peroxidation (TBARS) and protein-thiol groups (PTG) oxidation were quite surprising: **antioxidants declined but remained at a substantial level, peroxidation levels increased steeply, in a linear fashion up to an hour after the end of the session while protein-thiol groups declined during the same period.** All values returned to the baseline 24 hours thereafter and, in spite of the peroxidation increase, **no haemolysis was noted at any time.** Interestingly, even the exposure to oxygen alone induced a similar trend, although modifications of plasma levels were less marked.

We also investigated whether the plasma levels of three representative markers changed after the O_2-O_3 exposure. **Levels of IL-8 significantly increased 30 min after exposure.** Conversely, levels of myeloperoxidase (MPO) and transforming growth factor (TGF-beta1) either did not change or tended to decrease.

Plasma levels of hepatic enzymes and creatinine remained within the normal range.The interested reader can examine the diagrams and numerical data in the original publication (Bocci et al., 1999). All subjects tolerated the exposure to either gases or oxygen alone without reporting immediate or subsequent adverse effects. Oral intake of water was allowed at the end of the sauna. Four subjects enjoyed the sauna, but two reported that they would find it difficult to tolerate a period longer than 20 min in the cabin.

Although this preliminary study had some pitfalls, it was informative but it will be useful to examine the relevance of hyperthermia. Moreover we could have examined the effect of the hyperthermia alone, which in itself is quite interesting. We enjoyed reading a recent review on the "Benefits and risks of sauna bathing" (Hannuksela and Ellahham, 2001). Unlike the Turkish bath (45-48°C and humidity at about 100%), the sauna has a temperature ranging between 80-100 °C at the level of the bather's face and 30 °C at floor level and a relative humidity of about 20%. One good point of our study was to control the same subjects with oxygen alone. **Insufflation of oxygen alone is indeed able to increase peroxidation during the hyperthermic period.** The results are even more surprising if we consider that the gas flow was only 1 L/min and thus the total volume of 20 L of oxygen was diluted in about 440 L of air contained in the cabin. This suggests that **the heating per se must overwhelm the effect of oxygen alone.** However, **ozone clearly accounts for the significant linear increase of peroxidation values measured up to an hour after the session.**

Let us examine the risks: *firstly, ozone toxicity for the respiratory tract.* There must be neither contamination of environmental air with ozone nor any ozone inhalation and we took precautions to avoid that. The cabin must be tightly closed, the room must be well ventilated, the gaseous contents of the cabin must be quickly aspirated before it is opened and a monitor sensing the ozone level must be switched on.

Secondly, ozone toxicity for the skin. Depletion of antioxidants and the increase of malonyldialdehyde (MDA) in the outer epidermal layers are well documented, but in our study the final ozone concentration in the cabin could reach at most 0.9 mcg/ml by the end of the 20 min session. The final ozone concentration increases slowly because we must take into account the large dilution with a slight loss because the cabin remains at normal pressure and the rapid ozone decay at about 40°C (about 18 min). Thus the final concentration is about 10-20 times lower than that used during the final topical applications in skin ulcers or decubitus (Werkmeister, 1995). In conclusion, **we did not observe any acute or chronic toxicity.**

Thirdly, systemic toxicity of ozone. We had no information about this **but we reasoned that ozone would decompose entirely on the cutaneous surface and only some of the generated ROS and LOPs might be absorbed and enter the circulation.** The scheme shown in Figure 8 gives an idea of the site of action and fate of ozone in the skin. However, we knew already that blood is quite resistant to ozone, and body tissues and fluids have a great reservoir of antioxidant compounds, as well as the ability to regenerate them. **We envisaged that dilution, metabolic breakdown and renal excretion would minimize the increase, if any, of LOPs present in the plasma pool. Contrary to our expectation, there was a very significant increase of circulating LOPs which continued long after the session, suggesting a steady inflow from the skin prevailing over catabolism.** It would be interesting to follow the kinetics at 1.5-2-3-4 hours to localize the peak and the pattern of decrease. PTG values showed a consistent decrease, while (reassuringly) TAS values declined only slightly and temporarily. **The induced oxidative stress had a brief lifetime and did not cause haemolysis or any modification of important blood parameters.** Hepatic enzymes and creatinine plasma levels remained unmodified. Plasma levels of myeloperoxidase, a sensitive marker of the activity of neutrophils (Boxer and Smolen, 1988), did not change. No toxicity after repeated BOEX has been noted for the skin but we use the precaution of protecting moles at risk with a cream rich in vitamin E. **None of our volunteers, nor several patients, have reported acute or late side effects.** For experimental reasons, the author has undergone many BOEX, at different times, and each time he has experienced a feeling of great energy and euphoria for the next couple of days. In fact, it would be pleasant to have the time to do it twice weekly! A similar sense of wellness has been claimed by a few patients, who have repeatedly tried this procedure.

Is there an explanation for this good feeling and is it due to ozone or the sauna or both? We can certainly say that AHT (rectal insufflation is less effective) also give a sense of well being, but in the case of BOEX the hyperthermia itself may contribute. For a long time, we have wanted to evaluate the hormonal changes related to ozonetherapy and such a study would probably clarify this issue and broaden our vision. We found that **the short-term hormonal changes during and after sauna bathing, particularly the increase of growth hormone and beta-endorphin, are quite interesting** (Hannuksela and Ellahham, 2001). It is intriguing that long-term sauna bathing helps to lower blood pressure in hypertensive patients in spite of transient activation of the renin-angiotensin-aldosterone system. As expected, these changes are brief and reversible, and the same may occur for ozonetherapy. Whether ozone potentiates the effects of the hyperthermia remains to be seen but both stimuli are likely responsible for triggering a psychoneuroimmunological effect via the release of a cascade of hormones, namely of CRH, ACTH, cortisol, DHEA, growth hormone and so forth. After Payne and Krueger's findings (1992) and Reichlin's postulation (1993), one cannot avoid thinking how deeply ozone therapy can influence the neuroendocrine-immune relationship and how relevant its contribution is to the therapeutic effect.

Fourthly, does ozone switch on a dangerous oxidative stress? Although we noted a remarkable systemic increase of peroxidation, it was transitory, since the levels returned to baseline after 24 hours. If the reader has gone through the previous pages, she/he likely realizes that **we purposefully want to induce an acute oxidative stress in patients, using AHT** (and perhaps even with rectal insufflation). Probably it can also be realized that **this stress must be adequate** (otherwise it is a placebo), **calculated** (i.e. neither below nor too much above threshold levels) and **transitory. This is important because we do not want to override the antioxidant defence system nor cause any toxicity but we want to give a precise, atoxic shock to an organism which for various reasons has gone astray.**

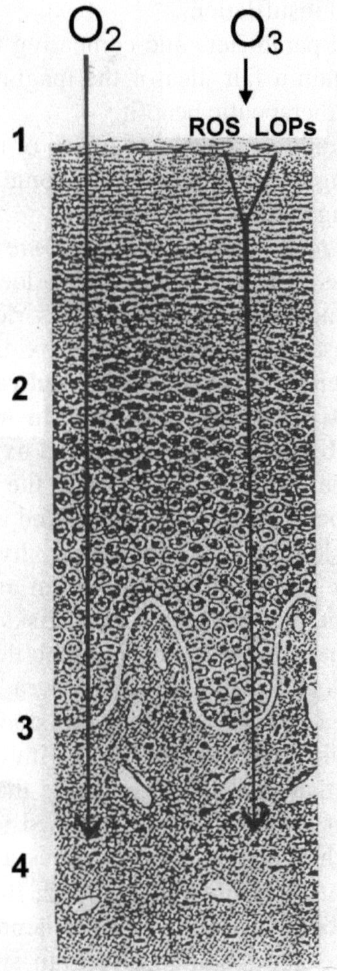

Figure 8. A schematic view of the fate of ozone in the skin during BOEX. Ozone dissolves in the water-sebum film overlaying the outer layer of the stratum corneum and reacts immediately with PUFAs, generating ROS (among which H_2O_2) and LOPs. These compounds can be partly absorbed and pass through the transcutaneous barrier, facilitated by the intense vasodilation induced by the hyperthermia. Both lymphatic and venous capillaries can rapidly transfer LOPs into the general circulation, thus inducing systemic effects. Numbers indicate: 1) The stratum corneum coated with a superficial hydrolipid film, in which ozone dissolves and generates ROS and LOPs. 2) Malpighi's layer.3) The basal cell layer. 4) The dermis.

One study we should do as soon as possible is to evaluate and compare the pharmacokinetics of LOPs (even if they are several and heterogeneous) in single patients during:

a) a standard AHT
b) a BOEX

c) a standard rectal insufflation.

By assessing several parameters and comparing them after each of these procedures, we could gain a fair idea of the magnitude of the biochemical modifications and their therapeutic benefit.

Another important study is to evaluate which of these three procedures is most effective in raising the adaptation to chronic oxidative stress (COS) and, in so doing, yielding clinical improvement.

Fifthly, does BOEX to oxygen-ozone have some advantages? During the treatment, there is a loss of 300-500 g of water due to intense perspiration, normal for sauna bathing. This loss of water is ridiculously advertised as greatly beneficial because the "body gets rid of oxidised toxins" in this way! **Transitory hyperoxygenation is also considered relevant, but it would be absurd to increase pO_2 levels through the skin when we could increase them far more simply by breathing humidified oxygen for one hour.**

The transitory thermal stress associated with the acute oxidative stress is possibly an advantage because it may enhance and accelerate the adaptation to COS. It is well known that **moderate hyperthermia positively modulates the immune system during infection and cancer.** On the other hand, excessive hyperthermia presents several risks (cardiovascular failure, etc.), induces a hypercatabolic state and immune depression; hence it must be avoided. An initial leukocytosis, followed by a modest leucopenia, was observed after exposure to oxygen-ozone in our study and was probably due to a transient release of IL-8. This agrees well with our previous data (Bocci et al., 1998b) showing that IL-8 is a chemokine that is released rapidly by leukocytes in blood that has been briefly exposed to oxygen-ozone. It may be useful in patients with infections, but it is necessary to explore further this finding and look for other cytokines such as Il-2, Il-12, IFNγ and GM-CSF. The simultaneous release of some pro-inflammatory cytokines may temporarily increase the hyperthermic effect. In spite of our approximate approach, we feel that our studies have some merit because they were the first to evaluate scientifically new ideas which have revived a stagnant field, restricted for three decades to AHT and RI.

What might be the practical usefulness of BOEX and does it have a future? If we listen to commercial advertising, which claims to cure cancer and AIDS, it will have a bright future. Yet we do not believe that the future of ozonetherapy lies in the claims of charlatans. However, we would like to compare the pros and cons of the current methods. If one uses the standard, optimised AHT method, one is able to slowly treat several ailments without any risk to the patient, but **one venous puncture is necessary.**

Rectal insufflation is extremely easy to do (once instructed by the ozonetherapist, the patient can do it at home by himself), very cheap and practically free of risk. Yet it is often objected to and the delivery of a precise dosage is always uncertain, although it may be beneficial in certain pathologies.

BOEX has distinct advantages: it is simple to perform, fairly inexpensive, non-invasive (no venous puncture) and does not involve the handling of potentially infectious blood, a point highly appreciated by medical personnel. We have noted some problems: the cabin must be well insulated and BOEX is best performed in a well-organised clinic or in a thermal resort with an entrance room, treatment room, adjacent room to allow a comfortable one-hour rest for the patient and another room with a shower. Whether this approach will truly become useful remains to be established by RCTs, but at this stage it seems to represent a promising tool to modify the biological response in some pathological states:

- **Chronic viral diseases (HBV, HCV, herpes I and II, HIV, HPV). It may be useful to treat chronic fatigue syndrome (CFS), even though it is probably not a viral disease.**
- **Metastatic cancer, to avoid palliative chemotherapy, which is usually useless and associated with a very poor quality of life. However, it could be even tried as an immunoadjuvant at earlier stages with polychemotherapy.**
- **Vasculopathies, particularly hind limb ischaemia due to atherosclerosis, Buerger disease and diabetes. Necrotic ulcers and dystrophic lesions must be simultaneously treated with topical therapy. Patients with severe coronary atherosclerosis, recent myocardial infarction or severe hypertension may undergo BOEX, but without hyperthermia, starting with a 10-min period and scaling up slowly. Patients with asthma and BPCO must also be treated cautiously.**
- **ARMD, particularly the atrophic form. Keeping the heating at a low level.**
- **Sclerodermia with Raynaud's phenomenon.**
- **Moderate burns, to prevent or reduce bacterial infections and enhance healing.**
- **Some muscular-tendinous lesions in athletes, to reduce muscle contraction and alleviate pain.**
- **Skin disease, such as infections, psoriasis, perhaps atopic dermatitis and eczema.**
- **Advanced lipodystrophies, such as Madelung disease. The lipodystrophy occurring during HAART may also be advantageously treated**

Our provisional protocol envisages a course of therapy twice weekly during the first and second weeks but it must take into account the patient's age, stage and type of disease. We always insist on the "start low, go slow" paradigm to allow for the adaptation to COS. The heating, hence the cabin's temperature should be gradually scaled up from 30°C to no more than 42°C, with periods from 10 min to a maximum of 20 min.

6. EXTRACORPOREAL BLOOD CIRCULATION AGAINST OXYGEN-OZONE (EBOO)

In both AHT and EBOO, blood is treated ex vivo either directly or with the intermediation of a membrane. The latter procedure resembles a classical dialysis with the substantial difference that the gas mixture: oxygen-ozone flows inside the hollow-fibres and blood flows, in the opposite direction, on the external side of the membrane. This approach has been developed with the enthusiastic collaboration of Prof. Nicola Di Paolo, who has been one of the very few clinicians interested in ozonetherapy. Our investigation started off on the wrong foot in 1991 but, unlike others, who also had the idea to realize a dialysis–like system, we have corrected our idea. Unfortunately, even today, unscrupulous quacks use the dialysis system exploiting cancer and HIV- infected patients with an ineffective and toxic technique in Kenya, India, Mexico and Malaysia.

Our results, detailed elsewhere (Bocci et al., 1999b; 2001c; Bocci and Di Paolo, 2004; Di Paolo et al., 2000; 2002), have clarified that **this** apparently obvious **method has in fact proved to be a formidable problem** which, only recently, has been **solved by using bio-compatible oxygenators (and not dialysis filters)** and continuously **monitoring biochemical results,** which is the only way of optimizing the method. The final system consists of a precise ozone generator, fed by therapeutic oxygen on line, able to deliver a constant flow of the gas mixture for hours. In the past we have assessed biochemical parameters and toxicity using ozone concentrations from 3 to 80 mcg/ml, but now, with very efficient gas exchangers, we can use ozone concentrations ranging from 0.2 to 1 mcg/ml throughout the session. I never get tired of repeating that ozone is a toxic gas and, when used as a drug, it must be used with great caution and within a defined therapeutic window.

During the last three years we tested several types of oxygenators, which are the ozone-resistant "lungs" of the system. This is essential to **prevent leakage of toxic compounds into the blood and this can happen with dialysis filters instead of oxygenators** currently used in cardiovascular surgery. These are made with microporous membranes made up with either polyethylene or polypropylene. They are hydrophobic, permeable only to gases and, unlike dialysis filters, do not form any ultrafiltrate. The exchange of oxygen, ozone and carbon dioxide occurs through the membrane without any bubble formation, thus excluding any risk of gas emboli. The gas exchange is proportional to the membrane surface that ranges from about 0.3 up to 1.6 square meters. Moreover, it varies according to the blood transit time, hydrostatic pressure, temperature, solubility and partial pressure of the gases on the opposite surfaces of the membrane. In the case of oxygen alone, elevated PvO_2 values are achieved implying full saturation of haemoglobin with a variable volume of oxygen physically dissolved in the plasmatic water. However I must immediately clarify that, by considering that the volume of blood exposed to gas per minute is about 1/60 of the blood volume circulating per minute, **the oxygenation per se has a minimal relevance.** On the other hand, ozone behaves quite differently from oxygen because firstly, it is ten-fold more soluble and secondly, owing to its strong oxidant potential ($E° = +2.076$ V), it reacts instantaneously with PUFA as well as reducing compounds present in plasma. Thus it is reasonable to assume that ozone reacts immediately at the gas-blood interface.

During cardiovascular surgery lasting several hours, oxygenators remain viable even though several studies have observed some undesirable immunologic modifications, particularly complement activation, a mild leukocyte activation and a decreased platelet count (Edmunds, 1998; Dernek et al., 1999; Stiller et al., 2001). These phenomena, though bothersome, have to be expected as hollow fibers present a foreign surface to blood components. However, when we initiated a preclinical study on sheep using standard oxygenators and heparin according to the standard procedure, we were disappointed to note that, in the presence of oxygen-ozone, the oxygenator decayed rapidly and the blood flow was blocked in about 5-10 min The oxygenator remained viable using oxygen alone, but as soon as even low ozone concentrations (3-5 mcg/ml) were added, it clogged rapidly and irreversibly. At first it was unclear whether ozone could switch on activation of coagulation factors or platelets. Two critical observations helped to clarify this problem. The first was that substitution of heparin with Na citrate (at full dose to chelate plasmatic Ca^{2+} level) allowed to normalize the extracorporeal circulation of blood in the sheep even using high ozone concentrations (up to 80 mcg/ml). Secondly, and most convincingly, was the observation made by using human platelet rich plasma, prp (Bocci et al., 1999a). On addition of ozone, prp in heparin showed a prompt aggregation while remained normal in citrate so that we could envisage the following sequence of events:

Heparinized prp + Ca^{2+} + O_2-O_3 → adhesion → aggregation →
→ degranulation → release of factors → coagulation

Other data by Iuliano et al. (1997) supported our observation that in the presence of physiological Ca^{++} ozone activates membrane receptors leading to irreversible damage.

We completed the preclinical study by using citrate and, by examining a number of biochemical parameters, we learnt how blood ex vivo behaves in the presence of progressive higher ozone concentrations (Bocci et al., 1999b). Total antioxidant status (TAS) and protein thiol groups (PTG) practically halved with an ozone concentration of 35 mcg/ml and the erythrocytic GSH content was also markedly reduced. Blood oxygenation, although remained at supraphysiological values, decreased at the outlet, showing that indeed more ozone is no better. On the whole, the results suggested that we could achieve better results with very low ozone concentrations. The citrate infusion also presented critical drawbacks such as the induction of severe hypocalcemia and acidosis that had to be continuously corrected by a simultaneous reinfusion of Ca^{++} and $NaHCO_3$.

However, it was reassuring to observe that biochemical parameters normalized quickly at the end of the EBOO and the sheep did not show any acute or chronic signs of toxicity confirming that the capacity of the antioxidant system is able to tame ozone activity. In spite of these encouraging results, we understood that the problem of platelet aggregation had to be solved because the use of citrate was unpractical and somewhat risky.

In order to clarify the problem we examined oxygenators perfused with heparinized swine blood in vitro. A control exposed to oxygen alone for 60 min showed only a minimal adhesion of platelets on the external surface of the fibers. In contrast, after 5 min exposure to ozone (even at a concentration of 5 mcg/ml), the polypropylene surface was coated with a thick layer mostly composed of platelets. This result indicated that **heparin-coated oxygenators,** which appear biocompatible in usual cardiopulmonary by-pass (Videm et al., 1999), **do not prevent platelet activation in the presence of ozone.**

Thus, initial diffusion of both gases was excellent: the pvO_2 raised up to 500 mmHg and TBARS values increased 3-6 folds from basal value. However, in a few minutes the pvO_2 fell progressively and peroxidation ceased completely. **After 10 min, pvO_2 levels became irrelevant because even oxygen diffusion was totally impeded. This was due to a coating of platelets, plus fibrin clots and blood cells thick enough to block any gas transfer.** We have postulated that gases are still exiting from the polypropylene micropores but, while oxygen remains trapped, ozone reacts with the adhering platelets leading to total occlusion.

Our work remained at a standstill for a couple of years until we could obtain the most technologically advanced oxygenators, where the external surface has been coated with various compounds. It is worth noting that heparin-coated oxygenators were unable to prevent platelet activation. On the other hand the new types of either albumin- or phosphorylcholine-coated oxygenators are biocompatible and display a better performance not only in cardiovascular surgery, but have allowed us to perform EBOO satisfactorily in heparinized patients. **The biocompatible layer on the polypropylene surface, in conjunction with the use of very low ozone concentrations, markedly delays platelet adhesion and allows the treatment to be performed in one hour.**

With the approval of the Ethical Committee of the University of Siena all the perfusions were and are performed in the Dialysis Unit of the University Polyclinic. The final EBOO system is schematically shown in Figure 9.

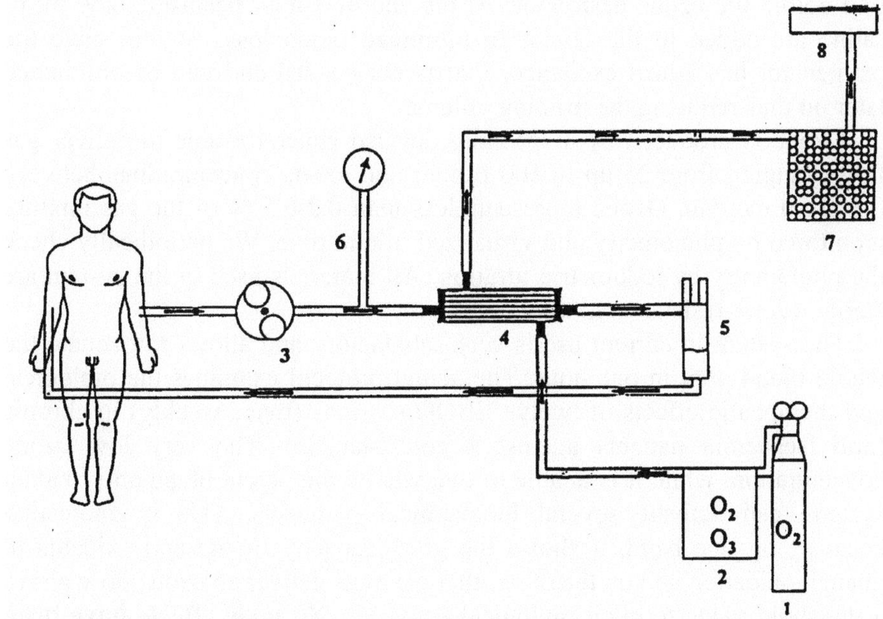

Figure 9. A schematic view of the simplified EBOO apparatus. 1) oxygen supply 2) ozone generator with photometer 3) roller blood pump 4) hollow-fibre oxygenator-ozonizer 5) two air traps with blood filters in series 6) blood pressure monitor 7) silica gel trap 8) ozone destructor

The oxygenator is made with thousands of externally coated polypropylene hollow fibers blocked with polyurethane (ozone-resistant) in a polycarbonate housing. The oxygenation membrane in current use has a surface area of 0.6 m^2. Extracorporeal circulation is carried out using a last generation apparatus normally used for haemodialysis. Just before perfusion the oxygenator and lines are routinely rinsed with 1 L saline before being connected to the catheters. The venovenous blood circulation is performed by means of standard arterial-venous fistula needle sets (usually G17), used with great care to maintain the venous access in good condition. One ml bolus of 5,000 IU unfractionated Na-heparin diluted with 10 ml saline is injected 5 min before starting the ozonation process and a subsequent slow delivery of a diluted heparin solution has proved unnecessary. The extracorporeal circulation is established by maintaining blood flow at 80-90 ml/min throughout the perfusion. This flow appears optimal because a slower one will yield excessive blood oxidation and a higher flow is not well tolerated by the venous access.

Although gas microemboli are not formed, for extreme safety, the line returning the arterial blood to the patient is intercalated with one air trap and a blood filter. Any trace of water vapour, possibly present in the exhaust gas containing CO_2 and the bulk of oxygen-ozone is retained by the silica gel trap before the ozone destructor. At the end of 1-hour perfusion, 250 ml of saline are added to the circuit to minimize blood loss. At this stage the oxygenator has a heat exchanger that is not needed and will be eliminated later on thus reducing the priming volume.

Ozone is produced by a specially devised generator able to deliver gas flows ranging from 25 up to 100 L/hour and ozone concentrations between 0.1 -10.0 mcg/ml. Ozone represents less than 0.2-0.5 % of the gas mixture monitored by photometry and visualized in real time. We periodically check the photometry by iodometric titration. All materials used in the system are sterile, ozone-resistant and used only once.

The system in current use is very satisfactory and allows to ozonate the whole blood pool in one hour. The actual protocol examines the biological and therapeutic effects of twelve EBOO sessions (twice weekly) in chronic limb ischaemia patients against a gold standard. The very low ozone concentration, while it is unable to overwhelm the potent blood anti-oxidant system, can activate several biochemical pathways. This is interesting because, on one hand, it shows the great capacity of the anti-oxidants to quench oxidation and on the other, that we must deliver an ozone dose above a threshold value to elicit biological activities. **No toxic effects have been noted.**

a) The extracorporeal circulation of blood against oxygen-ozone is a novelty and has become a reality. The main characteristic is that ozonation levels must be kept at very low levels because one treatment corresponds to about twenty conventional AHT performed simultaneously.

b) Technical and methodological aspects have been resolved satisfactorily and are susceptible to further improvements.

c) Owing to the improved efficiency of the oxygenator, up to 5 L of blood/hour can be exposed to very low ozone concentrations, just above the thresholds of the therapeutic window. To enhance ozone tolerance the first and second EBOOs last only 30 and 45 min, respectively.

d) As it occurs in the pulmonary circulation, the efficiency of the hollow fibers allows gas exchange in one minute. Needless to say that only a minor proportion of the two gases act on the flowing blood.

e) Both oxygenation and ozonation remain effective without any increase of venous pressure.

f) In arteriopathic patients (grade III and IV) subjective and objective clinical improvements have often been noted after the first treatment. Orthodox treatments usually do not provide such a rapid improvement. In fact this approach has been developed for the treatment of critical patients.

g) Neither metabolic derangement, nor changes in blood chemistry, nor any toxic effect has been observed during or months after the cycle.

h) It is necessary to prove objectively the clinical data and support them with laboratory data evaluating: 1) adaptation to chronic oxidative stress, by measuring levels of antioxidant enzymes, 2) various oxidative stress proteins, particularly heat stress protein (HSP32) or haeme-oxygenase (HO-1), 3) 2,3-diphosphoglycerate (2,3-DPG) values, 4) hormonal levels able to explain the feeling of wellness and disappearance of pain, 5) any modification of low and high density lipoproteins, cholesterol and fibrinogen levels, and 6) the immune status. It appears interesting to evaluate if one of the mechanisms switched on by ozonetherapy involves the release and activation of autoctonous staminal cells. If it does, it will represent a functional as well as an anatomical therapy. This could represent one of the most important effects and we will try to visualize a possible neoangiogenesis.

Some possible disadvantages must be taken into due consideration: 1) the cost of the disposable oxygenator, including ancillary materials is now near 700 Euro but it could decrease once the application will be used world-wide. 2) The cost of a qualified technician. 3) The potential deterioration of venous access. 4) The occasional need of inserting a catheter into a central vein to continue the cycle, with the related risk of infection (this recently occurred in two patients, who had to stop the treatment). The last problem may be reduced by using improved catheters impregnated with antibacterial substances (Wenzel and Edmond, 1999).

At this stage, we feel compelled to vigorously ascertain the therapeutic benefits of EBOO in the following areas:

a) Critical, inoperable ischemic limbs (stage III and IV, Leriche-Fontaine) when amputation remains the only option. Medical treatments (iloprost infusion, pentoxyphylline, electrical spinal-cord stimulation, anticoagulants, platelet anti-aggregation, anti-atherosclerotic drugs, etc.) help but are rarely successful (Bergqvist, 1999). The surgical procedure of distal venous arterialisation appears promising (Taylor et al., 1999) but it is also a complex and experimental procedure. However, a comparison of these approaches appears useful for further progress;

b) End-stage ischemic myocardiopathies, previously operated on with no success;

c) Acute cerebral ischaemia, to be treated with EBOO as soon as possible to reoxygenate the hypoischaemic (penumbra) and infarctuated areas, thus limiting neuronal death and favoring a more rapid recovery. Neurologists prefer the thrombolytic approach and are afraid of testing ozonetherapy.

d) Chronic HCV hepatitis in patients who are IFN-resistant or IFN-intolerant or because they refuse orthodox therapy;

e) Chronic renal failure, which is always accompanied by immunosuppression and a chronic oxidative stress (Witko-Sarsat et al., 1998; Morena et al., 2000) continuously aggravates the metabolic disorder (Bocci, 2002). In such a case, the oxygenator may be situated parallel to the dialysis filter and used following the dialysis session after a bolus infusion of antioxidants to reconstitute a sufficient antioxidant capacity depleted during dialysis (Chapter 9, Section IX).

In other diseases such as:

f) Metastatic, chemoresistant cancer and severe primary or secondary (to HIV-protease inhibitors treatment) lypodistrophies, the usefulness of EBOO remains to be considered against the validity and cost-benefit of this approach.

CONCLUSIONS: today ozonetherapy can be performed using six different modalities. Besides the old but still quite valid methods of major and minor autohaemotherapy and rectal insufflation, we have developed and evaluated other options such as the quasi-total body exposure to oxygen-ozone and the EBOO. In patients with precarious venous access, as a blood substitute, we are now using the gluco-peroxide solution, which represents a form of biooxidative therapy with a clear rationale and the advantage of being inexpensive and potentially useful to millions of people without medical assistance. Although all of these procedures must be controlled and supervised by physicians expert in ozonetherapy, a few of them are amenable to be used at home by the patient. Ozone must never be breathed but, if the dose is adapted to the potent antioxidant capacity of body fluids, the above described methods offer flexible and remarkable therapeutic advantages. Finally, when it was needed, I have successfully combined major and minor AHTs, RI, BOEX as well the gluco-peroxide infusion.

The central aim of ozonetherapy is to give a precise, atoxic shock to an organism which for various reasons has gone astray; the hope is that repeated, timely shocks will readjust several biological functions by means of many messengers (ROS, LOPs and autacoids generated by ozone) delivered by circulating blood to the whole body. **We have coined the term "therapeutic shock" to symbolize the possibility of reactivating the natural positive capabilities to restore health or, in better words, to stimulate the "vis medicatrix naturae".**

I believe that the simultaneous induction of an acute and precisely calculated oxidative stress on different areas such as blood, the skin and the gut mucosal system can result in a more comprehensive and perhaps synergistic response of the body defense system. Indeed chronic diseases must be attacked from different angles and we have evidence that the

stimulation of several biochemical pathways in different organs can be therapeutically beneficial.

Chapter 7

THE POTENTIAL TOXICITY OF OZONE. SIDE EFFECTS AND CONTRAINDICATIONS OF OZONETHERAPY

One good reason for the unpopularity of ozonetherapy in the medical field is that the toxicity of ozone is considered equal to that of ROS. In fact, there are substantial differences because **ozonetherapy is occasional and can be controlled whereas endogenous ROS formation goes on unperturbed throughout life** (Farber et al., 1990; Ames et al., 1993).

The topography of formation of ROS is also quite different: mitochondria, which convert 95% of the inhaled oxygen to harmless water, are the main source of ROS since at least 3% of oxygen is converted to superoxide, $O_2^{\bullet-}$ (Richter et al., 1988, 1995). Dismutation of superoxide by SODs (Fridovich, 1995; Carlsson et al., 1995) is the source of H_2O_2, whose reduction may generate the fearsome, non-specific hydroxyl radical, $OH^{\bullet}$. Halliwell (1994) estimated that a 70 Kg human produces no less than 0.147 moles or 5 g/day of superoxide, whereas **one AHT uses at most 20 mg of ozone, equivalent to less than 0.4% of the minimum daily production of superoxide!**

The huge formation of endogenous ROS in mitochondria, deeply immersed in the cell, explains the damage to mitochondrial DNA (Wiseman and Halliwell, 1996), which is oxidized about 10 times more than nuclear DNA (Richter et al., 1988) and remains persistently damaged (Yakes and Van Houten, 1997). Conversely, **ozone acts from the outside on the plasma, which has a huge reservoir of antioxidants.** Nonetheless, the ozone dose added to blood must reach a threshold level in order to generate sufficient H_2O_2, which passes from the plasma into the blood cell cytoplasm where it triggers several biological effects. We do not hide the fact that for ozone to act, we have to induce a calculated, transitory, acute oxidative stress that is rapidly corrected by the antioxidant system. Thus, there is no doubt regarding the formation of peroxyl radicals, hydroxyaldehydes and perhaps traces of $OH^{\bullet}$ and HOCl in the plasma. What is important to note is that all the vital cell compounds, such as enzymes, proteins, RNA and DNA

(Van der Zee et al., 1987; Stadtman and Oliver, 1991; Ames et al., 1993), are spared during the extracellular ozone decomposition.

Particularly in the USA, ozonetherapy is regarded as a *"barbaric"* therapy and unscrupulous ozonetherapists and quacks have done their best to reinforce this concept. However, it is now time to clarify this issue; without prejudices, we must evaluate the merits and demerits and **put an end to the confusion between the constant oxidative stress (COS) due to oxygen and the occasional acute stress due to ozone.**

Knowing the importance of oxidative DNA lesions in ageing and cancer, I am not surprised when often asked: **is ozone mutagenic? And does ozonetherapy accelerate ageing?**

I have already discussed in details (Bocci, 1996; 2002; 2004) a number of reports regarding these questions. Results have often been controversial because some Authors (Goldstein and Balchum, 1967; Freeman et al., 1979), working with saline-washed erythrocytes or with tissue cultures deprived of antioxidants, have observed a damage or mutagenic changes in cells exposed to ozone for a length of time. Once cells are washed in a protein-free saline solution, thus removing precious antioxidants, both oxygen and ozone become cytotoxic, as Halliwell (2003) and we (Larini et al., 2003; 2004) have re-emphasized. In a recent past, Galleano and Puntarulo (1995), Leist et al., (1996), Matos et al., (2000) and Dumaswala et al., (2000) have also shown that cell damage or genotoxicity induced by hydrogen peroxide or iron overload or prolonged storage can be checked if tissue culture media or plasma contain adequate amounts of antioxidants.

Victorin (1992), who has beautifully reviewed this topic, stated that "no cytogenetic effects have been reported for bone marrow cells or spermatocytes and the few experimental and epidemiological studies with human subjects do not allow a conclusion on the cytogenetic effects of ozone in human lymphocytes". The latest study by Diaz et al. (1995) is important because it was specifically carried out in lymphocytes of eight Retinitis pigmentosa patients before and after 15 treatments of AHT. The results showed no significant differences in sister chromatid exchanges (SCE), micronuclei frequencies and proliferation index values between control and ozone-treated lymphocytes. On the other hand, Diaz-Llera et al., (2002) demonstrated that one-hour exposure of **SALINE-DILUTED BLOOD** to 5 mM ozone induces genotoxic effects on human leukocytes. However, during AHT, **WHOLE BLOOD** is exposed for only a few minutes to ozone concentrations between 0.21 and 1.68 Mm that clearly explain why ozone is not mutagenic in practice. A careful study by Shinriki et al., (1998) has shown neither cell damage nor haemolysis of human blood exposed exactly with out technique to ozone concentrations up to 100 mcg/ml per ml of blood.

As far as induction of tumours is concerned, lung adenomas were induced in the sensitive strain A/J but not in Swiss-Webster male mice after

4.5 months of inhalation exposure to 0.8 ppm ozone (Last et al., 1987). Witschi et al. (1999) concluded that animal studies do not support the idea that ozone is a pulmonary carcinogen.

Trying to sum up this important topic, it appears that **the lack of natural antioxidants is critical in allowing mutagenic changes in cells exposed to ozone in vitro for a length of time.** After the removal of plasma, washing and resuspension in physiological media without or with only a small amount of antioxidants, erythrocytes and other cells (Larini and Bocci. 2004) become very sensitive to even very low ozone concentrations, as demonstrated by intense haemolysis or apoptosis. Instead of stigmatizing ozonetherapy as toxic, published papers (Goldstein and Bachum, 1967; Gooch et al., 1976; Freeman et al., 1979; Sato et al., 1999; Fukunaga et al., 1999) ought to have pointed out the importance of antioxidants in preventing damage.

Another blunder has been made by several cell biologists by keeping cell cultures under constant ozone exposure (Merz et al., 1975; Tarkington et al., 1994) at extremely low levels, but for several hours or days. The conclusion that ozone is toxic even at minimal levels is misleading: firstly, the level of antioxidants in tissue culture media is far lower than in plasma and, more seriously, **the authors have not taken into account the cumulative ozone dose.** Although I have already mentioned this point, it is appropriate to remind the reader that ozone solubility is very high: according to Henry's law, every second, ozone solubilizes into water, reacts and disappears, so that more ozone solubilizes and reacts, and this process goes on for days! Although minimal, all of these continuous reactions lead to increasing concentrations of H_2O_2, $OH^{\bullet}$, 4-HNE, etc., which go unquenched on account of the scarcity and consumption of antioxidants and thus become toxic. **Therefore, with time, even the lowest ozone concentration becomes toxic.**

In contrast, exposure of blood to oxygen-ozone is performed with ozone concentrations within the therapeutic window and is over after one min during EBOO and about 5 min during AHT. However, if the ozonetherapist uses either ozone concentrations above 100 mcg/ml or ozonated saline, he makes another blunder. A typical example is represented by the IV infusion of ozonated saline: Foksinski et al. (1999) infused into peripheral occlusive arterial disease (POAD) patients 500 ml of saline ozonated for 1 h (!), obviously without worrying about the high content of newly formed HOCl; they recorded a 450% increase of 8-hydroxy-2'deoxyguanosine (8-OHdG) in the lymphocyte DNA isolated from some of these unlucky patients. 8-OHdG is a marker indicating the occurrence of DNA oxidation. Thus Foksinski's results should absolutely preclude (as clarified in Chapter 6) the use of ozonated saline. An interesting, but not unexpected, result of this study was that only 3 of 6 patients showed the appearance of this marker, suggesting a possible genetic sensitivity to oxidative agents. Kleeberger et al. (1997) were the first to show that a susceptible strain of mice presents different ozone sensitivity (see also Cho et al., 2001). Unfortunately, the state of the ozonetherapeutic art is still too primordial to allow examination of the genetic pattern of antioxidant enzymes in putative patients. Nevertheless, it is necessary to check TAS levels in plasma and ascertain if patients have a G-6PD deficiency.

A reassuring fact is that after millions of AHT sessions performed in Germany, Austria, Switzerland and Italy, neither serious acute nor chronic side effects, nor an increased cancer incidence has been reported. Yet this does not absolve us from improving our controls by monitoring oxidative stress and lipid peroxidation in patients during and after ozonetherapy, e.g. by measuring F_2-isoprostanes (F_2-IsoPs), hydroperoxides and/or other parameters in plasma or urine. This is easier said than done, but I am hopeful that a specific and reliable assay for routine clinical use will soon become available. Furthermore, we must never lower our attention to the use of precise ozone generators and ozone doses that are biologically active but atoxic. If we work correctly, perhaps in due time the scientific community will accept the concept that ozonetherapy is not comparable to life-long endogenous ROS toxicity.

In conclusion, I cannot avoid saying that ozone is potentially toxic and mutagenic (like all cytotoxic drugs!) but so far, our experimental data and clinical evidence has not shown any risk. Jacobs (1982) has carefully examined all the possible negative effects of ozonetherapy. In spite of the famous "toxicity" of ozone, it appears that the incidence is only 0.0007%, one of the lowest in medicine. Four deaths due to direct IV injection of the gas were included in his data, but since 1982 other deaths due to malpractice have occurred, of which at least three in Italy. Thus Jacobs' data are valuable only with regard to side effects such as nausea, headache, tiredness and the like.

The reader will have to trust the Italian experience: at the Verona Congress (1999), Dr. Giuseppe Amato, who has always worked at the Hospital in Conegliano (Veneto) and is a very scrupulous ozonetherapist, reported only minor side effects and no sequelae in a thousand patients treated with AHT for several years. Our experience at the Siena University Hospital is also significant: since 1995, we have performed about 8000 AHT in ARMD patients and about 100 in patients with fibromyositis, as well as about 800 EBOO sessions, countless topical applications in chronic ulcers of the limbs, and either direct (intradisc) or indirect (chemical acupuncture with oxygen-ozone in the paravertebral muscles) applications in about 80 patients with backache.

Firstly, regarding side effects occurring during and after AHT, we have to distinguish about 5000 treatments performed between 1995 and June 2000, unfortunately using PVC autotransfusion bags. These contained 63 ml of CPD (up to 450 ml blood could be collected), but usually only 200-250 ml blood was withdrawn to treat ARMD patients. In order to avoid any contamination, the excess of CPD (about 30 ml) was not discarded and it was responsible for one of the following side effects. Moreover plastic autotransfusion bags had the following disadvantages:

a) Venous puncture was done with a venous fistula needle set (G17) and occasionally some patients fainted with fear. No case of lipothymia was observed, probably because, after blood collection during the ozonation process, a volume of about 100 ml saline was infused via the same needle.

b) Some patients (almost always women) reported a tingling sensation in the lips and tongue, most frequently towards the end of the reinfusion. This did not occur with very slow infusion, nor with the new atoxic system (sodium citrate solution well calibrated to the blood volume), nor with heparinized blood; hence **this symptom has been attributed to an excessively rapid reinfusion with a transitory slight hypocalcaemia due to the excess of citrate.**

c) During blood reinfusion, more frequently women (10-15%) have reported nausea, a feeling of stomach bloating and a strange metallic taste in their mouth, which could be due to Zn-stearate or Zn-2-ethyl hexanoate present as additives in PVC bags.

d) For about 1 day after the first 4-5 treatments, 20-30% of both male and female patients reported feeling tired. Another 10-20% had no symptoms, while 50% reported a feeling of wellness. It must be noted that in all of these patients (60-80 years old), the AHT were performed with a constant ozone concentration of 65-70 mcg/ml per ml of blood, without scaling up the dosage. In retrospect, this was a mistake and particularly in aged patients we must begin with a low ozone dose (20 mcg/ml) and slowly scale up to 40-50 mcg/ml. Since 2001, we have adopted the strategy: "start low (10-20 mcg/ml), go slow" (up to 40-80 mcg/ml, if necessary).

e) After 4-12 AHT sessions, four women patients (one with the history of an episode of anaphylactic shock to a wasp-sting) had a sudden appearance of a diffuse erythematous skin rash, with itching, nausea, hot flushes and slight hypotension, at the end of a blood reinfusion. Intravenous infusion of 1 g methyl-prednisolone Na-succinate relieved the symptoms in about 2 h. Interestingly, before undergoing ozonetherapy, one of these patients had participated as a control and had received 12 oxygenated (no ozone present) AHT without any problem. These cases of definitive intolerance were attributed to progressive sensitization to an immunogen due to phthalates bound to lipoproteins or to other PVC-additive components released after ozone addition.

From June 2000 until March 2004, we have been using the new atoxic system (glass, etc.), a precise volume of 3.8% Na Citrate to blood (1:9 v/v or 25-225 ml or 30-270 ml) and the slow scaling up of the ozone concentration (usually from 10 to 60 mcg/ml). ALL OF THE ABOVE-MENTIONED SIDE EFFECTS HAVE DISAPPEARED, AND NO OTHERS HAVE APPEARED. MOREOVER, NO ALLERGIC-LIKE INTOLERANCE HAS

BEEN OBSERVED. BECAUSE THE GLASS BOTTLE IS UNDER VACUUM, BLOOD IS DRAWN EASILY AND QUICKLY WITH A SMALLER NEEDLE (G19). IN ANY CASE **THE USE OF PVC BAGS HAS BEEN PROHIBITED BY THE MINISTRY OF HEALTH.**

Today ozone is widely used in orthopaedics, particularly in the case of low back ache (Chapter 9, Section XIII) and it has become fashionable to inject the gas mixture of oxygen-ozone into the trigger points detectable in the paravertebral muscles of patients. I defined this approach as "chemical acupuncture" (Bocci, 1998a) and a likely explanation is that ozone acts locally on nociceptors and evokes a rapid and effective (in about 2/3 of patients) antinociceptive response through chemical mediators. While direct intradisc injection of oxygen-ozone (to degrade the proteoglycans of the herniated disc) remains in the hands of orthopaedists and neurosurgeons, some physicians decide overnight to become ozonetherapists and, with the opportunistic encouragement of an ozone generator salesman, begin to practise the indirect method without knowing anything about ozone. This situation has some risks: in May 2001, one death in Naples was due to this therapy. Immediately after IM injection, ozone dissolves locally in the interstitial water and generates several ROS: if, at the first administration, the ozone concentration is 20-25 mcg/ml and the gas volume exceeds 10 ml, a very acute pain may occasionally cause vagal hypertone (inotropic and chronotropic negative effects), which may culminate in cardiac arrest. If the patient is lucky, he will recover or undergo only transitory lipothymia (bradycardia, hypotension, profuse perspiration, transitory loss of consciousness, etc.). Therefore, **it is advisable to practise "chemical acupuncture" with the usual precaution and by injecting the gas very slowly.** It is advisable to remind the patient of the aphorism "no pain, no gain" and that the pain will be bearable and will last only for a few minutes. In general, the improvement of backache outweighs the transitory therapeutic pain, so that the compliance is good. With a proper injection, the risk of oxygen embolism is nil and only one case of subcutaneous haematoma has been reported (Fabris et al., 2001).The direct intradisc injection may present very slight side effects and rare transitory cephalea. However, in the case of a herniated cervical disk in a young athlete, Alexandre et al. (1999) reported that the patient presented a bilateral amaurosis fugax after the injection, which fortunately reversed after one day. This serious complication can more likely be attributed to transitory ischaemia of the vertebral arteries due to an erroneous position of the head during ozonetherapy than to the ozone itself.

If ozonetherapy is performed correctly, it tends not to cause problems but the ozonetherapist must be able to overcome any emergency because a delayed intervention may end with death. He must know all the steps of basic life support (BLS) and have at hand the Ambu, medical oxygen, possibly an automated external defibrillator and

a few ampoules of epinephrine, atropine and corticosteroids (Cummins, 1994).

On the other hand ozonetherapy procures positive effects: about 2/3 of patients, particularly those that feel depressed and asthenic, report a feeling of well being and euphoria after a few treatments. Whether this is due to the "staging" of the procedure or to ozone or to oxygen, or to all these factors, remains unknown. For a long time, I have wished to perform a kinetic study of the hormonal pattern (CRH, ACTH, Cortisol, DHEA, GH, β-endorphin, somatostatin plasma levels) after these types of treatment. Needless to say, such a study must be performed with appropriate controls and this, unfortunately, will imply the collection of many blood samples. It will be more difficult to evaluate whether there is also a concomitant serotonin and/or dopamine upsurge.

An unresolved question is the optimal time of the day to perform the systemic approaches. On the basis of circadian rhythms of crucial hormones, I believe that the afternoon is the preferable period (Bocci, 1985b), but this is not always possible.

CAN OZONETHERAPY INTERFERE WITH CONVENTIONAL TREATMENT?

Before endeavouring ozonetherapy, the physician must know all the medical history of the patient and the drugs in current use. Mattassi et al., (unpublished), have observed a sudden marked hypotension upon rapid reinfusion of ozonated blood in patients treated with ACE inhibitors.This effect may be due to the activation of the kallikrein-kininogen cascade, as reported by Shiba et al (1997) and Abe et al (1998). However plasma bradykinin is degraded within minutes and a very slow infusion reduces this adverse effect. We have confirmed Mattassi's observations in two patients and I can suggest the following: firstly, warn the patient to omit taking the ACE inhibitor on the day of the AHT's treatment; secondly, slow down the infusion of ozonated blood and thirdly, keep ready a vasopressive drug.

ARE THERE CONTRAINDICATIONS FOR OZONETHERAPY?

This is particularly important for systemic therapy and the risk of ozonetherapy must be weighed against the clinical condition of the patient. Moreover, the following situations preclude or limit its use:

a) Patients with a significant deficit of G-6PD. Favism is a haemolytic disease observed in some people lacking the enzyme. **This enzyme provides crucial reducing equivalents able to abolish excessive oxidation and intensive haemolysis** (Chapter 4).The problem of genetic susceptibility to ozone is surely appropriate (McDonnell, 1991; Prows et al., 1997; Kleeberger et al., 1997) and besides individual TAS levels, each patient has a different enzymatic profile, different absorption and metabolism of antioxidants and so on. However, we are still at a very rudimentary stage with regard to the resolution of these problems.

b) Pregnancy, particularly the early phase, to exclude any mutagenic risk, although it is unlikely.

c) Patients being treated with ACE inhibitors.

d) Abnormal situations with hyperthyroidism, thrombocytopenia and serious cardio-vascular instability.

e) Allergy to ozone has been claimed, but what is it? I reckon that the hypersensitivity of asthmatic patients breathing air polluted with ozone has created some confusion (McConnell et al., 2002).

DOES PROLONGED USE OF OZONETHERAPY GIVE RISE TO SEQUELAE SUCH AS TUMOURS, DEGENERATIVE DISEASE, ETC.?

The question is theoretically appropriate because ozone induces ROS and these are at least partly responsible for many ailments and ageing. This is the sixth time that I propose that all national Health Authorities oblige all ozonetherapists (who ought to be physicians with appropriate specific training) to keep a medical register in which they should record all pathological events appearing in patients during and after ozonetherapy.

The following form may be useful:

Surname and Name ....
Sex..............................Age............................
Address...
Type of employment..
Diagnosis...
Type of O_2-O_3 treatment....................................
Period of treatment: from..............to ....
Clinical evolution ....

Whenever possible, the patient should be followed during subsequent years and it should be noted if the disease improves or persists or worsens, as well as the possible appearance of new pathologies related to oxidative stress.

Great attention should be given to:

agranulocytosis, asthma, atherosclerosis, bone marrow dysplasia or atrophy, cataract, degenerative diseases, emphysema, fibrosis (paravertebral muscles), gastrointestinal diseases, hepatitis, hypertension, leukemia and other haematological neoplasias, multiple sclerosis, neurodegenerative diseases (Parkinson, dementias), renal sclerosis, rheumatoid arthritis, scleroderma, skin carcinomas, SLE, solid tumours, others.

CONCLUSIONS: as other medical approaches using potent drugs, ozonetherapy may present some risks, which can be avoided if the ozonetherapist is theoretically and practically well prepared. The use of judicious ozone doses related to the antioxidant capacity of tissues and body fluids excludes the risk of citotoxicity and mutagenicity. Adverse effects, noted with the use of PVC bags and an excess of citrate, are now

totally avoided with the use of the optimized method using ozone-resistant glass bottles. Great care must be exercised when injecting the gas mixture directly into the paravertebral muscles: if this is done correctly, most patients comply well with the therapy. There are a few cases when ozonetherapy is contraindicated and, whenever possible, we must follow the patients during subsequent years and note any possible toxicity or new pathologies.

Chapter 8

IS OZONE REALLY A "WONDER DRUG"?

Even if the reader has only browsed through the previous chapters, he ought to have received my feeling that ozone has an enormous therapeutic potential that, so far, has been either disregarded, if not obstructed by world medical authorities. Reasons for delaying the use of ozone are multiple: while quacks and inexpert ozonetherapists are at fault for poor work, other aspects such as commercial interests, prejudice, lack of knowledge and a myopic medical vision have done their best to block a substantial progress.

Before examining the usefulness of ozone in various diseases (Chapter 9), I would like to summarize the number of biological effects induced by this gas on the body after stimulation of blood, skin, subcutis, muscles and gut lumen. Blood is obviously the best vehicle for transmitting the messages generated by ozone but other tissues have a cooperative relevance.

My words should not be misunderstood in the sense that I always give principal importance to orthodox medicine integrated, when necessary, by ozonetherapy. We shall see that there are vascular diseases such as chronic ulcers and never-healing wounds, where ozone therapy is essential, while in other diseases it has a useful but only a complementary role.

Vasodilation caused by an increased release of NO, nitrosothiols (Joyner and Dietz, 1997; Kashiba et al., 1999) and autacoids can save ischaemic areas in the limbs, heart, brain, kidneys and lungs. An increased supply and release of oxygen and nutrients is crucial for recovering moribund cells, so that a timely intervention can avoid irreversible damages and possibly death.

Release of an array of growth factors from platelets and endothelial cells, while almost impossible to describe in pharmacological and kinetic details, shows its importance by examining every day the extraordinarily rapid healing of necrotic ulcers, particularly enhanced by the topical application of ozonated water and oil.

At least everyone agrees on the disinfectant properties of ozone over the majority of pathogens but, in Western countries, the mental aptitude to profitably use ozone, particularly in chronic infections (large abscesses, peritonitis, osteomyelitis etc.,) is still primordial. How many thousands of patients with septic and toxic shock could have been saved if physicians had accepted my advice to treat them vigorously with ozone therapy?

In spite of the fact that my first interest in ozone was borne out by the finding that oxidants can induce release of cytokines such as TNF alpha (Bocci and Paulesu, 1990), much work remains to be done to fully envisage the activating or/and modulating effect of ozone on the immune system after several months of therapy. Nonetheless, we have gained some evidence that ozonetherapy can be a useful adjuvant for patients with HCV and HIV infections. In this regard, all the hype made by charlatans about the direct intravenous administration of ozone as a route able to "cure" AIDS is highly deplorable, mostly because it served to exploit the good faith of desperate patients. Truly enough, this does not happen only in this field because, in the last decade, too much noise was also made by official medicine regarding gene therapy of tumours (Wadhwa et al., 2002; Noguchi, 2003) and more recently antiangiogenesis. A similar sad story is repeated every day when performing a raving, high-intensity chemotherapy, which often destroys the last resources of the patient. In the cancer section (VI, Chapter 9), I will expand the concept that, while an initial, well-focused chemotherapy can be profitable for getting rid of the bulk or residual tumour, to stubbornly continue the administration of palliative cytotoxic drugs (owing to chemoresistance) is wrong, because **the prolongation of a few months survival is paid dearly by the suffering patients.**

When I name ozone "the wonder drug" of the XXI century, I am not making an overstatement as a foolish retaliation to an unjustified scepticism, but because I have good reasons to believe that prolonged ozonetherapy can allow four extraordinary phenomena:

A) the induction of oxidative shock proteins (OSP),

B) the upregulation of antioxidant enzymes,

C) hence, the reduction, if not the normalization of the oxidative stress and

D) the enhanced release of bone marrow staminal cells (BMSC).

Everyone aware of the current biological trends will agree that these are not farfetched ideas.

Regarding points A and B), the teleological significance of the OSP appears well demonstrated in bacteria, fungi, plants and mammals. These results are truly fascinating.

Any change of the external environment or internal "milieu" disturbs cell homeostasis, but if the stress is tolerable, or graduated in intensity, the cell can adapt to it and survive. If it is too violent, the cell programmes its own death, or apoptosis (Jacobson, 1996). The great number of stresses includes hyperthermia, hyperoxia, hypoxia, ischaemia, excessive ROS and LOPs production, heavy metals, ethanol, hypoglycemia, pH modifications, viral, bacterial and parasitic infections, antibiotics, malignancy, radiation, metabolic inhibitors, amino acid analogs and most likely mental stress and hormonal derangement. Obviously, **OZONE HAS TO BE INCLUDED:**

heat stress proteins (HSP70) is expressed after ozone inhalation (Su and Gordon, 1997) and an attenuation of ozone-induced inflammation has been recorded after repeated daily exposure (Christian et al., 1998). In relation to the variety of stresses, the cell either upregulates or synthesizes probably a hundred or more new proteins like HSPs, glucose-regulated proteins (GRPs) and OSPs, which allow the cell to resist against new and even more intensive stresses. As it has been observed in the cytokine field, also in this case there is an apparent redundancy, with the final aim of establishing "stress tolerance" and insuring cell survival. Already Paracelsus (1493-1541) had this intuition and in the "Nature of Disease" wrote that "the body possesses the high art of wrecking but also restoring health". The Romans, twenty centuries ago, already guessed the power of the "vis medicatrix naturae", or in other words, of the natural ability of the organism to heal itself when appropriately stimulated. The modern pharmacological approach, although useful, can often have a too narrow aim.

I believe that the future of ozonetherapy rests in part on the pedestal of OSP, but it will be necessary to demonstrate how best it can be obtained, its relevance and amplitude.The concept is old and it has been named in different ways only because it has been observed in different pathological conditions: Murry et al. (1986) pioneered the concept of "ischaemic preconditioning" for the heart, which after undergoing a brief, non-lethal period of ischaemia can become resistant to infarction from a subsequent ischaemic insult. Goldman (1996) has introduced the term **"hormesis"** for explaining "the beneficial effect of a low level exposure to an agent that is harmful at high levels", e.g. very low doses of radiation induce an adaptive response to a high dose in human lymphocytes (Olivieri et al., 1984; Wolff, 1996). Calabrese and Baldwin (2001) and Calabrese (2002) have presented numerous examples of stimulatory responses following stimuli below the toxicological threshold. This concept echoes Aristotle's thought (384-322B.C.): "Principium quantitate minimum, potestate autem maximum" ie, a minimal amount of a drug (ozone!) displays potent effects.

"Oxidative preconditioning" has been achieved by warm ischaemia or hyperthermia (Kume et al., 1996; Yamamoto et al., 2000), transitory limb ischaemia (Sun et al., 1999), AHT (Bocci, 1996a, c) and RI of ozone (Leon et al., 1998; Barber et al., 1999; Peralta et al., 1999, 2000; Borrego et al., 2004; Gonzalez, 2004). However, ***when ozone is used, the term "ozone tolerance" or "adaptation to COS" seems more appropriate because it specifies the inducing agent. We face a real paradox, since ozone, the "toxic gas", can be turned into a useful drug able to readjust an otherwise irreversible state of chronic oxidative stress.***

There are several pathologies, such as atherosclerosis, diabetes, ischemia, hyperhomocysteinaemia, neurodegeneration, nephropaties, chronic viral infections, autoimmune diseases and cancer where a vicious imbalance between oxidants and antioxidants becomes firmly established, leading more or less rapidly to death. Today we are also concerned about the *obesity* epidemic as a serious health-risk factor.

How can modern medicine correct this?

Let us first consider **the orthodox strategies to reduce oxidative stress in these diseases.** Owing to the great variety of metabolic disorders, approaches aims to:

1)**Inhibit xanthine oxidase** to reduce formation of superoxide using allopurinol (Farquharson et al., 2002).

2)**Inhibit NAD(P)H oxidase** (Lambeth, 2004). A direct action remains an unsolved pharmacological problem.

3)**Inhibit the renin-angiotensin system.** Angiotensin-converting enzyme (ACE) inhibitors and Ang-II receptor antagonists are broadly used drugs for effectively reducing blood pressure and interestingly they can also reduce oxidative stress by inhibiting NAD(P)H oxidase. On the other hand, Ca^{2+} channel blockers, beta blockers and alpha receptor blockers are antihypertensive but do not improve the antioxidant status in patients (Baykal et al., 2003). Administration of diuretics is helpful but is transitory.

4)**Inhibit 3-hydroxy-3-methylglutaryl coenzyme A (HMG-CoA) reductase**, which is the key enzyme of cholesterol biosynthesis. There are now a variety of lipophilic and hydrophilic statins able to lower serum cholesterol levels, increase the number of hepatic LDL receptors and modulate pathophysiologic processes in patients with acute coronary syndromes (Spencer et al., 2004). Statins have proved to be much more than simple lipid lowering agents (Liao, 2002) because, by blocking the synthesis of critical isoprenoid intermediates, they express several other effects such as: the inhibition of NAD(P)H oxidase, the increased expression of endothelial NO synthase and of tissue-type plasminogen activator, while the expression of plasminogen activator inhibitor and endothelin-1 are inhibited. Thus the multiplicity of hepatic and extrahepatic effects, by reducing inflammation, tumour progression (Katano et al., 2004) and an excessive immune reactivity (Vollmer et al., 2004) have raised statins at the level of a "miracle drug" comparable to the old penicillin (Roberts, 1996). Statins seem also able to mobilize bone marrow-derived endothelial progenitor cells (Llevadot et al., 2001) and practically every month a new beneficial effect is discovered. One problem with wonderful statins is their cost, which limits their use to an unacceptable minority of patients (Topol, 2004).

5)**Inhibit the excess of oxidants production by administration of either antioxidant vitamins** or of a "healthy diet" enriched with polyphenols and flavonoids (red wine, olive oil, etc). It is also known that administration of thiol-containing compounds (NAC and alpha lipoic acid) can inhibit LDL oxidation. This seems an easy solution but does ADMINISTRATION OF ANTIOXIDANTS really work? This is a recurrent and fashionable theme, often discussed by vitaminologists and by charlatans, who may intoxicate patients with megadoses of selenium, zinc, iron and vitamins A, C and E. Authoritative scientists have often posed the question as to whether supplementation with antioxidants (Antioxidant therapy, AT) reduces oxidative damage in humans. The conclusion is that an equilibrated dose may be essential during growth and useful in oxidative stress-related conditions, but there is little evidence that it can be a definitive remedy (Hennekens et al., 1994; Packer et al., 1997; Zino et al., 1997; Clinton, 1998; Halliwell, 1999 a,b; McCall and Frei, 1999; Pryor, 2000; Polidori et al., 2001, 2004; Bender, 2002; Vivekananthan et al., 2003; Seifried et al., 2003; Ames, 2004). An excessive amount may modulate the synthesis of HSPs and actually reduce the synthesis of HO-1 (Peng et al., 2000). If we wish to tackle this problem realistically, we must consider:

a) the uncertainty of intestinal absorption;
b) the individual variability of metabolism and excretion;
c) the variable and often reduced uptake of antioxidants by the cell;
d) the possible reduced synthesis of GSH (observed in HIV infection);
e) the potential toxicity of excessive doses;
f) the inability of antioxidants to stimulate the synthesis of antioxidant enzymes;
g) if not, to inhibit this process.

Thus the problem of antioxidant supplementation must be seriously considered and, while it is certainly useful to administer a correct and equilibrated amount, it cannot do miracles.

6)**Inhibit production of superoxide by long-term administration of L-arginine** (Enwonwu, 1989; Morris et al., 2000), which is the substrate for NO synthesis.

7)**Inhibit the excessive production of superoxide by SOD mimetics** (Fontana et al., 1999), because the administration of an exogenous enzyme, unable to enter into the cell, has shown to be useless. Induction of SOD by gene transfer is fashionable but, until we are able to control transgene expression and the homogenous distribution of the vector all over the vascular system, it remains a theoretical possibility difficult to realize.

8)**Inhibit the increase of homocysteine levels in the plasma** because the auto-oxidation of its sulfhydryl group generates superoxide and hydrogen peroxide that can become cytotoxic for the endothelium, Hyperhomocysteinaemia can be kept under control by the daily administration of folic acid plus vitamins B6 and B12 (Das, 2003) and by increasing the plasma level of adenosine (Riksen et al., 2003).

9)**Inhibit platelet aggregation** with aspirin, ticlopidine and the like.

10) **Inhibit the synthesis of pro-inflammatory autacoids** by the daily administration (2 g) of n-3 PUFAs present in fish oil, which enhance the generation of 3-series PGs and 5-series LTs, which are anti-inflammatory (Belluzzi et al., 1996: Mori et al., 2003).

11) **Inhibit hyperglycaemia** by carefully regulating caloric intake with abundance of fresh vegetables and adopt a correct life style without smoking

and find time for at least 30 min of a moderate physical exercise (Fontana et al., 2004).

I have just summarized the most relevant therapeutic strategies that orthodox medicine offers for reducing the chronic oxidative stress: with the exception of the statin and antihypertensive agents, the use of them separately makes little sense and cannot solve the problem. **Even if it implies taking daily six or more tablets, this long-term cocktail-type therapy is recommended in spite of the cost.** If the patient is compliant, the actual evidence is that the morbidity and mortality of seriously-ill patients diminish markedly, suggesting that **this multiform treatment can slow down the involution.**

Is any point in suggesting ozone therapy? Ozone cannot remove the primary causes of these diseases, but is able to reverse the chronic oxidative stress (Figure 10).

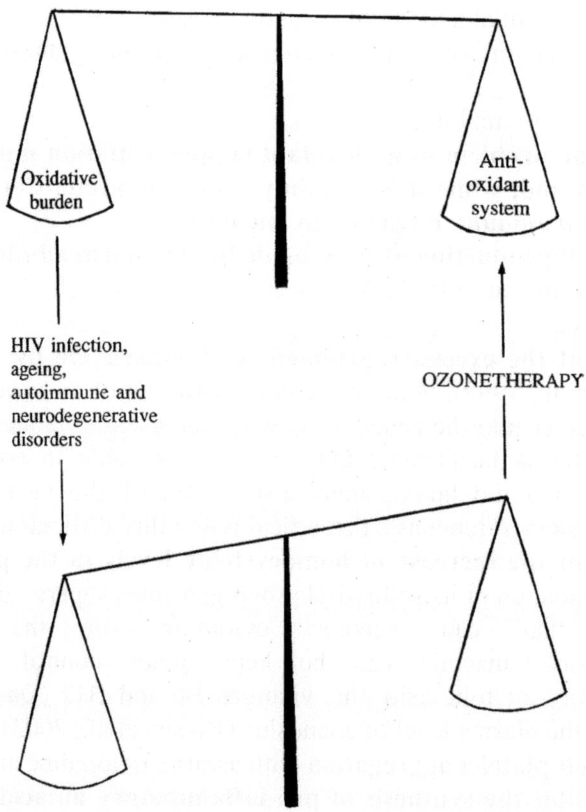

Figure 10. The normal and pathological redox balance. The scheme suggests that, by upregulating the expression of OSP and antioxidant enzymes, ozonetherapy may favour normalization of the impaired redox balance.

Can ozone alone do as much as the above listed eleven treatments? *I envisage the ozone treatment as a transitory and calculated oxidative stress resulting in a sort of "therapeutic shock" for the ailing organism.* ***Ozone realizes this shock because it generates a number of messengers that can reach all cells in the organism.*** How can this happen? First of all, it is necessary to distinguish local from parenteral treatments. Among the latter, major AHT, the "gluco-peroxide" infusion and EBOO are reasonably precise, and both the infused hydrogen peroxide, but especially LOPs with a long half-life, are the most important putative agents. BOEX and RI are somewhat imprecise approaches, but nonetheless likely to put LOPs, generated on the cutaneous and mucosal gut surface, into the circulation. Thus, during and immediately after one of these treatments, cells throughout the body will suddenly receive a pulse of LOPs and newly generated autacoids. As it was already mentioned in Chapter 4, these compounds are heterogeneous and undergo dilution and metabolism (Vasiliou et al., 2000). Over a certain level they are cytotoxic, while below 1 μM they can act as physiological messengers after binding to cell receptors and this is a very good reason starting ozonetherapy with low ozone doses scaled up slowly and cautiously. One possible way to interrupt the cell anergy, due to a chronic oxidative stress, may be an adequate and atoxic stimulation of the cell membrane receptors via a few LOP molecules. If the cell is still able to transduce the message to the nucleus, via phosphorylation of protein kinases and the like, it may represent the alarm signal able to reactivate gene expression, leading to the synthesis of OSPs and antioxidant enzymes. While a too high LOPs concentration or a too advanced disease will end with the cell death, a very low and gradual stimulation may favour a re-equilibration of the oxidant-antioxidant balance, as shown in Figure 10. **If the idea is correct, ozonetherapy should start at concentrations just above the threshold level, which is in line with the old concept "start low, go slow".** Experiments in laboratory animals (Leon et al., 1998; Barber et al., 1999; Peralta et al., 1999, 2000; Borrego et al., 2004) treated daily with RI of ozone have shown a really surprising adaptation to COS, with consequent resistance to prolonged ischaemia or toxic compounds, within two weeks (10 treatments). Of course experimental results in rats are not the last word because in a healthy volunteer (Fig.11), as well as in HIV patients, I found that it took from 2 to 4 weeks (5 to 9 AHT; twice weekly) to detect an increased plasma level of SOD and a concomitant decrease of the TBARS level.

Which proteins and enzymes are important in correcting the COS? This problem has been extensively investigated in the last 15 years and it has been shown that hyperoxia and ROS can induce increased levels of SODs, GSH-Pxs, GSSGR and catalase (Heng et al., 1987; Rahman et al., 1991; Shull et al., 1991; Doroshow, 1995; Hernandez et al. 1995; Bocci, 1996a; Tacchini et al., 1996; Sagara et al. 1998; Wang et al., 1998; Barber et al.,

1999; Chen et al., 2000; Csonka et al., 2000). All of these data have been extremely encouraging to evaluate the effects of ozonetherapy.

We are continuing to investigate the levels of antioxidant enzymes, G-6PD (Puskas et al., 2000) and some OSPs inducible by hydrogen peroxide and ozone (Jornot et al., 1991; Cardile et al., 1995; Kiang and Tsokos, 1998), before, during and after ozonetherapy. We are particularly interested in analysing the pattern of HO-1 (or HSP-32) because even a gentle exposure of blood to ozone is likely to release traces of haeme and its breakdown generates beneficial molecules, such as CO and bilirubin (Abraham et al., 1996), as well as free Fe^{2+} which, if not promptly chelated, may act as a pro-oxidant (Dong et al., 2000; Ryter and Tyrrell, 2000; Snyder and Baranano, 2001). On the whole, **HO-1 is becoming a most interesting enzyme** (Galbraith, 1999; Zuckerbraun and Billiar, 2003), involved in protecting the skin (Reeve and Tyrrell, 1999), in avoiding acute haeme toxicity and iron overload (Nath et al., 2000), in suppressing endothelial cell apoptosis (Brouard et al., 2000), in blocking the growth of vascular smooth muscle cells (Durante, 2003), in rejection of mouse to rat cardiac transplants (Sato et al., 2001) and in protecting heart, liver, kidneys and lungs against ischaemia/reperfusion and hyperoxia injury (Csonka et al., 1999; Amersi et al., 1999; Otterbein, 1999; Miyazono et al., 2002; Choi et al., 2003; Wagner et al., 2003).

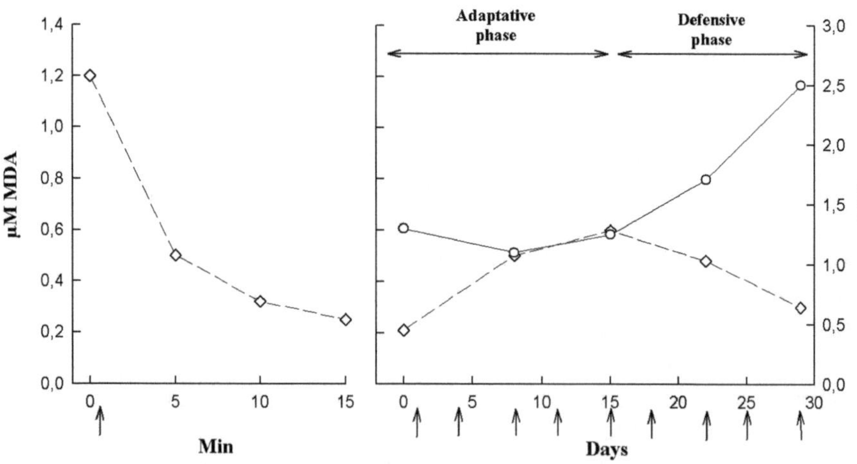

Figure 11. An AMRD patient's response to a single (left side) or intermittent (right side) infusion of ozonated AHT (300 g blood treated with an ozone dose of 21 mg per session). MDA, malonyldialdehyde (◊) and Mn-SOD (U/ml plasma O) are reported on the ordinate. Arrows indicate the time of blood reinfusion.

Thus, **there is already supporting evidence that the adaptation to COS can be realized with ozonetherapy.** Before starting ozonetherapy, we should at least once determine the TAS of each patient. If this is not possible and if the patient is in a critical condition (cachexia, anorexia, great pain, etc.), I feel it is necessary to give a daily well-balanced and reliable supplementation of antioxidants one week before ozonetherapy, calibrated at a correct level. Moreover, in the case of a demanding approach, such as EBOO performed in critical patients, we can start with short treatment periods (20 min only, followed by 30, 40, 50 and finally 60 minutes,corresponding to the 1st, 2nd, 3rd, 4th and 5th treatment, respectively). **We regularly prescribe the following daily oral supplementation:**

- **0.5 g of vitamin C (morning). This dose saturates the body (Levine et al., 1996). I do not see any need of a megadose that may be only partly absorbed, may quench ozone activity, may act as an oxidant and, most likely, be rapidly eliminated with highly acidified urine.In critical patients this dosage can be doubled (Polidori et al., 2004).**
- **0.6 g of NAC (either morning or evening) (Bridgeman et al., 1991; Hack et al., 1998) as the precursor of GSH. I would like to remind that exogenous (oral or/and IV) administration of GSH, with a few notable exceptions (hepatic poisoning etc.,) is a biochemical and pharmacological nonsense. In particular situations this dosage has been increased four-fold. (Hack et al., 1998; Tepel et al., 2003).**
- **an approved multivitamin complex (RD doses) including vitamin E, alpha lipoic acid and selenium;**
- **a rich dietary intake of fresh fruit and vegetables.**

This antioxidant regimen can be maintained throughout the therapy and will allow us to progressively increase the ozone dose without risk. **My belief is that, unless we are able to ACTIVELY increasing the intracellular antioxidant capacity, even if body fluids are flooded with exogenous antioxidants, there is no hope to rehabilitate the cell and to achieve a therapeutic result.**

I wish I could give a definitive answer whether ozone therapy can do as well or even better than the eleven treatments previously discussed and it is pointless to debate this issue unless we can compare them in a randomized clinical trial. This is certainly an impossible task for our means and orthodox medicine will never entertain it because **statins alone represent a colossal "business".** For the time being and the sake of the patient, I can only suggest accepting orthodox therapy associated with the least invasive ozone therapy for obtaining the maximal effect with a minimal discomfort.

The final point D) regards the exciting possibility to improve the oxygenation of ischaemic tissues by promoting angiogenesis. It has been shown already that autologous bone marrow stem cells (BMSC) or/ and endothelial progenitor cells (EPC) can play a role in accelerating angiogenesis of the human myocardium thus improving the perfusion of the infarct zone leading to regeneration (Strauer et al., 2001; Orlic et al., 2001; Schwartz and Curfman, 2002; Aicher et al., 2003).

First of all let us consider how conventional medicine has tried to solve this problem. Two main approaches have been used: The first consists in collecting autologous BMSC and transplanting them via intracoronary or transendocardial routes. The invasiveness of this method may limit its clinical application. The second exploits the release of SC in the circulation after administration of granulocyte-colony stimulating factor (G-CSF). After collection of enriched haemopoietic stem cells (using CD34 as a marker of SC) from the circulation, these have been infused via intracoronary route. This method is fairly practical but there is a risk of in-stent restenosis (Kang et al., 2004). Thus, although both approaches can improve myocardial perfusion, they don't seem ideal procedures.

Ozone therapy could be advantageous because it rapidly improves the oxygenation and the metabolism of ischaemic tissues and could itself mobilise endogenous SC, thus avoiding the need to collect and to transfuse cells. The hypothesis that ozonetherapy may enhance the release of SC from bone marrow was put forward some time ago (Bocci, 2002) for explaining the surprisingly long lasting remission in two of seven cardiopathic patients after EBOO's treatment, when the usual therapeutic effect lasts only a few months. It became obvious to imagine that a sort of myocardial repair could have occurred if BMSC have homed in the infarct zone and regenerated the necrotic myocardium but, regrettably, an appropriate evaluation could not be performed. It is also possible, although less likely, that *in situ* cardiomyocyte replication allowed replacement of the myocardial scar tissue. In a brilliant review, von Harsdorf et al., (2004) have discussed this possibility as the "newt" approach that has been clearly shown in amphibians.

Even if the location of SC remains elusive, it seems that every organ (liver, brain, skeletal muscle, skin and now even cancer) is gifted with these cells but the real trove seems to be the bone marrow that contains about 1 % of haematopoietic and some 0.05 % of mesenchymal stem cells (MSC). It has been demonstrated (Barakat et al., 2004) that in rats, after intraperitoneal injection of ozone at variable concentrations (4.0, 40.0 and 75.0 mcg/ml), an induction of neoangiogenesis can be achieved in both skeletal and cardiac muscle with the medium ozone concentration. **If this happens during prolonged ozone therapy, it remains undetermined, but it is one of the most exciting avenues of research.** After all, almost every day, we notice a far more rapid healing of cutaneous ulcers in patients with chronic limb ischaemia undergoing ozonetherapy, so why couldn't the skin reconstruction mirror the heart repair!

The idea that ozone therapy could mobilize BMSC is supported by sound biochemical data: four years ago, we demonstrated that LOPs present in human ozonized plasma induced NO synthase (NOs) in human endothelial cells and we measured a significant release of NO and nitrosothiols (Valacchi and Bocci, 2000). These compounds are of fundamental importance in the physiology of the vascular bed because they enhance vasodilation and inhibit platelet-leukocyte aggregation-adhesion and muscle cell proliferation (Joyner and Dietz, 1997; Kashiba et al., 1999; Stamler, 2004). Aicher et al., (2003) have added the crucial finding that the induction of endothelial NOs is essential for neovascularization because NO activates matrix metalloproteinase-9 (MMP-9) indispensable for SC mobilization.

In conclusion this process can be distinguished in four phases:

i. MOBILIZATION or RELEASE of BMSC, MSC and EPC. Reinfusion of ozonated blood represents an acute, precisely calculated stress able to stimulate the bone marrow by means of LOPs and possibly autacoids, growth factors and cytokines. ***The sudden homeostatic change in the bone marrow microenvironment caused by these messengers (particularly NO) may well be an effective way for enhancing the output of stem cells.***

ii. THE JOURNEY TO THE TARGET: circulatory BMSC, MSC and EPC do not get lost in the vast expanse of the vascular bed and eventually **home in an injuried site that likely is an ischaemic and/or an infarcted area.**

iii. HOMING may be determined by chemoattractive mechanisms as a damaged tissue may release chemoattracting factors or express new receptors where SC can dock.

iv. INCORPORATION and TISSUE REPAIR, given due time, can occur via proliferation and appropriate differentiation of SC, thanks to improved oxygenation and presence of growth factors in the microenvironment. If this is correct, **even a small number of SC can be eventually sufficient to reconstruct the infarcted zone.**

Although humans have not the power to regenerate the organs, except the liver, the present state of the art is encouraging for the heart and can also help to spare amputations of limbs in some patients. An astonishing result observed in one of our patients at the 4th stage of POAD after ozone therapy has led us to believe that only the new formation of an efficient circulatory network could have allowed the recovery from an apparently irreversible damage. However, **highly compromised patients with advanced dysmetabolic syndrome appear unable to recover.** There can be little doubts that, besides a correct timing and efficacy of the therapy, genetic, metabolic and neuro-endocrine factors play an important role in the final outcome because only a minority of patients have a positive response. Results obtained with prostanoids' infusion are inferior to ozone therapy suggesting that ozone deserves to be thouroughly examined. It will not be easy but we will try our best to investigate with refined instrumental analysis if this repair process really occurs in vasculopathic patients treated with ozone therapy. If ozonetherapy really offers an advantage over the more elaborate administration of staminal cells via special routes (Strauer and

Kornowsky, 2003), it ought to be seriously investigated because we could easily and inexpensively help a far larger number of critical patients.

A final remark regards the duration of an ozonetherapeutic treatment and if it allows to "*cure*" a disease. Around 80 A.D. Tacitus wrote "nature infirmitatis humanae tardiora sunt remedia quam mala" or, on the basis of the nature of human frailty, remedies work more slowly than illnesses. This remains true today for both orthodox medicine and ozonetherapy. With this complementary approach it takes some time to notice a real improvement and this depends very much on the state of the patient, age, type of disease, the quality of the treatment and also on the capacity of the ozonetherapist. Moreover ozonetherapy only rarely can "*cure*" a disease **but it can correct or block its progression and the benefit can often be conserved with a maintenance therapy.**

CONCLUSIONS: this chapter was written for outlining the number of potential benefits obtainable with ozonetherapy. There is a real possibility that, by combining the use of the best medical drugs (statin and hypertension inhibitors) with ozonetherapy, we can really defeat the infamous chronic oxidative stress (COS) with all its negative consequences. As slowly we move on and observe the validity of ozone therapy in new diseases, we are surprised of the breadth of action of this approach and its atoxicity against the blackest predictions. It is regrettable that for lack of organization and resources (practically nothing in comparison to orthodox medicine!), basic and clinical researches progress at a snail's pace. Yet they allow putting forward new exciting scientific ideas that would indicate the ozone capability of restoring health, if we can prove their exactness.

Chapter 9

THE CLINICAL APPLICATION OF OZONETHERAPY

The reader may be eager to examine in which diseases ozonetherapy can be proficiently used and she/he will be amazed by the versatility of this complementary approach (Table 5). The fact that the medical applications are numerous exposes the ozonetherapist to medical derision because superficial observers or sarcastic sceptics consider ozonetherapy as the modern panacea. This is so because ozone, like oxygen, is a molecule able to act simultaneously on several blood components with different functions. The ozone messengers ROS and LOPs can act either locally or systemically in practically all cells of an organism. **In contrast to the dogma that "ozone is always toxic", three decades of clinical experience**, although mostly acquired in private clinics in millions of patients, **have shown that ozone can act as a disinfectant, an oxygen donor, an immunomodulator, a paradoxical inducer of antioxidant enzymes, a metabolic enhancer, an inducer of endothelial nitric oxide synthase and possibly an activator of stem cells with consequent neovascularization and tissue reconstruction**.

Table 5. Ozone therapy can be used in the following medical specialities

Angiology	Gynaecology	Pneumology
Cardiology	Hepatology	Rheumatology
Cosmetology	Infectivology	Stomatology
Dentistry	Intensive therapy	Surgery
Dermatology	Neurology	Urology
Gastroenterology	Oncology	
Gerontology	Orthopaedics	

Figure 2 (Chapter 4) has tried to give a comprehensive idea of how ozonated blood cells and LOPs interact with a number of organs after the initial reaction of ozone with plasma components. One of the substantial differences between classical pharmacology and **ozonetherapy** is that this approach **generates a heterogeneous number of compounds, which, in**

submicromolar concentrations, can trigger a variety of functional activities, hence multiple therapeutic responses rarely obtainable with a single drug. We know that chronic diseases are the result of a number of dysfunctions and the use of a reductionist approach can be disadvantageous. Indeed atherosclerotic patients often complain that during the day they must remember to take six or seven drugs such as a statin, folic acid, antioxidants, an antiplatelet agent, an anticoagulant, an ACE-inhibitor etc., to keep the disease at bay. This example is mentioned not for disregarding conventional medicine but to point out a reality that presents some problems with compliance and eventual outcome. Actually statins produce pleiotropic effects thus resembling ozone because, by inhibiting 3-hydroxyl-3-methylglutaryl coenzyme A reductase, an enzyme crucial to cholesterol and nonsteroidal isoprenoid compounds biosynthesis, they have antiatherosclerotic and surprising immunosuppressive effects (Mach, 2003; Vollmer et al., 2004; McCarey et al., 2004).

On the other hand also **ozonetherapy has drawbacks**: **ozone is a gas intrinsically toxic that cannot be breathed, cannot be stored and must be used with caution and competence.** Thus ozonetherapy can be performed only by physicians after an appropriate training in ozonetherapy using a precise ozone generator equipped with a well-calibrated photometer. It is disgraceful that it is also performed with unprecise ozonators by charlatans and speculators without a medical qualification and this very fact compromises the credibility of ozonetherapy in the medical field. Hopefully this drawback will be overcome when ozonetherapy will become part of official medicine and all public hospitals will have an appropriate service. In the future, with medical supervision and a suitable ozonator, it will be possible to do, at least in part, some automedication using either rectal insufflation or/and body exposure (BOEX). This will represent a big step ahead because chronic patients will treat themselves comfortably at home with the result of maintaining a good quality of life.

The main problem remains the scarcity of clinical trials and the difficulty of knowing and organizing reliable clinical results obtained by individual ozonetherapist. As a consequence, referees have been keen to suggest doing first animal studies. This suggestion is unrealistic because, beside rectal insufflation or intraperitoneal administration of gas (with obvious problems), laboratory animals are not suitable for examining the value of prolonged AHT. Moreover as millions of AHTs carried in humans have already proved their efficacy and atoxicity, why should we waste time with animal models? Too often it has happened that, **even extremely successful results with human tumour transplanted in mice (see the clamour of "tumour infiltrating lymphocytes" and the media frenzing unleashed by the New York Times' article reporting the antiangiogenetic effect of "endostatin") have not been reproduced in the clinical setting!**

While I admire some important and clear-cut results achieved with randomized clinical trials, all of us have to consider that behind such studies there are thousands of biochemists, immunologists, pharmacologists, clinical scientists, statisticians and, even more important, giant pharmaceutical industries providing huge fundings for the research. Even so, because of always being in a hurry to sell the new drug, they commit mistakes and recently a statin had to be withdrawn because of deadly effects. Thus, it is unfair when referees disregard our almost "heroic" efforts to do a clinical study without a sponsor and no other professional help. **From the height of their chairs, they disdain to read or to have the referees' comments regarding papers dealings with ozonetherapy, by solemnly declaring that "the topic is under-researched, the quality is very poor and the theme is not of wide interest to an international readership".** Nothing could be more false than those statements because we address critical issues where official medicine fails to be satisfactory, such as chronic limb and heart ischaemia, ARMD and chronic cutaneous wounds and ulcers that never heal. While it is true that some ozonetherapists treat the trivial cellulite for earning a living, this topic is certainly irrelevant but it is unfair drawing negative conclusions on the whole approach.

Italian Health Authorities, well supported by conventional clinicians, who do not know anything about ozonetherapy, are also disdainful of our work: during the last 12 years they have done their best in rejecting our efforts by obstructing to perform ozonetherapy in public hospitals with the usual excuse that ozone is toxic or that is used badly owing to the lack of precise regulations. Thus, **not only there is no financial support but what is more indecent is that these supreme judges are full of prejudices and refuse to understand even the simplest concepts of this therapy.**

However I am not begging indulgence because any complementary approach must accept and undergo the regulations endorsed by conventional medicine and must clarify whether a treatment is really effective and atoxic. I will then describe the results so far achieved by either presenting either data of any available clinical trial, or a "best case series" or anecdotal, yet reliable results. It will be shown that in many diseases, conventional medicine is quite adequate and ozonetherapy does not necessarily represent the first choice treatment. Indeed the competent physician-ozonetherapist must know all the conventional "gold standard" therapies and use them. **Only when the best standard treatment is not satisfactory, the ozonetherapist may propose the option of ozonetherapy, only if he is sure of its efficacy.** It is also possible that in some diseases, ozonetherapy may complement the conventional treatment and accelerate the resolution of the disease.

I would also state that **the term "alternative medicine" must be rejected because ozonetherapy is still an experimental approach and cannot be antithetical but only complementary.** In spite of important progresses, conventional medicine is still unable to provide a significant improvement in some diseases. Thus **it is ethically correct to take advantage of ozonetherapy when the best orthodox treatment has failed.**

In the next eighteen sections, the biological and clinical effects of oxygen-ozone therapy will be discussed and it will become apparent that this approach can be of critical importance in some diseases, useful if combined to orthodox medicine in others and, so far, useless in a few.

1. INFECTIOUS DISEASES (BACTERIAL, VIRAL, FUNGAL, PARASITIC)

There is no doubt that ozone can have an important therapeutic role in various types of infections because it generates ROS ($O2^{\bullet -}$, $OH^{\bullet}$, H_2O_2, $NO^{\bullet}$ and HOCl), also produced by granulocytes and macrophages during an infectious process (Badwey and Karnowsky, 1980; Chanock et al., 1994; Anderson et al., 1997; Saran et al., 1999; Titheradge, 1999; Babior, 2000). Moreover, neutrophils have a wealth of antimicrobial proteins in their granules and release proinflammatory cytokines which, by exerting a variety of effects, cause tissue damage as well (Witko-Sarsat et al., 2000). Nieva and Wentworth (2004) have entertained the possibility that ozone may be produced *in vivo* via the antibody-catalyzed water-oxidation pathway through a postulated dihydrogentrioxide (H_2O_3) intermediate. To our surprise, it appears that **Nature is able to generate gaseous and reactive molecules (CO, $NO^{\bullet}$ and O_3), which, in trace amounts, may display critical physiological roles, while, during inflammation, excessive amounts cause a continuously damaging oxidative stress.** This reality strengthens my conviction that ozone, used in appropriate doses, can be therapeutically useful.

We observe that, owing to diffuse antibiotic-resistant bacteria, rich countries continue to use expensive and often ineffective antibiotics, while poor countries use ozone which is quite active and has not yet induced resistance. Ozone is profitably employed either as a gas mixture composed of oxygen and ozone, which must be well contained in an ozone-resistant bag and saturated with water vapour, or better as ozonated bidistilled water and oils (to be used only topically), for the treatment of war wounds, anaerobic infections, trophic ulcers and burns (Miroshin and Kontorshikova, 1995). Cellulitis, abscesses, anal fissures, decubitus (bed sores), fistulae, fungal diseases, furunculosis, gingivitis, inveterate osteomyelitis, peritonitis, sinusitis, stomatitis, vulvovaginitis and wound healing disturbances have been shown to improve rapidly because **ozonated solutions display a cleansing effect and act as a powerful disinfectant, which kills even antibiotic-resistant or anaerobic bacteria.** On the whole, ozonated solutions control the bleeding, improve the metabolism and reduce the infection (Payr, 1935; Aubourg, 1940; Rokitansky, 1982; Werkmeister, 1995; Shaschova et al., 1995; Filippi and Kirschner, 1995; Wasser, 1995a; Bulinin et al., 1995; Kudravcev et al., 1995; Kasumjan et al., 1995; Steinhart et al., 1999).

In poor countries, by sheer necessity, physicians have had to devise all sorts of ways to employ the gas, or more easily the ozonated water, to avoid environmental contamination. **In Western countries, we still need to create the mental attitude to profitably use ozone.** Yet, I am convinced that, once medical personnel realize the advantages, it will be put into general use, for the benefit of patients. Moreover, with the current increase in medical costs, ozonetherapy deserves attention because it reduces hospital assistance and is extremely cheap. Obviously we will need to explain how ozone works and show what ozone concentrations are appropriate for the particular infection or lesion. The scheme reported in Figure 4 (Chapter 5) shows that a concentration of 80 mcg/ml (as gas) can be used only during the first phase, in which there is pus, bacteria and necrotic tissue. The wound must be cleaned and exposed to the gas for only 10-15 min. Bidistilled water ozonated with 80 mcg/ml has an effective content of about 20 mcg/ml ozone and is far more practical for cleansing the wound and changing the compress throughout the day. Ozonated oil can be applied any time and certainly for the night. As the infection regresses, ozone concentrations must be lowered to 2-5 mcg/ml to avoid cytotoxicity and to activate local metabolism, cell proliferation and synthesis of cytokines (PDGF, bFGF, TGFβ1, EGF, KGF), so as to promote the synthesis of the intercellular matrix and the healing process (Beck et al., 1993; Pierce et al., 1995; Sporn and Roberts, 1993; Schmid et al., 1993; Slavin, 1996; Martin, 1997). Topical treatment is easy to perform because daily observation of the wound is a good guide; however, it helps to know that time, patience and compliance are good allies.

The problem is more complex in systemic infections (peritonitis, large abscesses, pleural empyema), possibly complicated with toxic and septic shock. In the United States, about half a million patients per year develop sepsis and mortality reports vary between 30 and 70 %. The pathophysiology of severe sepsis is highly complex and includes the activation of the innate immune system, a profound alteration of endothelial cell functions and of the haemostatic system with abnormal release of inflammatory mediators and multiple organ failure (Cohen, 2002; Aird, 3003). Once again, very successful results from animal models of sepsis have not been translated into the clinical setting and the history of therapeutic interventions has been referred to as the "graveyard for pharmaceutical companies". New approaches appear promising and particularly the benefits and risks of activated protein C (drotrecogin alfa) have been recently discussed (Warren et al., 2002; Riedemann et al., 2003). In the past, owing to the lack of an effective treatment, I repeatedly tried to evaluate if ozonetherapy performed in the intensive therapy unit could be of any value (Section XV) but my proposals have been always rejected because, in the case of patient's death, the ozonetherapist is afraid to be considered responsible and penally pursued. There are good reasons for justifying the application of ozonetherapy: removal of purulent material and rapid washing with ozonated water can be useful, particularly combined with AHT which, during the acute phase, can be carried out 2-4 times a day at low ozone concentrations (20-25 mcg/ml per ml of blood). Ozonated AHT is intended to improve tissue perfusion, oxygenation and metabolism but not to increase production of pro-inflammatory cytokines, which are already superinduced by bacterial toxins. It is also clear that it cannot sterilize blood: although most pathogens suspended in water are sensitive to ozone, they become fairly resistant in plasma because of the protection exerted by endogenous antioxidants. Direct IV injection of gas, similar to the sterilization of drinking water in an aqueduct, is simply a mad idea and is proscribed.

In the case of septic ulcers and wounds, topical treatment must be coupled to AHTs because there is a synergism leading to more rapid healing. ***The problem of ulcers which never heal due to diabetes, atherosclerosis, old age and paralysis is one of the most distressing of our times and there are millions of patients suffering with only a faint of hope of solving it. The cost is huge as well and if medical authorities will endorse and develop ozonetherapy, they will assist to a real revolution in the medical treatment of this affection.***

Chronic osteomyelitis, although less frequent, is a disease with severe complications. So far we have treated five patients, three women (age: 51, 81 and 83 old with cleft spine and paraplegia, uremia and uremia plus diabetes, respectively) and two men (39 and 63 old with either an initial dental abscess or multiple myeloma, respectively).

All of these cases had a fistula releasing a foul-smelling secretion, septic fever and two were cachectic and lethargic. They were treated for periods from 6 to 10 weeks with several wide spectrum antibiotics with no improvement. These patients were lucky because they were eventually treated with ozonetherapy as follows:

in a well-ventilated room, direct insufflation of 20 ml of gas (ozone concentration: 70 mcg/ml), via a polyethylene catheter deeply inserted into the fistula, was performed every 4-5 min for one hour, twice daily for the first 6 to 9 days, followed by instillation of ozonated olive oil, that remained all night long. During the first week the topical treatment was combined with one daily AHT (depending upon body weight, 200-300 ml of blood were ozonated with increasing ozone concentrations from 40 to 70 mcg/ml per ml of blood). Supportive therapy with antipyretics and antioxidants without any antibiotics was performed. On the average, after one week, the purulent secretion ceased and fever also receded. **The topical treatment continued once daily for 1 to 3 months and is believed to have been most**

important. During this period, AHT was performed three times weekly and was likely responsible for improving the general conditions. When the ozonated water was available, it was also used intermittently with the gas. The above schedule may appear approximate and is up to the judgement of the ozonetherapist to increase or decrease the frequency and intensity of the treatment that, in any case, must aim firstly, to eradicate the local infections supported by antibiotic-resistant bacteria and secondly, to stimulate the immune system.

We have had another two patients where we could evaluate **the validity and effectiveness of the combination of AHT and topical therapy.** The first case was a patient with a chronic (one year and two months) **empyema** developed after surgical resection of the left lung for a neoplasm. All the best orthodox medications proved to be of no avail and ozonetherapy was tried as the last resort. AHTs (225 ml of blood with 25 ml of sodium citrate 3.8 %, plus 225 ml of gas with increasing ozone concentrations from 20 up to 70 mcg/ml per ml of blood) were performed three times weekly for a month and then twice for the second month and seemed useful to reinvigorate the patient. However **the topical therapy was crucial in slowly eliminating the secretions:** firstly, via the fistula, by using a polypropylene catheter, we washed the pleural cavity with freshly prepared ozonated water (the initial ozone concentration was at 20 mcg/ml but it was progressively reduced down to 3-4 mcg/ml) and, after draining the water, we insufflated daily for two weeks some 800 ml of gas (oxygen-ozone) at progressively lower concentrations (from 60 mcg/ml down to 5 mcg/ml) every day for the first two weeks and then every other day. The pneumothorax was open via the fistula. Near the end of the second month, the patient was practically cured and topical application of ozonated oil enhanced the healing of the fistula.

The second striking result was achieved in a 67 year-old woman, who had undergone dialysis for several years. **The initial infection started with a bed sore** in the coccygeal area but, in spite of intensive conventional therapy, the infection spread to both legs **evolving towards a necrotizing fascitiis.** A dermatologist took care of the patient and, after a microbiological analysis, administered antibiotics as well as topical antibiotic therapy. However the patient progressively worsened with septic fever and a semicomatose state. After the relatives signed an informed consent, we could perform both parenteral (EBOO) and topical ozonetherapy. The latter was carried out by applying continuously compresses soaked with ozonated water during the day and ozonated oil at night time. Once again this therapeutic combination cured (Figure 12) the patient in about two months (Di Paolo et al., 2002).

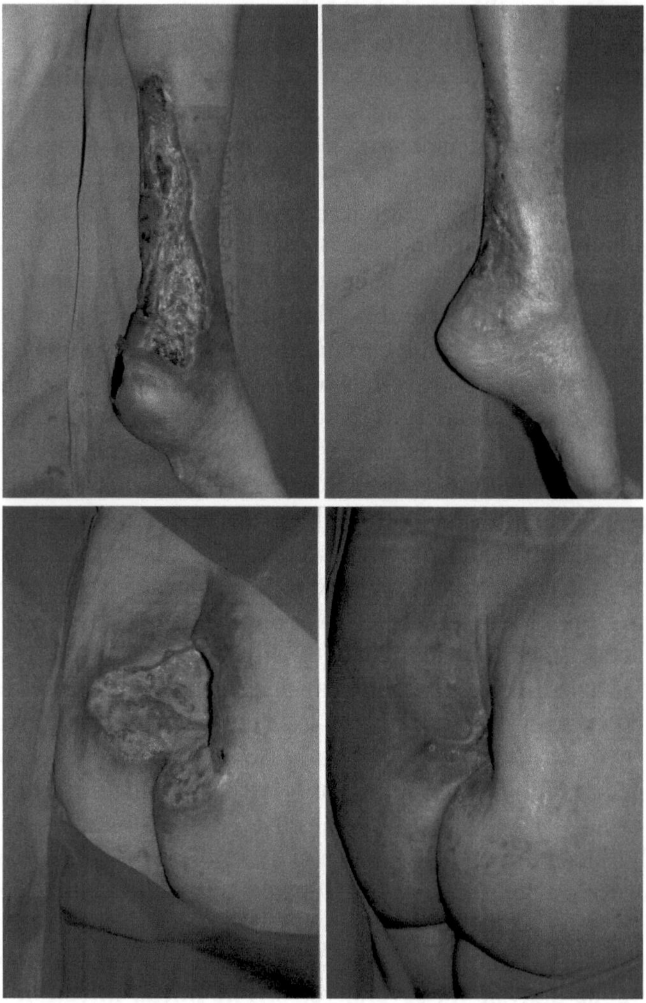

Figure12. The amazing results obtained in one patient with necrotizing fasciitis treated with parenteral (EBOO) and topical (ozonized water and oil) treatments. Extensive necrotic lesions were present between the buttocks, on the legs and heels. Before (left) and after (right) the treatments.

Another **infection** that recently has attracted great attention is maintained **by Helicobacter Pylori** (Hp). This is a gram negative, microaerophilic bacterium which, acquired in childhood, infects the stomach of about 50-80% of children and remains for life (Rowland, 2000). About 50 % of subjects may later on present ulcer disease, chronic gastritis and possibly gastric adenocarcinoma and gastric B cell lymphoma. Surprisingly Hp thrives in the acid environment of the stomach by activating its own cytoplasmic urease which, by converting urea into carbon dioxide and ammonia, neutralizes the acidity of the gastric juice and allows the bacterium colonisation.

Official medicine has elaborated a good therapeutic approach aiming at eradicating the infection. The combination of two antibiotics chosen between clarithromycin, amoxicillin and metronidazole plus a protonic pump inhibitor (omeprazole) is markedly effective but, owing to poor compliance or bacterial resistance, only about 80% of patients are cured. Although the usefulness of drinking idoneous concentrations of ozonated water or/and a dilute solution of hydrogen peroxide is known, no serious study has been yet performed for this chronic infection. Hp bacilli are localized in the deep portion of the mucus gel layer and in between this layer and the apical surface of the gastric epithelial cells. Hp is known to be sensitive to ozone (Baker et al., 2002) and to the generated ROS and therefore, in case of antibiotic resistant bacteria, one can envisage the use of ozone along the line of previous experiments performed at the Cuban Centre of ozonetherapy on Cryptosporidiosis and Giardiasis. However, in order to create a hostile environment to Hp, we must be concerned with safety because the gastric mucosa contains normally a protective mucous layer that may be discontinous in pathological states and allow an oxidative insult to the mucosa (Das et al., 1997). It may suffice to ingest on an empty stomach in the morning, 200-300 ml of freshly ozonated water (final ozone concentration should not exceed 10 mcg/ml) one hour before breakfast The treatment can continue for four weeks before repeating the tests (Hahn et al., 2000) for evaluating the possible eradication of the infection. A serious disadvantage is the need of daily prepared ozonated water and the problem in poor countries, where Hp infection is widespread, may only be solved by developing an effective vaccine.

Fungal, parasitic and protozoan infections, more frequent in hot-humid countries, are seen less frequently in Europe, either as opportunistic infections or after a trip to the tropics. Chagas' disease (American trypanosomiasis) caused by *Trypanosoma cruzi* and African trypanosomiasis (sleeping sickness) caused by *Trypanosoma gambiense and T: rhodesiense* are almost forgotten infections affecting millions of African and Latin America people. Although only an effective vaccine may reduce the problem, I am wondering if ozone therapy could be of any use.

Among fungal infections, those that have been treated with ozone are onychomycosis (*tinea pedis* or athlete's foot) and *candidiasis*. As necessity is the mother of invention, scientists and physicians in Havana have used ozone successfully, showing that it is an effective low-cost antimycotic drug. In a controlled, randomized phase III trial (200 patients) treatment of *tinea pedis* with 1-2 drops of ozonated sunflower oil for six weeks led to a complete and stable cure in 75% of patients (the remaining showed marked improvement). Similarly, 81% patients of the group treated for the same period with ketoconazole cream 2% twice a day were cured (Menendez et al., 2002). Using ozonated olive oil topically, we have achieved incredible results in a variety of chronic infections, particularly relevant in diabetics and invalid patients; in fact, we now believe that **this simple preparation is effective not only because it is a good disinfectant but because is able to stimulate the healing process.** I would like to emphasize that, as soon as prejudices disappear and physicians become aware of this fact and try ozonated oil with good results, **it will become widely used worldwide with great satisfaction for the patient.**

Vaginal infections sustained by *Chlamidia, Candida* and *Trichomonas* have become frequent in young women and can be treated with systemic and topical antifungals. However, if they cannot be eradicated, vaginal washing with ozonated water and oil applied as a pessary have equal, if not superior, effectiveness.

Giardiasis is a parasitic infection caused by the protozoan Giardia lamblia, common in areas with poor sanitation and present even in the United States.

Cryptosporidiosis is also a diarrhoeal disease, caused by protozoa of the genus Cryptosporidium. Good drugs like metronidazole are effective but have some side effects. In Cuba, at first they used to drink ozonated water, at least four of five glasses per day on an empty stomach for repeated periods of 10 days separated by a 1 week interval. According to Sardina et al. (1991), up to 48% of patients became asymptomatic after the second cycle. Ingestion of ozonated oil seems more effective, but it is hard to swallow. An improved administration is represented by capsules (possibly gastro-resistant) filled with ozonated oil. A 10-day cycle "cured" 79% of children, while the remaining 21% showed a marked improvement of symptoms but still had cysts or trophozoites in the faeces (Menendez et al., 1995). No side effects were reported.

There is no need to report other studies because the therapeutic modality is the same. However**, it is certainly worth keeping this approach in mind for use in poor countries of Africa, Asia and South America affected by several fungal and parasitic diseases.** Areas lacking electricity cannot produce ozone and ozonated water**. Thus the World Health Organization (WHO) ought to promote a standard and very economical production of ozonated oil (which keeps well) and distribute it where needed.** I am trying to promote this enterprise, although it may have little value unless we can reduce the rate of infection by improving sanitation in all directions.

Just a few words about **malaria**, which remains another scourge of our time, exacting a toll of more than 1 million deaths each year. Unfortunately, the anopheles mosquitoes have become resistant to insecticides but now it is hoped that the protozoan *Plasmodium falciparum* will remain sensitive to the artemisinin-based combination therapy. Almost 20 years ago, Dockrell and Playfair showed in mice that hydrogen peroxide is able to kill Plasmodium yoelii. At the XV IOA Congress (London, September 10-15, 2001), Viebahn-Hansler et al. reported that parasite growth can be inhibited by ozone at a concentration of 80 mcg/ml after ozonation of a blood cell suspension. In contrast to the sarcastic opinion of many scientists that ozone is a panacea, I doubt that ozonetherapy would ever be useful because parasites are well protected by the plasma and cellular antioxidant system, as well as being hidden in the spleen and other sanctuaries. Moreover the treatment of blood with ozone is a demanding approach and would be difficult to organize in tropical countries for the treatment of millions of people. **One possible solution may be the use of the gluco-peroxide solution because it is reasonably simple to prepare and there is no need for electric power. However I feel pessimistic about wasting our meagre resources on diseases such as HIV and malaria for which the administration of oral drugs or a long-sought vaccine appear rational and could be more useful on a large scale.**

1.1 Viral infections.

It is likely that today there are a billion people affected by chronic viral infections and the potent disinfectant action of ozone comes to mind as a possible helpful solution. While most lipid-enveloped viruses in aqueous media are ozone-sensitive because ozone easily oxidizes glycoproteins and lipoproteins of the external envelope (Akey and Walton, 1985; Shinriki et al., 1988; Vaughn et al., 1990; Wells et al., 1991; Carpendale and Freeberg, 1991), **the virucidal activity becomes uncertain when viruses are in biological fluids or, even worse, when they are intracellular (hepatocytes, epithelia, CD4+ lymphocytes, monocytes, glial and neuronal cells) because, ironically, the potent antioxidant system protects viral integrity.** *This emphasizes once again the irrationality of direct IV injection of gas performed even today in countries lacking medical control.* Quacks exploit anguished patients and spread false and sensational news that this method cures patients and in this way they compromise the progress and acceptance of ozonetherapy.

In order to explore if ozonetherapy can be useful in viral diseases, since 1990 (Bocci and Paulesu, 1990) we examined the possibility that ozone may act in vivo. The following mechanisms may have some relevance:

a) ***A prolonged ozonetherapeutic treatment appears able to induce an adaptation to COS,*** hence a re-equilibration of the cellular redox state, which is a fundamental process for inhibition of HIV, HBV and HCV replication (De Maria et al.,1996, Romero et al.,1998, Akaike et al., 1998; Morisco et al., 2004). As an example, by means of some viral components, e.g. HIV-1 trans-activator of transcription (Tat protein), HIV is able to inhibit or downregulate the synthesis of antioxidant enzymes such as SOD and GSH-Px. This induces an intracellular chronic oxidative stress (increase of $O2^{\bullet -}$, $OH^{\bullet}$), which favours viral replication and, by accelerating cell death, enhances expansion of the disease (Ho, 1997). There are unequivocal experimental data (Westendorp et al., 1995; De Maria et al., 1996; Ranjbar and Holmes, 1996; Schwarz, 1996: Akaike et al., 1998; Larrea et al., 1998; Romero et al., 1998; Rubartelli et al., 1998) that fully agree with the fact that an excess of NAC, GSH and cystamine suppresses in vitro HIV replication (Roederer et al., 1990; Kalebic et al., 1991; Bergamini et al., 1994), while a GSH deficiency impairs survival (Herzenberg et al., 1997). The increased release of extracellular Tat, associated with circulating IFNα, also suppresses immune cell activation and inhibits the production of C-C chemokines, leading to immune collapse (Zagury et al., 1998).

b) ***The induction of cytokine synthesis, such as IFNs and ILs,*** in ozonated blood has been shown to be possible. *Although ozone is a weak inducer, the reinfused lymphocytes and monocytes, by migrating through the lymphoid system, can activate other cells that, in time, will lead to a stimulation of the immune system.* This may represent an important process because it is known that an acute viral disease becomes chronic either because the virus is particularly virulent, or because the heterogenous viral population evolves rapidly and escapes immune control, or because the immune system becomes tolerant to viral antigens and becomes unable to counteract the infection. Moreover, besides the induction of HO-1, a very protective enzyme, the release of some heat shock proteins (HSP) such as HSP60, HSP70 and HSP90 is in order. These proteins are potent activators of the innate immune system, able to induce the synthesis of proinflammatory cytokines by the monocyte-macrophage system and the activation of antigen-presenting cells.

c) ***Oxygen-ozone therapy certainly improves oxygenation and hepatic metabolism*** and indeed we have always found that fibrinogen and prothrombin plasma levels tend to normalize in infected patients, suggesting an improvement of the hepatic protein synthesis. It has not yet been clarified whether ozonetherapy is able to enhance the release of hepatocyte growth factors or of TGF alpha, which may improve liver regeneration.

d) During blood ozonation ex vivo for the minor AHT, using ozone concentrations near 90 mcg/ml per ml of blood, ***it may be feasible to induce the oxidation of free viral components,*** which could represent an inactivated and immunogenic vaccine.

e) It is very likely that ***ozonetherapy activates the psychosomatic system***, thus allowing the release of the growth hormone, ACTH-cortisol and possibly neurotonic hormones and neurotransmitters. If we could demonstrate this point, we would clarify why so often infected patients report a feeling of euphoria and wellness during therapy. Obviously the disappearance of asthenia and depression, a reduction of the wasting syndrome, associated to the lack of side effects, represent positive results.

f) ***In the HIV infection, ozonetherapy may able to correct hyperlipidemia and the acquired lipodystrophy*** that accompanies metabolic and cardiovascular complications (Kotler, 2003; Garg, 2004).

I will then make a few comments for each type of viral infection.

1.1.1 HIV-1 infection

Since 1993, owing to false claims by charlatans, the mass media have misinformed the public, boasting that ozonetherapy or hyperbaric oxygenation could cure HIV infection. The spreading of sensational news is a typical but reprehensible propensity of practitioners of complementary medicine including ozonetherapy. During the period 1991-1995, the epidemic was mounting, AZT monotherapy was hardly useful and only one study using ozonetherapy had been surprisingly accepted and published in AIDS (Garber et al., 1991). This work, poorly conceived, neither showed efficacy, nor toxicity. I leave to the reader to decide about its scientific validity because only 10 ml of infectious blood was treated with an unknown ozone concentration plus heat, plus irradiation with UV before being reinjected intramuscularly as a sort of minor autohaemotherapy.

In 1995 many patients refused AZT because more toxic than effective and solicited me to perform ozonetherapy mostly because news from Germany claimed excellent results with major AHT. Distinguished virologists and clinicians warned me that ozonetherapy, being an oxidative approach, could worsen the disease that by itself was inducing a hyperoxidative state. One hope, that in hindsight has proved to be correct, was ozonetherapy may slowly reverse the unbalance and normalize the redox state, thus limiting the viral replication. The trial accrued ten patients, went on for about 7 months and three patients underwent as many as 54 AHTs, receiving an overall ozone dose of 1080 mg evenly distributed in 16.2 L of blood (Bocci et al., 1998c).

Although the study analysed a limited number of patients, repeated measurements of relevant virological markers indicated that **ozonetherapy carried out with an accurate method (that, very unfortunately at that time used PVC bags for autotransfusion that released immunosuppressive plastic microparticles and phthalates)** neither improved nor worsened the dynamics of HIV-1 replication. $CD4^+$ lymphocytes slightly increased ($p=0.066$) from 272±99 to 341±133. Therapy was stopped in one patient after two months because the viral load in plasma showed a marked increase. Plasma HIV-1 DNA remained stable (~57,000 copies/106 CD4) and HIV-1 RNA levels also remained practically unvaried, except in one case. Serum β2-micro-globulin increased significantly, possibly as a result of ozonetherapy-mediated immunological enhancement. Analysis of the three long-term ozone-treated patients at week 24 confirmed sustained CD4 counts and a stable viral load. While in the lay press there have been many undocumented claims that ozonetherapy is effective in HIV-1 infection, we could not document any substantial advantage (was this due to the use of PVC bags?) even though no patient reported side effects, haematology parameters remained stable and some patients reported a feeling of well-being and a decreased incidence of oral candidiasis and herpes labialis. In any event, against the most pessimistic predictions, ozonetherapy did not harm the patients and it is possible that the documented adaptation to COS countered the oxidative stress established by the virus. Indeed in two patients, we measured a significant increase of erythrocytic SOD after 4 and 5 AHTs (Bocci, 1996a).

Even in these days, I continue to ask myself if I was wrong in selecting the ozone concentration (~68 μg/ml per ml blood), or the schedule, or the use of PVC bags or what else? I also regret that I was unable to retrace these patients and see how they fared.

After the enlightening vision by Ho (1997) and the long overdue introduction of the far more rational highly active anti-retroviral therapy (HAART), the viral replication is usually so well inhibited that levels of free virus in plasma become undetectable in about two-thirds of patients and morbidity and mortality have markedly decreased. In spite of this great progress, it is not yet possible to eradicate the virus (Chun and Fauci, 1999), continuous HAART is toxic (Hruz et al., 2001), difficult to adhere and expensive (although it has the advantage of the selfadministration) and therefore official medicine has proposed to follow the "structured intermittent therapy" (Ruiz et al., 2001) with the possible SC administration of IL-2 for stimulating the lymphocyte proliferation and immune system recovery.

Thus a question often posed is:

does it make any sense today to think that ozonetherapy could help HIV patients? My answer remains: yes and no! No, if we want to substitute HAART with ozone. The former is in continuous evolution and frequently we receive even more potent and less toxic drugs, thus reducing treatment failures due to the induction of resistance or poor compliance (Lalezari et al., 2003). Despite the anecdotes I receive from quacks, I am convinced that ozone cannot match HAART in removing HIV from the plasma, when we know that blocking viral replication is a fundamental step.

However, ozonetherapy may be useful as a complementary therapy for the following reasons:

a) Now, with the new option of BOEX (or at least RI), we have a practical, inexpensive and above all ***non-invasive*** *approach (no venous puncture or risk of infection).*

b) Using a gradual increase of ozone concentrations (from low to medium: 20-40 mcg/ml), we may achieve:

b1) adaptation to COS, hence a re-equilibration of the cellular redox state, which is a fundamental process for inhibition of HIV replication.

b2) correction of hyperlipidemia and peripheral lipodystrophy.

b3) a correction of the wasting syndrome instead of administering recombinant GH and DHEA (Murphy and Longo, 2000).

b4) a feeling of euphoria, counteracting asthenia and depression.

The same objectives can be achieved using AHT but this approach is technically more complex, invasive, more expensive and objected by medical personnel. If we want to assess whether ozonetherapy has any value, we must conduct appropriate studies in collaboration with expert infectivologists but, **in order to satisfy the supreme interest of the patient, we must first use the best of official medicine possibly helped by ozonetherapy.**

1.1.2 Chronic Hepatitis B and C

Chronic HBV and HCV infections affect either 350 or 300 million people worldwide, respectively. The numbers vary in different countries and, as an example, Italy and Egypt have about 2 and 10 million patients, respectively.There is also a different geographic distribution of the known six HCV genotypes and more than 50 subtypes: in Europe and USA, genotype 1, is the most virulent and frequent while genotypes 2 and 3 have a low prevalence. Genotypes 4 and 5 are dominant in Africa and genotype 6 prevails in Asia. Genotype differences deeply influence the susceptibility to antiviral therapy (Hui et al., 2003; Zeuzem, 2004).

Chronic hepatitis diseases are less dramatic than HIV but are certainly very serious ailments from a socio-economic point of view. Not all patients report an aggressive disease and the majority has a mild infection that can perdure for 20-30 years. Nonetheless sooner or later, depending upon sex, ethnicity, age, genotype, viral load, diet, alcoholism, obesity and quality of life, a number of patients develop liver cirrhosis, ascites, hepatocellular carcinoma and eventually end-stage liver disease. Moreover hepatitis may become complicated by cryoglobulinemia, vasculitis, membraneproliferative glomerulonephritis and arthritis (Johnson et al., 1994).

Is there any clinical evidence that ozonetherapy is useful in chronic hepatitis? Until recently we had only anecdotal and insignificant communications and a publication by Knock et al. (1987) who reported "more than satisfactory results" (?) in patients with chronic HBV infection treated with ozone via RI. In collaboration with Dr G. Amato, one of the most reliable Italian ozonetherapist, we carried out two pilot studies: the first one, in 1997, administered 40 AHTs, treated with an ozone concentration no higher than 40 mcg/ml per ml of blood to nine patients in five months. **It was a failure with no results, probably owing to the low ozone concentration and to use of PVC bags for autotransfusion.** The second trial in 14 patients started in 1999 and we used the atoxic glass bottles for the AHT and a constant ozone concentration of 70 mcg/ml per ml of blood.

Unfortunately the schedule, adjusted to the hospital possibilities, was unsuitable: three AHTs per week for three weeks followed by one AHT every month for one year. **All three hepatic enzymes (SGOT, SGPT and GGT) decreased progressively and were within the normal range ($p<0.01$) after 12 months but the viral tests remained positive** (Amato et al., 2000). Although the results were encouraging, the schedule was poor and one monthly treatment appeared absolutely insufficient being aware that even a tight yearly schedule with IFN is partly effective. We realized that lack of funding and the impossibility of performing domiciliary treatments hampered the research.

Since the last seminar congress in Munich (May 23rd-25th 2003), **the outlook has changed thanks to the clinical trial performed by Prof. Nabil Mawsouf et al., (2004) in Egypt.** The study has included 60 patients (45 men and 15 women, age 34-65 years) with chronic HCV infection (genotype 4), treated with AHT and RI during the first two months three months weekly and then twice weekly during the following four months. Each AHT included 150 ml of blood and an equivalent volume of gas, of which the ozone concentration was correctly upgraded from 25 up to 60 mcg/ml per ml of blood. RI was performed with ozone concentrations from 20 up to 40 mcg/ml and gas volumes from 300 up to 350 ml. As I discussed before***, although the RI approach is very approximate, the association with AHT is meaningful and it may display a synergistic effect.***

Extensive tests performed after 8 and 24 weeks showed a highly significant decline of the viral load (up to 95%) and a marked correction of transaminases plasma levels. No side effects were reported and the preliminary conclusion was that ozonetherapy proved to be effective, inexpensive and safe. This is a first serious study but the Authors concluded that it will be necessary to follow up these patients and to programme a randomized double-blind placebo study lasting 12 months. By considering the complexity of the procedures and the need for a total compliance, I am wondering how ethical, although scientifically correct, the evaluation of the placebo (oxygen only) is. At this stage I am unable to evaluate how many of these 60 patients had a total and durable response for making a comparison with the actual gold standard.

Since the early 80s, IFN alpha has been considered the treatment of choice although up to the end of last century, its therapeutic activity was not impressive. Even after intensive (half-one year) therapy, up to 50% of patients showed a good clinical response, but about half of them, particularly those including genotypes 1 and 4, soon relapsed. Side effects, ie., the typical flu-like syndrome, were most frequent during the first month of therapy and elderly patients showed a worrisome depressive state (Bocci, 1988a; Musselman et al., 2001), occasionally leading to suicide. About 20% of patients refused to continue the therapy and those with thrombocytopenia, anaemia and liver fibrosis needed to be cautiously treated. The last breakthrough has come with the more rational introduction of the combination: pegylated (Peg) IFN alpha 2a or 2b (1.5-2.5 mcg/Kg), plus ribavirin (0.8-1.2 g a day) for at least six months. Peg IFN alpha is a "retard" IFN with a very long half-life with the great advantage that the patient can do a self-injection once a week. Remarkably, the response rate is now up to 30-43 % after six months therapy for genotypes 1 and 4 and up to 62% for genotypes 2 and 3. However also the Peg IFN induces adverse effects similar to those with the unpegylated counterpart and ribavirin, an oral purine nucleoside analogue, occasionally induces haemolytic anemia (Zeuzem, 2004).

Owing to effective vaccination, **chronic HBV infection is becoming less frequent but the risk of developing cirrhosis and liver cancer remains high.** Orthodox medicine is providing new effective therapeutic strategies based on IFN, which has antiviral and immunomodulatory properties, and several nucleoside/nucleotide analogues, namely lamivudine, famcyclovir, adefovir dipivoxil, etc., which inhibit HBV polymerase. Vaccines and antisense oligonucleotides complete the armamentarium, which is promising particularly because it combines drugs with different mechanisms of action (Boni et al., 1998; Dianzani, 1999; Pianko and McHutchison, 1999). Usually an intensive, six months therapy elicits a positive response in about 40% of patients and this is a remarkable result.

Almost every day Italian patients call me and ask my advice. I always suggest going to the nearest hepatology centre and starting IFN therapy. Some patients are afraid of side effects and some say that they are intolerant or unresponsive to it. This digression has two purposes: the first is to inform the ozonetherapist of the state of the art, because **she/he has the duty to inform the patient thoroughly about IFN therapy.** The second is to point out that orthodox medicine receives plenty of funding from national agencies and multinational pharmaceutical industries, which are interested in developing drugs to recover their investment and making a profit. In comparison, ozonetherapy is like an ant to an elephant: no funding, no laboratories, no clinics and total disorganization. Moreover, there is another huge disadvantage: although IFN therapy is expensive, the National Health Service pays for the drug and the patient, once instructed, can do it easily at home and visit the hospital every three months for a check up. In contrast, except for the very empirical RI still doubtful on its own, the AHT has to be performed privately and the patient must bear the financial burden out of his own pocket. Moreover, medical personnel are hostile to both ozone and the handling of infected blood. Thus, **although ozonetherapy is relatively inexpensive we cannot evaluate the cost/benefit ratio because the benefits have not yet been definitively demonstrated.** In Italy it is not possible but there is a hope that other countries like Egypt can do further studies. **I still believe that we should clarify whether ozonetherapy has some merits.** This can only be done by randomized clinical trials, comparing ozonetherapy against the orthodox gold standard. The evaluation of the oxygen alone will be important because the relevance of spontaneous remissions must be clarified but the ethic aspect is difficult to accept and, in any case, we must insure that the patients will be properly treated in a second phase.

The most suitable and practical methods are 1) AHT alone or combined with RI and then we could test BOEX in patients with poor venous access. Among chronic hepatitis diseases, we could examine hepatitis C with defined HCV genotype, possibly without any previous treatment because of refusal of IFN. Patients should be of both sexes, between 30 and 50 years old. Informed consent is needed. The most practical schedule seems twice a week (M and Th or Tu and F). **A possible protocol is the following:**

225 ml blood in 3.8% Na citrate (25 ml) plus 225 ml oxygen alone or oxygen-ozone. Use of citrate instead of heparin may reduce ozone's effectiveness but avoids possible complications due to dyscoagulation and potential formation of miniclots.

1st week: 20 mcg/ml for a total ozone dose of 4.50 mg per treatment,
2nd week: 30 mcg/ml for a total ozone dose of 6.75 mg per treatment,
3rd week: 40 mcg/ml for a total ozone dose of 9.00 mg per treatment,
4th week: 50 mcg/ml for a total ozone dose of 11.25 mg per treatment,
5th week: 60 mcg/ml for a total ozone dose of 13.50 mg per treatment,
6th week: 70 mcg/ml for a total ozone dose of 15.75 mg per treatment,

to be continued for 24 weeks (48 sessions) unless a problem arises. We must always apply the strategy of "start low, go slow" for achieving the adaptation to the acute oxidative stress imposed by ozone. Therapy may be continued once a week during the second semester depending on the results. Possible schedules for RI and BOEX have been indicated in the relative sections. Patients should take the usual daily oral antioxidant supplement. Evaluation of therapeutic effectiveness should consider the following end-points:

a) Permanent serum HCV RNA clearance, tested with the most precise system. Viral load should be assessed before treatment, after 3 and 6 months therapy and then after a further 3 months.

b) Normalization of hepatic biochemistry (SGOT, SGPT, GGT, bilirubin levels). Test as in (a).

c) Liver histological results, whenever possible before and 3 months after the 6-month course. If liver biopsy is refused, a surrogate test to indirectly evaluate liver fibrosis may be used. Moreover, in addition to all the routine biochemical tests, TAS, TBARS and PTG should be measured every 3 months. Of particular interest is the evaluation of cholesterol, LDL, HDL, albumin, fibrinogen, prothrombin and CRP.

Patients with HIV, autoantibodies, autoimmune hepatitis, hypergammaglobulemia, haemochromatosis, liver metastasis, incipient cirrhosis, extrahepatic manifestation of HCV infection should be excluded.

Treatment must be obviously cost-free and control patients have the right to be treated with ozone therapy after the first semester. This switch-over might actually be interesting to clearly demonstrate the role of ozone. It would be very important to have the results of this study and I would be glad to collaborate with anyone seriously interested in conducting it. **If they show that at least 40% of patients are good responders, ozonetherapy could be useful in patients who do not tolerate IFN, in elderly patients particularly sensitive to psychotic effects, in hepatitis C patients with normal serum aminotransferase levels but with viremia (Hirsch and Wright, 2000), in patients after liver transplantation and in patients who cannot afford the cost of IFN.**

As the current best conventional combination (Peg IFNα-2a with ribavirin) is good but not entirely satisfactory, it could be supplemented with one AHT treatment per week, which may reduce the severity of adverse effects and enhance immunoactivation.

Moreover, on the basis of our experience clearly showing that a short course of ozonetherapy cannot reduce the viral load, we could test a hybrid approach: firstly, knock down the viral load with a short (2 weeks) intensive treatment with IFNα (Neumann et al., 1998) or IFN-β (Ikeda et al., 2000) followed by AHT according to the schedule described above.

In conclusion I would like to thank Prof Mawsouf and Collaborators for their study showing, for the first time, a serious possibility of using ozonetherapy proficiently so that today we can say that ozonetherapy might be useful in complementing the orthodox therapy to achieve a favourable outcome. I would like to make a plea to all hepatogists to abandon absurd prejudices in order to intensify the research on behalf of too many patients waiting for an appropriate treatment.

1.1.3 Herpetic infections and *Herpes Zoster*

Herpes simplex viruses (HSV-1 and HSV-2, *cold sore virus*) cause human infections involving mucocutaneous surfaces, the CNS and possibly visceral organs in immunosuppressed patients. HSV-1 is mostly responsible for causing oral-facial herpes, but it can spread to give a herpetic eye infection that may lead to corneal blindness. HSV-2 is frequently responsible for lesions on the genitalia, and it recurs periodically. HSV infection of the finger (herpetic whitlow) usually represents a complication of oral or genital herpes.

Although these infections are usually limited, their frequent recurrence compromises the patient's quality of life (Arvin and Prober, 1997). Effective antiviral chemotherapy is prevalently based on systemic (oral and/or IV) administration of nucleoside analogues: acyclovir, famcyclovir, and valacyclovir (Kimberlin and Rouse, 2004). Ganciclovir has been found particularly effective in inhibiting cytomegalovirus (CMV) replication before the development of CMV pneumonia and CMV retinitis in immunocompromised patients. (Crumpacker, 2004). Occasionally, owing to acyclovir-resistant strains, these drugs can be less effective.

Control of HSV infection may be achieved by a vaccine, which has been late in coming and has showed effectiveness only in women previously infected with HSV-1. A promising therapy for genital herpes is the local use of a gel containing an immune response modifier called resiquimod, which is able to stimulate antibody and cytokine production (Bishop et al., 2001).

Herpetic cheratitis can be treated with ophthalmic IFNα or IFNβ plus acyclovir.

Herpes zoster (HZ), or shingles, or Saint Anthony's fire is a distressing disease affecting about 1% of the over-60 population. It is caused by the varicella-zoster virus, which remains in a quiescent state in the nerve root ganglia after recovery from chicken pox. The virus may be reactivated during an immunosuppressive state caused by ageing, chemotherapy, chronic infections or use of steroids. It causes a unilateral dermatomal, vesicular rash associated with severe pain. The frequency of location is: trigeminal (16%), thoracic (50%), cervical (14%) and lumbar (12%) dermatomers. If the disease goes untreated, the pain can last for months and can be complicated by **post-herpetic neuralgia (PHN).** This complication is rare in young and middle-age patients (30-50 years) but is frequent in elderly patients. PHN should be prevented by intensive therapy as early as possible. The sooner an appropriate treatment is started, the better. Unfortunately, the incidence of this complication increases with age and with immune depression. It seems that microinfusion of anaesthetics via the peridural route, initiated no later than 1 week from the appearance of the cutaneous exanthema, may reduce the incidence and minimize the pain. By blocking the axonplasmatic transport, local anaesthetics can prevent diffusion of the HZ virus to neurones in the spinal cord, thus reducing neuronal death and the consequent allodynia and abnormal sensations. The anti-epileptic, gabapentin, is widely used, but is not always effective. Prophylaxis in patients over 60 and at risk has been partially accomplished by the administration of specific zoster immune globulin (ZIG) or by shingles vaccine (NIAID, Bethesda, USA, 1999). Antiviral chemotherapy is based on acyclovir, valacyclovir or, probably even better, famcyclovir with or without prednisolone (Wood et al., 1994), but they have little effect on the healing of skin lesions or pain. The use of corticosteroids is controversial: although they reduce inflammation, they inhibit healing and enhance immunosuppression, which is exactly what favours the virus. Administration of amitryptline (25 mg for 3 months) seems to reduce the pain (Dworkin, 1999). Taking antiviral drugs continuously can reduce or suppress herpetic infections, but it is expensive, may cause adverse effects and induce viral resistance.

This is what official medicine offers today, but it cannot necessarily satisfy all patients. Although this disease is not deadly, it is painful and can become serious in immunosuppressed patients. It appears that ozonetherapy can on its own be helpful or, it can beneficially complement orthodox treatments.

Mattassi et al., (1985) treated 20 patients, of which 11 presented herpes simplex and 9 had HZ. I believe the patients were treated with 5 to 12 IV injections (!) of oxygen-ozone. After a few injections, all patients overcame the infectious episode and only a few had a recurrence over several years. None of the patients had side effects. It was stated that results were incredibly rapid and that to be successful the therapy should be started as soon as the lesion appears. Dr. J. Delgado, of the Centre of Medical and Surgical Research in Havana, treated 15 patients suffering from HZ with daily IM infections of gas and topical applications of ozonated sunflower oil. He noted a marked improvement after a few days and all patients were cured after two weeks, without showing any relapse. He concluded that "the low cost, the easy availability and simple application made ozonetherapy the treatment of choice". Konrad, working in Sao Paulo (Brazil) has reported (1995, 2001) that AHT was effective in both herpetic infections and was able to minimize the complication of PHN evaluated in 55 patients. The work of Dr .G Amato in treating PHN patients performed at the Hospital "DeGironcoli" at Conegliano (Veneto) in Italy during the last decade is outstanding (Personal communication). Although this is an open study, it is praiseworthy and regards 180 patients (84 men and 96 women) between 40 and 85 years of age:

Age 40-50: 30 patients (16.7%).
Age 51-70: 60 patients (33.3%).
Age 71-80: 54 patients (30%).
Age 81-85: 36 patients (20%).

The location of HZ was as follows:

Ocular region: 18 patients.
Head, neck and arms: 30 patients.
Thorax: 30 patients.
Lumbar region: 48 patients.
Limbs: 54 patients.

Unfortunately patients always arrived at the hospital with some delay when previous physicians felt unable to deal with the intense pain of acute HZ infection. Evaluation of pain was carried out with the visual analogue scale (VAS). On the basis of previous experience, Amato decided to abandon all conventional medication and examine ozonetherapy associated with the microinfusion of anaesthetics (usually 12 ml of marcaine at 0.25% daily) mostly via the epidural route to block the sympathetic system in relation to the dermatome presenting the cutaneous rash. The concomitant use of two therapies or the lack of a control is open to criticism, but in the case of PHN it was done for ethical reasons in order to reduce the pain.

Amato proceeded systematically to perform:

a) **AHT** (150 ml of blood collected in Na citrate and a total ozone dose of 10.5 mg or 70 mcg/ml) every day for 4 consecutive days and then every other day for 2 weeks (at least 10 treatments).

b) **Local treatment using compresses moistened with ozonated water during the day and application of ozonated oil at night**. The topical treatment does alleviate the pain and enhance healing.

c) **Sympathicolysis of the stellate ganglion or other ganglia at various levels.**

Owing to the fact that patients below 50 years rarely develop PHN, they underwent only ozonetherapy. Pain disappeared after 2-3 days (i.e. after 2-3 AHT) and the exanthema also improved very rapidly. Three patients (out of 30) developed PHN after 2 months and they were promptly treated with anaesthetics. However, in the subjects over 50 (150 patients), Amato believed it ethically correct to practise both therapies on a prophylactic basis, because they were at a real risk of developing PHN.

Anaesthetic treatment was performed daily for no more than 10 days at the level of the stellate ganglion and for no more than 20 days in other locations.

On average, after 3-4 days the pain disappeared in about 90% of patients and, although further treatment seemed unnecessary, it was continued for up to 20 days in order to prevent PHN later on. All patients were followed up for 2 to 5 years: of 99 patients older than 50 and treated as indicated above for the first week, only 12 developed mild PHN that was successfully treated with both therapies. Of the remaining 51 patients treated with a delay longer than one week, the percentage increased and was in relation to the delay. **In conclusion, it appears that the combination of ozonetherapy with anaesthetic intervention is most effective in preventing PHN in patients over 50. Remarkably, patients did not take any antiviral drug.**

In view of the difficulty of managing PHN, the results appear impressive. By sheer necessity, they lack controls (oxygen only) and, in this regard, I must report another surprising study. Olwin et al. (1997) found that **minor AHT (10 ml of blood NOT treated with oxygen-ozone or oxygen alone) was effective in eliminating clinical sequelae in 8 of 12 (66%) patients with thoracic HZ, in 9 of 9 (100%) patients with ophthalmic HZ and in 1 with lumbar-thigh HZ.** They claimed (data not presented) that IFNα, IFNγ and IL-4 levels were increased in the patients within 24 hours after the IM blood injection. They also mentioned that another 25 cases of herpes infections of various types yielded favourable results, noting that the rate of success depends on early intervention. A delay of 2-13 months between the first symptoms and treatment yields negative results. As this report originates from reliable institutions (Rush Presbyterian St. Luke's Medical Center and Life Sciences Department, ITT Research Institute, Chicago, USA), the data ought to be reliable. If they are, they partly support Amato's data; yet they refute the value of oxygen-ozone. Moreover, ***if they are true, Health Authorities and official Medicine have the obligation to verify them:*** irrespective of the skepticism toward ozone, it appears ridiculous to use expensive drugs when a few trivial injections of autologous blood into the patient's buttock could relieve awful pain in 2 to 8 days. However, authoritative scientists and clinicians obviously do not bother to believe, or to read, papers published in the Journal of Alternative and Complementary Medicine and prefer to administer antiviral drugs. I would like to remind that the minor AHT is an old medical practice (Maddox and Back, 1935; Hardwick, 1940; Martindale and Capper, 1952); even I performed it in 1953 when I was an intern in Clinical Medicine! My interpretation is that the minor AHT, without or better with ozone, act as a vaccine and I am convinced that the ozonation enhance the immunogenicity of virus particles.

In Chapter 6, the approaches of the so-called "major" and "minor" AHT have been extensively described. If venous access is lacking, we can use the option of RI or BOEX. **Minor AHT, without or with ozone, is an interesting immunoenhancer approach and it is easy, simple, inexpensive and rapid to perform.** Starting with a low dose and gradually increasing it, we can ozonize 5 ml blood (70-90 mcg/ml and upward) followed by IM injection three times a week, and then slow down as soon as the lesions are healed and the pain is gone. **Local treatment is also important and effective when combined with AHT.** It can be performed easily by applying and repeatedly changing a compress moistened with ozonated water and ozonated oil at night. Vaginal or rectal suppositories of ozonated oil can be employed in genital-anorectal herpes. We must try to start the treatment as soon as prodromic cutaneous-mucosal symptoms appear; the viral reactivation should be suppressed as soon as possible because it reduces the PHN complication. It appears necessary to alert all GPs to send HZ patients to the special PHN unit at the hospital as soon as they make the diagnosis.

CONCLUSIONS: Herpetic infections are painful, depressing diseases and particularly those due to HSV-I and HSV-II are recurrent. They cannot be underestimated because they procure a very poor quality of life. It appears that both herpetic infections and the fearful HZ with the possible combination of PHN can be treated reasonably well with either antiviral drugs or ozonetherapy. However, for the many patients, who suffer more or less frequently of these affections, this is an unsatisfactory information because they only want to know which is the most rapid and effective treatment. It would be a great advancement if we could programme a comparative study including three arms: antivirals, ozonetherapy and both. Such a huge study involves hundreds

of patients, many clinicians and great resources for various analyses and it is beyond our possibility. Only imaginative public-health leaders could undertake this endeavour but do they exist?

Meantime the solution that may yield the best and lasting result (if not the cure) can be obtained by COMBINING the orthodox antiviral agents with ozonated major plus minor autohaemotherapy and topical application of ozonated oil. Genital herpes is the infection that often causes severe psychological effects and the majority of patients feel devastated when they learn the diagnosis. This is the reason why I strongly recommend a combination therapy carried out for a prolonged period and likely to reduce recurrency and the risk of transmission.

1.1.4 Papillomavirus Infections (HPV)

HPV infects the epithelium of skin or mucous membranes and may produce warts, or benign and malignant neoplasias. Common warts *(Verrucae vulgaris and plana)* may be present in children, while plantar warts *(Verrucae plantaris)* are painful and fairly common in young adults. The incidence of venereal warts *(Condyloma acuminatum)* has risen, particularly in women, and represents a common sexually transmitted disease (Cannistra and Niloff, 1996). Viral genotypes 6 and 11 carry a low risk and may cause modest dysplasia of the uterine cervical epithelium, known as cervical intra-epithelial neoplasia (CIN I). Viral genotypes 16, 18, 31, 33 and 35 are more carcinogenic and can induce a CIN II or the more severe form, CIN III (Liaw et al., 1999). Laryngeal papillomas are typical of children and may produce life-threatening airway obstruction. Anogenital warts (venereal warts) can reach monstrous proportions and may be associated with cervical cancer.

Effective conventional therapies include cryosurgery, surgical excision and ablation with a laser. Topical treatments with antimetabolites and podophyllum preparations are scarcely resolutive because the virus is widespread in the basal cell layer and persists if the immune system is unable to destroy infected cells. The use of both IFNalpha and IFNbeta have been successful for laryngeal papillomatosis and partly useful (30-40% response) in preventing venereal HPV recurrences even after prolonged treatment (Friedman-Kien et al., 1988; Kirby et al., 1988; Weck et al., 1988; Bocci et al., 1990). Both the cost and the adverse effects of IFNs reduce the compliance. The fact that HPV infection is an important risk factor for carcinoma is well known and several HPV vaccines are undergoing trials. However, **ozonetherapy could be useful as a complementary therapy**. To the best of my knowledge, reliable data are still lacking, but it may be worthwhile evaluating a protocol in the hope of eradicating cervical-vaginal infections. Therapy should combine parenteral approaches, such as major or minor AHT, RI or BOEX, with local treatment. **After the basic surgical**

treatment, always important to remove the bulk of infected tissue, there are several possibilities: one is the intralesional injection of small volumes of O_2-O_3 (from 10 to 20 mcg/ml). The infiltrating injections of gas must be done slowly and with great care, possibly at the base of the wart; as reported for IFN beta, they are painful and the patient may get discouraged. Intravaginal insufflation of gas (concentration: 30-50 mcg/ml) for a few seconds is more acceptable, as noted during treatment of bacterial and fungal vaginitis. Instillation of ozonated water (final ozone concentration ~ 20 mcg/ml) for 5-10 min can be done at home and application of an ozonated oil pessary every night is practical and certainly far less expensive than an IFN beta gel. The application of an ozonated oil pessary before intercourse may prevent the transmission of sexually transmitted diseases.

The benefit of ozonetherapy remains to be ascertained, but there is no risk, no side effects and certainly a low cost. The possibility of minimizing viral shedding, thus reducing the potential of transmission to sexual partners is not a trivial advantage.

1.1.5 The common cold

This viral infection affects at least once a year the majority of the population. The well-known manifestations of the common cold, i.e. rhinorrhea, nasal congestion, lachrymation and sneezing, are often accompanied by sore throat, malaise and headache. Although the common cold resolves without sequelae in 4 to 9 days, it is a very bothersome infection. Normal individuals do not need particular treatment, but immunosuppressed patients are at risk of pulmonary infections and can be **prophylactically** treated with IFNs. Inhibitors of the viral binding to mucosal cells are not yet available but will be expensive and scarcely effective.

Enormous funds have been spent in the hope that a few applications of IFNs sprayed at the appearance of the first symptoms would abort the infection. As a matter of fact, the applications are always too late: in order to establish the antiviral state, the IFN should bind to the cell receptors a few hours before the viral invasion. The IFN approach has been a financial blunder because the local adverse effects of IFN are worse than the infection itself. Ozone as a gas is toxic for the nasal and respiratory mucosa and must not be used. However, in our lab, during the last five years we have prepared a lot of ozonated bidistilled water every day. It is ready after 5 min of bubbling ozone (concentration 80 mcg/ml) in water. The final ozone concentration is about 20 mcg/ml and, if it is stored in a glass bottle with a tight Teflon tap, it keeps for two days, even though the ozone concentration progressively decreases. If anyone thinks he has caught a cold, he/she can aspirate the ozonated water into the nostrils 3-4 times a day and can take the bottle home for further use. Water passes into the rhinopharynx and is eliminated, but it is not harmful if swallowed. It also helps to gargle the ozonized water at the same time. Although nasal aspiration of ozonized water causes transient irritation (10-15 sec), it is unbelievable how rapidly the nasal congestion, sinus oedema and pharyngodynia disappear rapidly for 3-5 hours, after which it is necessary to repeat the procedure. Inhalation of ozonated oil helps for the night. The infection resolves in 3-4 days, but it is far more tolerable than if it went untreated. Whenever possible, a daily major AHT during the initial 3-4 days does alleviate the asthenia.

This approach is quite empirical and, by considering the instability of ozonated water, it is difficult to develop a practical system.

CONCLUSIONS: it can be said that ozone, in spite of its potent disinfectant activity in vitro, is NOT as active in vivo because pathogens are normally protected by the plasma and cellular antioxidants. This point must be emphasized to prevent the direct intravenous administration of gas into patients practised by quacks, which often leads to deadly oxygen embolism. Nevertheless *ozone can be useful in infectious diseases by activating ancillary mechanisms.* Luckily orthodox medicine has made available a number of antivirals, which, when used in COMBINATION, are effective (but not always curative) in rapidly clearing viruses from the plasma and cells. Unfortunately the hope to eradicate the HIV has not come true and, at this point, ozonetherapy can become useful because it is able to activate several biochemical and immunological pathways that eventually may further reduce the morbidity. This is a realistic vision that regretfully is not shared by orthodox infectivologists but it is hoped that the tendency of treating chronic and complex diseases with reductionist approaches will vanish when clinicians will become convinced of the effectiveness and atoxicity of ozonetherapy. What is at stake is not the validity of one or the other approach but the wellbeing of the patient!

2. ISCHAEMIC DISEASES (HIND-LIMB ISCHAEMIA, CEREBRAL AND HEART ISCHAEMIA, VENOUS STASIS)

After having tested ozonetherapy in a variety of diseases, **the best clinical results have been achieved in ischaemic diseases.** A partial obstruction of limb arteries due to atherosclerosis (Lusis, 2000) or diabetes, or Buerger's disease (thromboangiitis obliterans) leads to a progressive reduction of blood flow to the feet. Lack of perfusion leads to tissue ischaemia and possibly, cell death. Any minor trauma, normally irrelevant, facilitates the formation of an ulcer, which will not heal because oxygen, nutrients and soluble mediators involved in the repair process are lacking.

Acute limb ischaemia, frequently caused by acute thrombotic occlusion of a pre-existing stenosis or by an embolus, requires immediate surgical or medical attention. Atherosclerosis, diabetes, smoking and a stressfull life are factors responsible for an increase of chronic limb ischaemia, which represents a serious socio-economical burden. In Europe, on the basis of the Leriche Fointaine's classification we distinguish four stages,

· **stage I**: Feeling of cold or numbness in the foot and toes. Skin temperature is reduced. The foot is pale and frequently becomes cyanotic.

· **stage II**: Paresthesia and hypoesthesia, firstly localized and successively diffused to the whole foot. Hyporeflexia. This is the phase with incipient neurological defect. Intermittent claudication. Pain may cease with rest.

· **stage III**: Pain at rest with nocturnal exacerbation. Cyanosis becomes well evident in one or several toes, with an incipient trophic lesion or a frank ulcer. (Rate of amputation is ~ 15%).

· **stage IV**: Partial or total necrosis of one or several toes. Pain often becomes unbearable (Rate of amputation is ~ 50%).

The aims of orthodox therapies for vasculopathies are the following:

a) Prevention of critical limb ischaemia.

b) Reduction or disappearance of pain elicited by hypoxia or/and nociceptors's stimulation.

c) Improvement of trophism with enhanced healing and of the quality of life, and

d) Reduction of the amputation rate.

The angiologist has several, precise non-invasive techniques to objectively assess the severity of peripheral occlusive arterial disease (POAD). Extensive epidemiological studies have shown that these patients have practically the same relative risk of death from cardiovascular causes as do patients with a history of cerebrovascular or coronary disease. So far POAD has been the most amenable to be evaluated with ozonetherapy. The following parameters are being used for evaluating the therapeutic efficacy:

1) **"Claudication"** distance in meters.
2) Timing (seconds) for covering a certain distance.
3) An important predictive value is given by **the ankle-brachial index (ABI).** It is assessed by using a 5- to 7- MHz handheld Doppler ultrasound stethoscope. The normal range of values is 0.91-1.30, which decreases to 0.90-0.41 in mild to moderate POAD and to below 0.40 in severe POAD. Patients with ABI below 0.40 are at high risk of a cardiovascular event and present an annual mortality of about 25%.
4) Toe systolic pressure lower than 30 mm Hg.
5) **Arterial stenosis** (as a percentage) at one or more levels.
6) **Conventional angiography.** This is a very useful test but it cannot be repeated frequently and it is first necessary to evaluate the renal function.

7) By using a polarographic needle electrode, **it is possible to measure the pO_2 and the pH at the muscle level.**
8) Doppler waveform analysis and exercise Doppler stress testing.
9) **Thermometric evaluation**
10) **Clinical and photographic evaluation of throphic lesions** with measurements of the size and depth of the lesion.

Before entertaining ozone therapy, the patient must be evaluated for any possible revascularization and there are several operative procedures attempting to achieve the vessel desostruction or the recanalization by either a stenting or a bypass. However lumbar sympathectomy is no longer performed because it does not increase blood flow to the muscle. Spinal-cord stimulation also does not prevent amputation (Klomp et al., 1999). At the extreme, Taylor et al. (1999) have shown that distal venous arterialisation is a unique procedure with promising possibilities for salvage of critically ischaemic, inoperable limbs (stage IV). **Besides surgery, orthodox medicine offers several therapeutic options, including useful supportive measures,** such as quitting smoking, proper hypocaloric and antiatherosclerotic (with n-3PUFA) diet, exercise (Davies, 2000) **and pharmacological treatments as follows:**

1) **Vasodilatators** must be able to improve collateral blood flow and avoid "stealing" blood away from underperfused muscle. **Pentoxyfylline** may enhance oxygenation in ischaemic tissues by increasing blood flow to the microcirculation. It may improve blood rheology by decreasing blood viscosity and enhance erythrocyte flexibility. However, a recent double-blind RCT showed no significant difference in healing rates of pure venous ulcers between patients taking pentoxyfilline and those taking placebo (Dale et al., 1999). In 1999, the FDA approved cilostazol, an inhibitor of phosphodiesterase type 3, which by increasing the concentration of cAMP causes vasodilatation and reduces claudication. **PGE_1 and a stable prostacyclin analogue (iloprost) have been infused in patients with critical leg ischemia** (Wigley et al., 1994). **These compounds, termed prostanoids,** can inhibit the synthesis of thromboxanes, improve the deformability of erythrocytes, reduce blood viscosity, inhibit the production and release of ROS, proteinases and leukotriene (LTB_4) from neutrophils, induce vasodilation for increased production of NO and increase fibrinolysis by activation of tissue plasminogen or/and urochinase. Moreover it seems that the increased consumption of glucose is accompanied by a reduced production of lactate and by a stimulated protein synthesis of the skeletal muscle. **Interestingly, ozonetherapy exerts similar mechanisms of action.** Although both cilostazol and iloprost improve POAD, they cause frequent headaches, palpitations and dizziness and should not be used with patients, who also have heart failure.

2) **Progression of atherosclerosis may be delayed** by treatment of hypercholesterolemia with statins (HMG-CoA reductase inhibitors, 40 mg

pro die) and fibrates (Spencer et al., 2004). These two drugs should not be used simultaneously because of the risk of fatal rabdomyolisis (Lane and Phillips, 2003). **Platelet aggregation inhibitors** (aspirin, ticlopidine, clopidogrel), represent a therapeutic pillar while thrombolytic intervention does not really help POAD patients. **Propionyl levocarnitine** improves muscle metabolism and seems useful in improving the quality of life, but certainly does not solve the central problem. Needless to say, **diabetic patients must be kept under strict control**, **the homocysteine serum concentration must be lowered** and **hypertension controlled with ACE inhibitors or Angiotensin II soluble receptors, beta-blockers and diuretics**, if necessary.

The prognosis of severe POAD patients is dim, with progressive deterioration that limits their ability to perform daily activities. These patients often complain the need to take 6-9 tablets daily and the compliance tends to be poor with time. Although the therapy tries to stabilize the progression of the disease, new approaches have been proposed for generating new vessels and correcting the ischaemia: **neoangiogenetic and growth factors therapy using VEGF, bFGF and HGF** injected either systemically or injected into the ischaemic areas have been tested with some improvement (Lederman et al., 2002). As these factors have a brief life-time, a more durable effect will be probably achieved with gene therapy hoping to eventually find an ideal vector (Laitinen et al., 1998). The latest biotechnological attempt is being performed with **staminal cells obtained mostly from the patient's bone marrow and perhaps from embryonic cells in the future** (Tateishi-Yuyama et al., 2002). An interesting possibility discussed in Chapter 8 is that ozone therapy may be able to mobilize BMSC and allows the neovascularization of the ischaemic limb provided that the patient is not too seriously compromised by the dysmetabolic syndrome (diabetes, uremia, advanced atherosclerosis, etc.)

It is not surprising that patients search for other treatments that may improve their condition. In the field of complementary medicine, several approaches such as acupuncture, homeopathy, herbal therapy, meditation and Chinese medicine have been tried with modest, placebo-like effects. **Oxygen-ozone therapy deserves attention even though orthodox angiologists regard it with scepticism.** My feeling is that patients must first follow the traditional multiform medical therapy and, if results are unsatisfactory, they can undergo ozonetherapy because the so far achieved results indicate a significant advantage. In this field until recently, the work by Rokitansky (1981, 1982), was revered as the best: he evaluated two groups of patients of which the first (232) were treated with the direct intra-arterial (IA) administration of oxygen-ozone into the femoral artery and a second group (140) received conventional vasodilation therapy. The most marked improvement was determined in chronic limb ischaemia, stage II, patients (80 % versus 43.8 %, respectively). The rate of amputation declined from 15 to 10% for stage III and from 50 to 27% for patients, stage IV, treated with intra-arterial (IA) ozone plus topical bagging. Mattassi et al., made a step ahead when in 1987, they compared IA gas injection with the classical AHT and proved that the last method yielded better results without any local complication. A similar study with analogous results was published by Romero et al., in 1988. **The IA administration of gas is now proscribed because it is less effective and risky**. A randomized, double-blind, placebo-controlled crossover study was carried out by Kraft et al., (1998), who examined the effect of AHT on 17 patients with mild hypertension. Although the methodology was exemplary, they investigated the wrong disease and they could only show a transitory (about four months) decrease of the blood pressure that could have been easily achieved by conventional remedies, if not simply by a low-salt diet!

An interesting observation was made by Amato (2000) on the effect of AHT as a unique therapy for angina abdominis (AA). This is a rare, painful abdominal syndrome that manifests itself after a meal, probably owing to a localized transitory ischaemia of the gut. Surgical vascular correction normally solves the problem, but in the three elderly patients studied by Amato it was not feasible. A cycle of 10 AHTs (150 ml of blood treated with ozone: 20-40 mcg/ml per ml of blood) followed by maintenance therapy (one treatment every month) resolved the problem very well and patients, no longer afraid to have a meal, showed a marked improvement without any side effects. The oldest patient, a woman of 87 years, has undergone this therapy since 1994 proving, beyond any doubt, the atoxicity of ozone.

Recently **important studies, analysed with modern techniques, have appeared clearly indicating that ozonetherapy can produce a significant improvement in blood flow and oxygenation in ischaemic tissues:** it appears that the more poorly oxygenated muscles benefited most from the therapy even though this result had been achieved after only two AHTs (Clavo et al., 2003). In comparison to the baseline, common carotid blood-flow was increased by 75% after the third AHT (Clavo et al., 2004). Giunta et al., (2001) treated 27 POAD's patients (clinical stage II and III) with ozonated AHT and detected an improvement of haemorheological parameters and an increased oxygen delivery to ischaemic tissues. Two papers by a Polish group have shown, in comparison to an oxygen-control group, the clinical efficacy of ozonated AHT in haemodialyzed patients with intermittent claudication, without any side effects (Tylicki et al., 2001; 2003; 2004; Biedunkiewicz et al., 2004). ***These results fully confirm the postulation that ozone, via ROS and LOPs, activates several biochemical pathways increasing the vascular flow in the ischaemic areas. Besides the vasodilation due to S-nitrosothiols and the enhanced oxygen delivery, release of growth factors (PDGF, TGF-beta and VEGF) from activated platelets greatly influences tissue regeneration.*** Anecdotal informations suggest that AHT can also be proficient in patients with Raynaud's disease

(Cooke et al., 1997), where infusion of iloprost has proved to be effective for short-term palliation (Block and Sequeira, 2001).

Once again I would like to emphasize the extraordinary efficacy of combining AHT with topical therapy (gas, or better, ozonated water and oil) to allow healing of severe decubitus or necrotic ulcers in the limbs. Regarding ulcers on limbs, irrespective of the aetiology (atherosclerosis, Buerger's disease, diabetes, Raynaud's phenomenon), they do heal, even in exceptional cases (Figure 13) described by De Monte et al. (2004). In the first, a woman was initially treated with a percutaneous chemical (phenol) lumbar sympathectomy, supplemented with a continuous infusion (0.5 ml/hour) of bupivacaine 0.15% via an epidural catheter; this treatment only controlled the pain. The second case was a man with painful bilateral leg ulcerations due to a vasculitis. A lumbar epidural catheter delivering 0.5 ml/hour of bupivacaine 0.20% and 0.125 mg/hour of morphine (3 mg/day) barely controlled the pain and the ulcers worsened. In both cases, healing was achieved by removing the catheters and performing AHT but, because topical therapy with ozonated oil was **not** performed, patients underwent an exceedingly high number of AHTs.

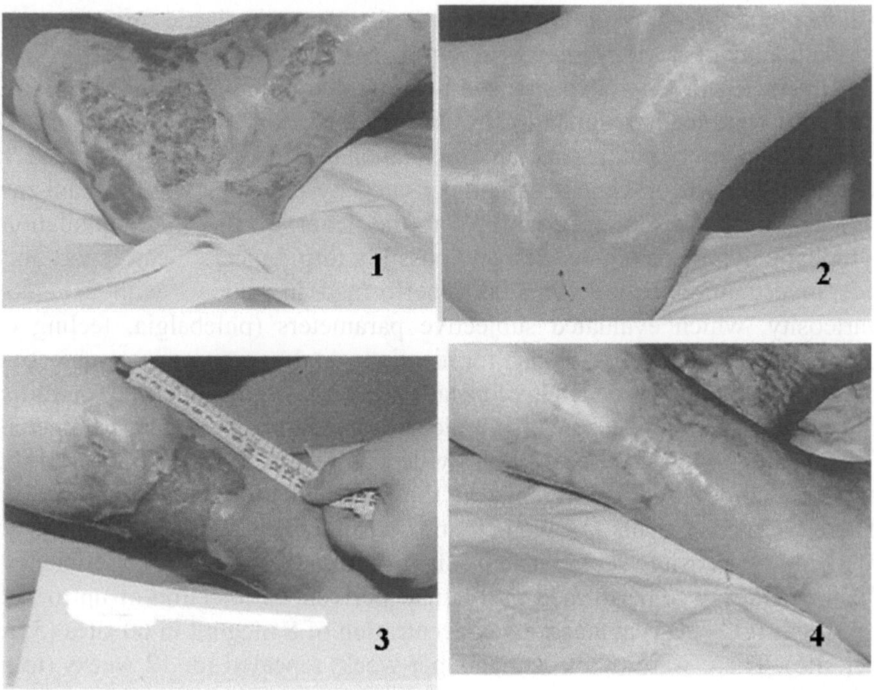

Figure 13. Ozonated AHT in a vasculitis patients before (1) and after (2) 70 treatments. Diabetic patient before (3) and after(4) 40 treatments. Both had a complete healing of the ulcerations

Owing to the high frequency of chronic limb ischemia, **I have been frequently asked which, between prostanoids (the orthodox gold standard) and ozonetherapy, is the treatment that I would select.** There is no doubt regarding the use of the basic conventional drugs (aspirin, statin, hypertension inhibitors etc.,) but then, on the basis of my clinical experience, I prefer ozonetherapy. **Both procedures need a venipuncture** but prostanoids need to be infused for a long time, at least 12 hours at the time, against 15-20 min for the ozonated blood. **Prostanoids often procure side effects** such as pain, edema and erythema after an intra-arterial infusion of the limb. Upon intravenous infusion, patients often complain of cephalea, hot flushes, nausea, vomiting, diarrhea, occasional hypotension, dyspnea and more rarely hyperthermia. **Ozonetherapy, not only is absolutely adverse effects-free, but the majority of patients feel euphoric and energetic in the next few days.** Moreover **ozonetherapy is today the only approach able to normalize a chronic oxidative stress due to ageing, atherosclerosis and diabetes.** A consistent therapeutic cycle (16-20 treatments) followed by two treatments monthly for years can indeed change a dim future. **The cost of prostanoids ranges among 1300-2000 Euro per year, while the cost of the disposable materials for 30 treatments is about 300 Euro.** While patients are always enthusiastic to repeat an ozonetherapy cycle, they are reticent with prostanoids.

Ulcers from venous stasis have been treated and they also heal rapidly with the combined treatment. However, phlebopathies have attracted less interest than arteriopathies and are often amenable to be surgically treated. If venous hypertension cannot be compensated by physiological mechanisms, it leads to increased permeability at the level of the microcirculation, lymphatic hypertension, oedema and possibly torpid ulcers. I can report only one investigation (Lo Prete, 2000) performed in patients with extended varicosity, which evaluated subjective parameters (phlebalgia, feeling of orthostatic weight or pain, formication and paresthesia), objective parameters (evening oedema, constant oedema, haemosiderinic dermitis, fibrous hypodermitis, eczema, skin ulcerations) and instrumental parameters (plethysmography in reflected light, videocapillaroscopy with optical probe and evaluation of circumference at the calf and at the ankle-malleolus). There were 15 patients (14 women and 1 man), from 20-60 years old, with marked varicosity complicated by chronic venous insufficiency. Ozonetherapy was performed by SC and perivenous injection of up to 300 ml of gas (O2 + O3) at an ozone concentration of 8 mcg/ml in 60 sites (5 ml per site). There were two treatments per week, repeated for 12 weeks (total 24 sessions).

There was a marked reduction of the peripheral venous stasis, likely due to improved microcirculation. The SC and perivenous administration of gas caused modest but transitory pain. No more than 5 ml per site ought to be injected. There are no other adverse effects. Simultaneous topical treatment enhances the healing of torpid ulcers. The association with AHT may further improve the treatment.

As **ozone therapy is a valid approach** in treating vascular disorders of the limbs, there is little doubt that it could also be useful **for myocardial and cerebral ischaemia**, because it can: a) increase oxygen, glucose and ATP delivery by several mechanisms discussed in Chapters 4 and 8; b) enhance neoangiogenesis **and possibly the implantation of BMSC**; c) induce the preconditioning phenomena by upregulating the expression of antioxidant enzymes, HO-1 and HSPs and d) trigger a neuro-humoral response for improving the quality of life.

Owing to the systemic nature of atherosclerosis, both the heart and CNS are at high risk in POAD's patients, and indeed there is a rather high incidence of myocardial infarction or ischaemic stroke in these patients. That is why **we are testing the validity of EBOO in end-stage cardiopathic patients when either transplantation or surgical revascularization is not feasible.** Our methodological studies (Bocci et al., 1999 b; 2001 c; Bocci and Di Paolo, 2004) and a preliminary study (Di Paolo et al., 2000) on several patients yielded results that, although encouraging, are regarded as anecdotal because angiocardiographic examination could not be repeated after the treatment. We are still baffled by the prolonged improvement of two of these patients, which might have been caused by an effective cardiac neoangiogenesis induced by ozone therapy. This project is still in progress and has been delayed by technical difficulties due to the need for perfecting the gas-exchanger and the ozone generator. For the time being, there are two studies to be regarded as simply indicative: the Russian trial was carried out in 39 patients with advanced coronary atherosclerosis. They underwent five daily infusions (for 20 days) of ozonated saline solution. I believe that ozonation was carried out at a very low ozone concentration (perhaps 2-3 mcg/ml), so that levels of HOCl were not too high and thus not too caustic! I must say that I am dead against this procedure. However, Zhulina et al. (1993) concluded that the treatment was effective because angina attacks decreased from an average of 6 to about 2 per day. There were no controls with either oxygenated saline or simple saline, which might have shown the relevance of a placebo effect. *Instead of using the ozonated saline, I would strongly suggest to infuse the "gluco-peroxide solution"*, which has a precise rationale and is atoxic. The second study was by Hernandez et al. (1995), who performed AHTs, five days per week for a total of 15 treatments, in 22 cardiopathic patients. They found a significant decrease in plasma cholesterol and LDL levels (we shall see if we can confirm this finding after EBOO) and an increase of erythrocytic GSH-Px and G-6PD, which is in line with the phenomenon of adaptation to COS paradoxically induced by ozone. While the increase of antioxidant enzymes is a good result, a possible advantage for the coronary circulation remains unclear.

Since 2002, **the great hope of modern medicine is to use either angiopoietins or gene therapy or stem cells to elicit therapeutic angiogenesis in patients with chronic myocardial ischaemia for correcting the progressive degeneration** (Patterson and Runge, 2000; Jackson et al., 2001). However, while these new approaches mature (Stamm et al., 2003; Tse et al., 2003; Murry et al., 2004), I do not see anything wrong in evaluating ozone therapy because it may also enhance the mobilisation of endogenous stem cells.

In about 80% of patients, **ischaemic stroke** results from atherothrombotic or thromboembolic processes. Stroke is a major public-health burden worldwide and can strike relatively young persons at the peak of their intellectual activity and, if not fatal, can be highly debilitating (Warlow et al., 2003). Fortunately, Handel and Pasteur, to cite a few, were able to make great contributions to music and science in spite of suffering a stroke. Modern medicine has developed prophylactic measures, previously discussed, able to reduce the risk of transient ischaemic attacks (TIAs) or of stroke in prone individuals by 20-30% (Gubitz and Sandercock, 2000). Moreover, anti-atherosclerotic drugs and, if necessary, carotid endarcterectomy appear beneficial. In the case of an acute stroke, therapy must begin within the shortest possible time (from 0.5 to 2 hours) to reperfuse the ischaemic penumbra surrounding the core of a cerebral infarction. Time delays are predominantly in the pre-hospital phase and can be fatal or cause a permanent invalidity. **Hypoxia induces a cascade of metabolic disorders**, such as tissue acidosis, reduction of ATP levels, Ca^{2+} overload, activation of glutamate receptors, N-methyl-D-aspartate (NMDA) channel opening, release of ROS and proteinases, **leading to neuronal death** (Small et al., 1999; Rosenberg, 1999).

Since the 1990s, the lysis of the clot using recombinant tissue plasminogen activator (Tpa), with due caution to avoid cerebral haemorrhage, has been applied to shorten the time of reperfusion and reduce neuronal damage. Intravenous thrombolysis is a sort of endovascular surgery operated by the enzyme and is one of the remarkable discoveries of modern medicine. Six clinical studies performed in 300 hospitals of 18 countries and including 2775 patients have confirmed that the best results are obtained if the thrombolytic therapy is carried out as soon as possible (1-2 hours) from the stroke first symptoms (Lancet: **363,** 768-774, 2004).

With regard to ozone therapy, a preliminary study has been reported by Dr. G. Wasser, a German ozonetherapist, who has treated stroke patients privately, with all possible inherent disadvantages. He reported at the XII IOA Congress (Lille, 1995, b) that he had treated several patients some time after they suffered an acute stroke. In spite of this limitation, the use of AHT every day seems to have improved the outcome, in the sense that no patient died and they apparently recovered very rapidly. In Cuba, where there is a lack of Tpa, many hospital emergency units have ozone generators at hand

and patients with stroke are luckily treated as soon as possible with oxygen-ozone therapy. Consequently, in 1998, Rodriguez et al., (personal communication), by using multidimensional evaluation test, examined 150 patients before and after ozone therapy phase observing an astonishing improvement in 86% of the patients with, as expected, better results when treatments were performed in the early phase. Unfortunately these results do not get any credit in Western countries

At my University, I have found great disinterest; neurologists do not want to risk what they consider a conventional valid treatment (thrombolysis) for the uncertainty of ozonetherapy. This aptitude is quite correct because the patient's life is the most important issue. If a rapid and intensive ozonated autohaemotherapy would do BETTER than thrombolysis remains unknown and this dilemma will be probably answered in a poor country lacking the expensive drug. I believe that a controlled study using either Tpa or ozonetherapy, or **even better a combination of the two,** performed at the earliest possible time after a stroke, would be informative and could help to save lives, reduce the disability and hospital costs.

CONCLUSIONS: a combination of the basic orthodox medicine and a life-long prolongation of ozone therapy is potentially able to correct the chronic oxidative stress underlying the vascular disease and restore health in seriously ill-patients. This is due to the multiform and simultaneous effects elicited by ozone therapy, a virtue not shared by other approaches. Patients are very much interested to know which will be the best and simple course for taking full advantage of ozone therapy. Among the described approaches (Chapter 6): AHT, RI, BOEX and the "gluco-peroxide"infusion are the least invasive, well tolerated and absolutely atoxic in the long term. RI is the least expensive and the instructed patient can do it at home. In such a case, ozone concentrations may range from an initial 5 mcg up to 20 mcg/ml, increasing the gas volume progressively from 150 ml to 450 ml in 2 weeks The other methods, depending upon the stage of the disease, require two cycles (of 14-20 treatments each) per year with at least one monthly treatment as maintenance in between. Chronic limb ischaemia is often accompanied by an ulcer that will never heal unless we normalize the delivery of oxygen and stimulate tissue regeneration. In this disease, ozone delivers its best messages and behaves really as a wonderful drug when the ozonetherapist combines the ozonated AHT with domiciliary topical therapy with ozonated oil. The local induction and release of growth factors in a sterile and revascularized tissue has a fundamental importance for the healing process. The disappointing clinical outcome from growth factor trials (Bennett et al., 2003) is due to the fact that exogenous hormones applied on an infected and ischaemic tissue are useless. Almost needless to say that the patient must continue

the basic conventional therapy that aims to block the progression of the disease.

3. RETINAL DEGENERATIVE DISORDERS

There are some degenerative disorders of the retina and optic nerve that are progressive and irreversible, for which no therapy has proven effective. These are:

1) Age-related macular degeneration (ARMD),
2) Degenerative myopia,
3) Retinal vascular disorders due to diabetes,
4) Retinal inherited-degenerative disorders of which retinitis pigmentosa is typical,
5) Ischaemic optic neuropathies and
6) Glaucoma.

In spite of the fact that most opthalmologists are sceptical, since 1995, we have shown that ozonated AHT can improve the vision acuity, particularly for affections no. 1, 2 and 5. ***Although it cannot normalize the eye sight or "cure" the disease, it offers the chance for improving the quality of life***

First of all it appears worth while to briefly remember the anatomical aspect of the retina to fully understand the physiopathology and the rationale of the therapy. The retina is a transparent membrane lining the interior of the eye able to receive and process the visual stimuli arriving from the external world. Its outer face is in contact with Bruch's membrane, which separates the vascular choroids from the retinal pigment epithelium (RPE), which represents both the histological and functional connection with the photoreceptors (rods and cones) situated in the outer layer. The neurosensorial retina is one of the most complex structures of the body because is separated into ten layers, of which the photoreceptors are located in the outer layer while the axons of the ganglion cells (second-order neurons) are placed on the inner layer to form the optic nerve. The RPE is vital to the integrity of the photoreceptors. It exerts crucial functions such as the daily phagocytosis of about 10% of the tips of the outer segments of the photoreceptors, the recycling of vitamin A and the transfer of oxygen and nutrients from the choroids to the photoreceptors and outer retina.

The central area of the retina is referred to as the *macula lutea* (about 2 mm in diameter) and in its centre is the *foveola*, an area of about 0.35 mm in diameter where the retina is very thin (about 130 μm) and avascular (Kimura et al., 2003). The *foveola* has the highest concentration of cones and is responsible for the visual acuity, i.e., for the detection of the finest details of any object. For its metabolic requirements, the *foveola* depends entirely on the choriocapillaris circulation because there are no retinal vessels and, among the various tissues of the body, it has a far higher consumption of oxygen than the liver, the kidney and the brain. Thus it becomes understandable that the lack of oxygen rapidly leads to peripheral and/or central loss of vision by degeneration of the neurosensorial cells (D'Amico, 1994). The neuronal degeneration triggered by the ischaemia leads to cell death by the simultaneous participation of several deleterious processes such as bursts of free radicals, generation of peroxynitrite, Ca^{2+}-induced damage and glutamate toxicity. Seddon et al. (2004) have detected a good correlation between C-Reactive Protein (CRP) serum level and ARMD, implicating the role of inflammation in the pathogenesis of the disease. This is possible especially in patients who are obese, smoke and have been exposed to excessive light. On the other hand, ischaemia induces the production of an angiopoietin: the vascular endothelial growth factor (VEGF), which stimulates an imperfect neoangiogenesis, which is the formation of abnormal blood vessels growing from the choriocapillaris through the RPE. **Although this is the natural attempt to correct the hypoxia, it ends up in creating an abnormal vascular network that disrupts the delicate equilibrium between the RPE and the photoreceptors.** The simultaneous hyperpermeability favours leakage of plasma components responsible for the formation of an exudate (or even haemorrhage): this causes a serous retinal detachment able to definitively exclude the photoreceptors from the light stimuli. The complexity of the pathophysiological modifications does not exclude the possibility of defining new pharmacological approaches able to limite the neurotoxic injury. However these are not yet available and meantime oxygen-ozone therapy appears a workable treatment for the management of some ischaemic and neurodegenerative disorders.

An increasing percentage of the population is ageing and the maintenance of a good quality of life for elderly people imposes an ever increasing strain on the national health services. ARMD represents the main cause of irreversible visual loss in developed countries in people over 50 years of age affecting 20-30% of people over the age of 65 (Bressler et al., 1988; Pauleikoff and Koch,1995; Chopdar et al., 2003). Since this section of population is expected to increase during the next century, ARMD is becoming a serious public health problem. Owen et al., (2003) reported that in the United Kingdom there are 214000 partially sighted or blind patients, who, by the year 2011 will increase to 239000. In the USA, between 6 and 10 million Americans are blind from ARMD (Evans, 2001).

ARMD's aetiology remains uncertain, but could be due to a number of factors such as ageing (>55 years), genetic predisposition, smoking (Vingerling et al., 1996), excessive exposure to sunlight causing a photo-oxidative stress (Cruickshanks et al., 1993; Darzins et al., 1997), a blue iris, hyperopia, vascular diseases with hypertension and possibly a nutritional deficiency of zinc and antioxidants (lutein, zeaxanthin, etc,). Moreover in patients affected with the "dry" form the mutation of the ABCR protein appears relevant. This is a transporter protein detected in the outer segment of rod cells that, upon mutation, may favour the accumulation of degraded material able to interfere with the retinal metabolism. These factors, to some extent, reduce choroidal perfusion, cause vascular and haemoreological abnormalities and chronic oxidative stress ultimately leading to death both retinal ganglion cells and photoreceptors.

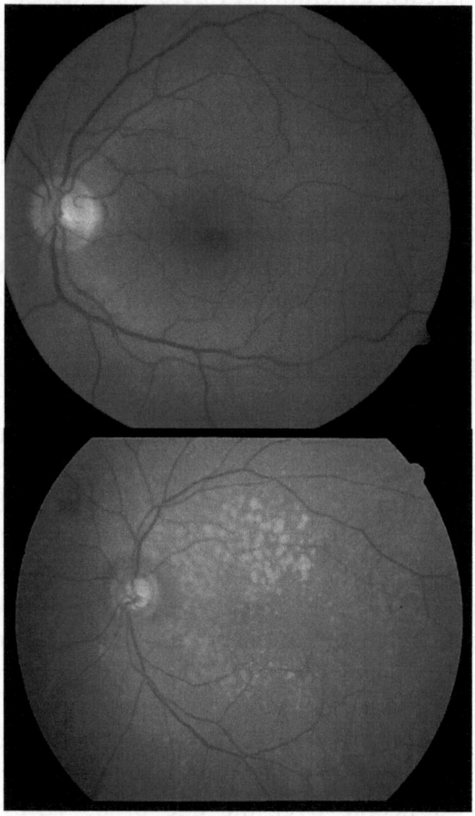

Figure 14. Image of a normal ocular fundus (top) with the macula lutea at the centre. On the bottom, the ocular fundus of a dry form of an ARMD's patient shows a conspicuous number of drusen.(By courtesy of Dr.R.Smettan, Kornwesthein, Germany).

Two main forms of ARMD have been described:

-**"dry"** (non-exudative, atrophic) **form**, characterized by the presence of drusen (hard, soft, mixed) and areolar (geographic) atrophy of the retinal pigment epithelium (RPE). This form occurs in 80-95% of patients and the visual deterioration is slow and becomes serious in only 5-10% of cases, in relation to to the location and the area of atrophy.

-**"wet"** (exudative-neovascular) **form**, characterized by choroidal neovascularization, detachment of the RPE and fibrovascular disciform scarring. It is fairly rare (5-20% of patients), but is associated with poor visual prognosis owing to the loss of central vision in about 75% of cases.

The most frequent signs of ARMD are:

1) ***disturbance of the RPE,*** which may appear disrupted into small areas of hypo- and hyperpigmentation (pigmentary changes) or may become absent, forming large areas of atrophy (areolar [geographic] atrophy). The RPE appears to normally release the pigment epithelium-derived growth factor (PEDF) that has antiangiogenetic and antivasopermeability effects and probably allows the normal proliferation of the RPE (King and Suzuma, 2000; Chader,2001; Rasmussen et al., 2001; Liu et al., 2004).

2) ***drusen.*** These lesions are ophthalmoscopically visible as pale yellow spots that may occur individually or in clusters throughout the macula as well as in the retinal periphery. They consist of an accumulation of amorphous material (hard, soft or mixed) between the RPE and Bruch's membrane, resulting in a microscopic elevation of the RPE. Although their exact origin remains unknown, current theories favour the accumulation of oxidised lipids, polysaccharides, glycosaminoglycans, lipofuscin and other cellular debris derived from cells of the RPE that are compromised by age or other factors. Crabb et al., (2002) have hypothesized that OXIDATIVE INJURY contributes to the pathogenesis of ARMD and oxidized proteins may have an important role in drusen formation. Drusen can then perpetuate a state of inflammation and of chronic oxidative stress (Pauleikhoff et al., 1990). This concept is in line with other age-related diseases such as neuro-degenerative diseases.

3) ***choroidal neovascularization***. In response to ischemia, choroidal vessels proliferate across Bruch's membrane under the RPE and, frequently continue their extension into the subretinal space, thus disrupting the crucial anatomical and functional relationships. Neovascularization is most likely stimulated and mediated by angiopoietins, probably produced by endothelial cells under hypoxic conditions. Copious leakage from these new and abnormal vessels results in *exudative detachment of the RPE or haemorrhages*, which may be confined to the area under the RPE or may extend under the retina. The natural course of this process is fibrotic evolution, with formation of a *disciform scar*. An obvious therapeutic aim is either to suppress the secretion of VEGF with antiangiogenic compounds or to stimulate the synthesis of PEDF (Liu et al., 2004).

The most frequent symptoms of the macula's alterations are:

- **decreased visual acuity** (loss of central vision, colour vision, ability to see fine details),
- **metamorphopsia** (distortion of the shape of objects in view),
- **paracentral-central scotoma**, that is a sort of a round black spot.

By examining Figure 15, the reader can rapidly determine if he/she suffers of any of these symptoms.

Loss of vision in ARMD is the result of photoreceptor death, occurring when RPE cells, with which they are associated, deteriorate and die. The loss of vision resulting from drusen and pigmentary changes (early stages of the disease) is highly variable: most patients are asymptomatic or experience only a small visual loss or metamorphopsia. With the progressive development of larger areas of atrophy of the RPE involving the *foveola*, visual acuity decreases abruptly and relative or absolute scotomas appear within the central 10 degrees of the visual field. Sudden substantial loss of central vision, over a larger area and often at an earlier age, is generally the result of choroidal neovascularization, with serous or haemorrhagic detachment of the RPE. The natural clinical course of ARMD is progressive and the final visual acuity is usually < 20/200 (Piguet et al., 1992; Sarks et al., 1988; Klein et al., 1993, 1997).

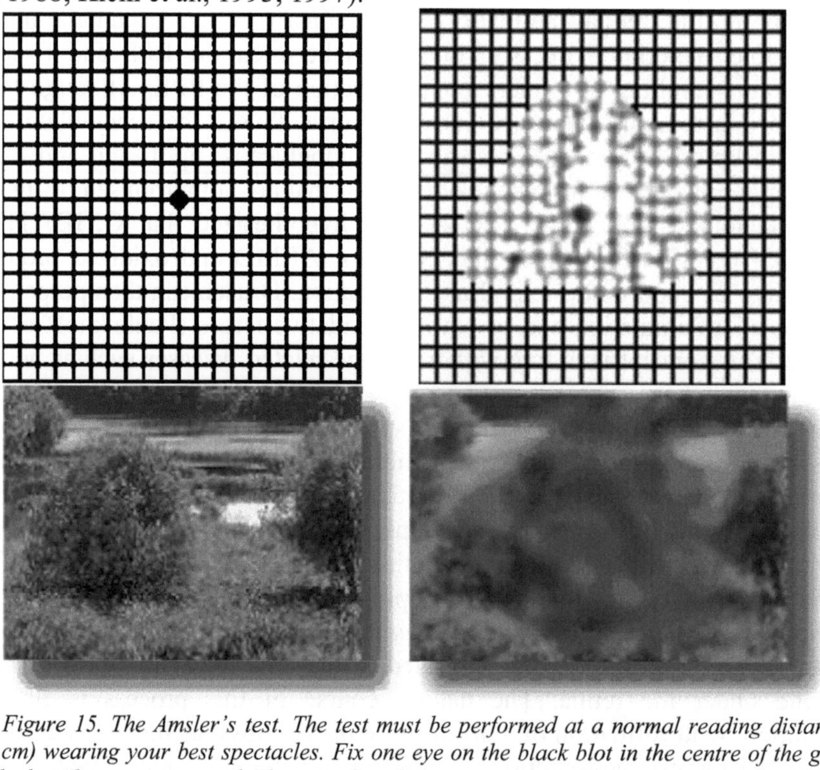

Figure 15. The Amsler's test. The test must be performed at a normal reading distance (30 cm) wearing your best spectacles. Fix one eye on the black blot in the centre of the grid while the other eye is covered. Then test the other eye. Lines should be seen perfectly lined up. If the lines or the images appear distorted or if a black or gray spot appears in the center of the image (scotoma), please consult urgently your ophthalmologist because you may be affected with macular degeneration. Left, normal vision; right, pathologic vision

Most **potential therapies have been addressed to the "wet" form of ARMD** in order to reduce the neovascularization: laser photocoagulation (Macular photocoagulation group, 1991; Figueroa et al., 1996) indicated only for selected patients with well defined extrafoveal and juxtafoveal neovascular membranes, the more selective photodynamic therapy using verteporfin that can be applied to subfoveal membranes (Verteporfin Study Group,2003; Chan et al.,2003), several medical approaches such as radiation therapy (Finger et al., 2003), antiangiogenic compounds such as IFN alpha2a (Fung 1991), endostatin and subretinal surgery, directed to remove the offending neovascular membrane. All these therapies aim to stop the natural course of the disease or at least to slow it down, but can't recover the lost visual acuity, and may have disturbing side-effects.

For the "dry" pre-angiogenic form of ARMD instead, there are no useful therapies at all, and a few postulated treatments remain controversial. On the basis of the role of oxidative stress in the pathogenesis of the disease, the protective effect of several food supplements such as zinc (Newsome et al., 1988), antioxidants like vitamins A, C and E (Sperduto et al., 1990; Seddon et al., 1994; West et al.,1994;), the today popular antioxidant carotenoids like lutein and zeaxanthin (Chopdar et al., 2003; Krinsky et al.,2003) as well as statins (van Leeuwen et al., 2003) have been investigated, and, according to the results of the Age-Related Eye Disease Study Research Group (2001), the only possibility for these patients to reduce the risk of progression of the disease, is the **continuous oral administration of antioxidants. Certainly these compounds do not harm but, although useful, do not improve vision.** In anaemic patients, administration of erythropoietin (EPO) may be beneficial because photoreceptors or retinal ganglion cells can degenerate in hypoxic conditions. This aspect, to my knowledge, has been not yet evaluated in a clinical trial.

Briefly we already know that ozonation of blood brings about several effects such as:

- Improvement of blood rheology.
- Improvement of the glycolytic pathway on erythrocytes.
- Activation of the hexose-monophosphate shunt on erythrocytes with a possible increase levels of 2,3-DPG levels.
- Increased oxygen availability and delivery to hypoxic tissues due to a shift to the right of HbO_2 dissociation curve.
- Increased concentration of ATP levels on erythrocytes with possible microrelease at hypoxic sites.
- Vasodilatation by increased release of nitric oxide or and prostacyclin.
- Release of growth factors from platelets and cytokines from leukocytes.
- Upregulation of the enzymatic antioxidant system and ozone tolerance.
- Enhanced activity of HO-1 with release of CO and bilirubin.

The fascinating aspect of ozonetherapy is the ability of activating the cooperation of a number of defence mechanisms against ischaemic and neurotoxic injury, thus preventing cell death. Since 1995, owing to the lack of an othodox therapy for the dry form of AMD, it was considered worthwhile to carry out an investigation in the Department of Ophthalmology and Neurosurgery of the University of Siena, using the classical method of AHT, in order to check the safety of the method and the clinical usefulness of this approach, comparing it with a control group. Within 6.5-7.5 weeks we evaluated the effect of a cycle of 12-13 ozonated AHT in 54 patients and of only oxygenated AHT (control) in 23 patients. In both the ozone therapy and control groups there was a slight prevalence of men with an age ranging from 63 to 81 years old. All of them presented the dry form, prevalently with soft, confluent drusen followed by the geographic atrophy form. Mean baseline visual acuity (logMar equivalents) was either 1.27±0.49 or 0.95±0.5 for the treated or control group, respectively. It must be emphasized that the type of treatment is the same except that blood was exposed only to oxygen. Orthodox medicine requires a control but today this appears unethical. Best corrected distance visual acuity (Snellen chart), and a complete biomicroscopic and ophthalmoscopic examination with intraocular pressure measured by applanation tonometry were recorded before the first treatment (baseline), after the last one (post-treatment) and then, when possible, every 3 months for up to 18 months; in addition, in order to check the safety of prolonged AHTs, general haematochemical parameters (blood cell counts, plasma proteins, plasma lipids, coagulation and fibrinolysis tests) were recorded at the baseline time and after the end of the cycle of treatments.

With regard to optalmological results, change in visual acuity from baseline at each follow-up examination was the primary parameter used in order to verify the response, if any, to AHT, compared with the control group. Mean distance best corrected visual acuity (logMar equivalents) was significantly improved in the treatment group of dry ARMD's patients, while in the control group only a modest improvement in mean distance visual acuity was observed, which was not statistically significant.

In the treatment group we observed an improvement in visual acuity more than 2 ETDRS lines in 36 patients (66.6%), equal or less than 2 ETDRS lines in 18 patients (33.3%); in the control group an improvement in visual acuity more than 2 ETDRS lines was observed in 7 patients (30.4%), equal or less than 2 ETDRS lines in 16 patients (68.5%). These differences were statistically significant (chi-square).

In the treatment group, to our surprise, the improvement remained reasonably stable during the first semester, and then slowly declined, but after 18 months only a minimal visual improvement remained in comparison to the acuity values assessed at the start of the study (Figure 16). On the basis of erythrocyte life-time (4 months) and of the usual short-biochemical memory, we did not expect this result. In the control group after 6 months visual acuity had returned to the pre-treatment values and the natural course of dry AMD progressed, with its continuous and rapid visual loss.

A number of laboratory data reported in Table 6 show that AHT does not cause significant modifications of critical parameters measured just before and at the end of the treatment. Typical liver enzymes levels were also unmodified. In some of the patients we ascertained that there was no increased peroxidation. 2,3-DPG levels remained practically constant but they increased only in a few patients who had a basic low level. SOD levels slightly increased after the first 5 sessions and then returned to normal values. The slight increase of G6PDH was also not significant. These data need to be investigated in a far larger group of patients and during a prolonged therapy.

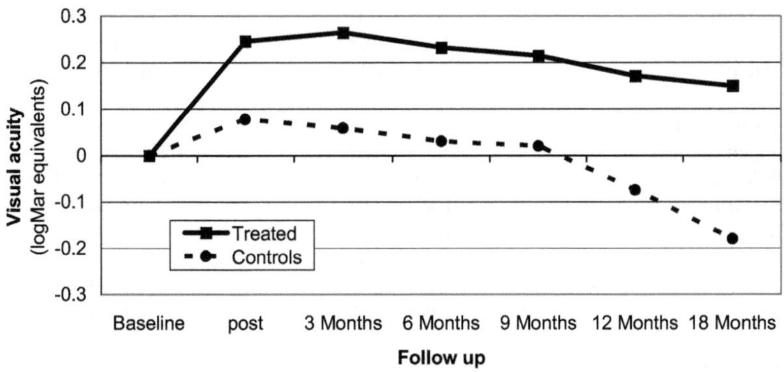

Figure 16. Visual Acuity changes from baseline observed during the trial.

We have not observed any side effects either during or after the treatment. Actually most of the patients reported an improvement of their general conditions particularly in terms of increased stamina, appetite, mental concentration and memory. The compliance of patients was excellent and moreover they accepted and followed enthusiastically the maintenance treatment. The only noticeable problem is that a few patients had a poor venous access, and this occasionally required more than one venipuncture. If absolutely necessary, this problem is now solved by the infusion of the "gluco-peroxide" solution with a very small needle.

At the present time there is NO other effective medical therapy for the atrophic form of ARMD. Most of these patients, still being physically and mentally active, are very concerned about the lack of an effective treatment and although there are new avenues of research, it will take time before they will be transferred to the pharmacist's shelf. On this basis **we feel that it is ethically correct to use ozone therapy not only because patients**

appreciate it but because this approach is based on precise biochemical reactions and is not toxic. We deeply regret that orthodox ophthalmologists, having been informed of this possibility and without knowing anything about ozone therapy, continue to be sceptic without any consideration for the patients.

Table 6. Laboratory tests carried out in 54 dry ARMD's patients before (pre) and after (post) 13-15 sessions of ozonated AHT.

	PRE	POST
Blood cells		
RBC (M/µl)	4.6 ± 0.7	4.5 ± 0.7
Hb (g/dl)	14.0 ± 0.2	13.6 ± 0.2
Hct (%)	41.7 ± 0.6	40.6 ± 0.7
MCV (fl)	91.0 ± 0.8	91.4 ± 0.8
MCH (pg)	30.5 ± 0.3	30.7 ± 0.3
MCHC (g/dl)	33.5 ± 0.1	33.5 ± 0.1
PLT (K/µl)	232.2 ± 9.2	237.4 ± 9.8
WBC (K/µl)	6.3 ± 0.3	6.4 ± 0.3
Coagulation tests		
Fibrinogen (mg/dl)	293.6 ± 12.5	327.6 ± 14.7
F.VIIIvV (%)	151.6 ± 12.8	153.6 ± 14.0
F.1+2 (nM/l)	1.42 ± 0.14	1.15 ± 0.11
AT III (%)	100.9 ± 3.6	100.9 ± 2.6
PT (%)	96.2 ± 3.1	96.1 ± 1.8
a PTT (sec)	31.7 ± 0.7	30.3 ± 0.6
TT (sec)	19.4 ± 0.7	19.6 ± 0.3
Fibrinolysis tests		
t-PA (ng/ml)(sec.)	11.2 ± 0.8	10.4 ± 0.9
PAI1 (IU/ml)	11.2 ± 1.4	13.1 ± 1.6
FM test (µg/ml)	10.1 ± 1.2	13.5 ± 3.0
FDP (µg/ml)	6.4 ± 0.8	7.6 ± 2.4
D dimer (ng/ml)	111.3 ± 5.5	114.5 ± 8.4
Lp (a) (mg/dl)	43.8 ± 10.4	35.7 ± 8.2
Platelets tests		
PF4 (IU/ml)	4.6 ± 0.8	3.8 ± 0.5
β-TG (IU/ml)	27.1 ± 2.0	29.2 ± 2.9
Plasma proteins		
Proteinemia (g/dl)	6.8 ± 0.4	6.9 ± 0.5
Plasminogen (g/l)	0.12 ± 0.4	0.12 ± 0.6
Fibronectin (mg/dl)	43.3 ± 1.5	45.4 ± 2.2
Plasma lipids		
HDL (mgf/dl)	60.2 ± 2.6	54.9 ± 2.8
Cholesterol (mg/dl)	285.5 ± 8.9	278.9 ± 8.5
Triglycerides (mg/dl)	119.9 ± 13.9	114.4 ± 10.5

Results presented in Figure 16 suggest that, after a cycle of 14-16 treatments, therapy should not be discontinued for as long as one semester and actually should be continued all the time with a decreased frequency. At the moment we are evaluating whether a **maintenance therapy** of one monthly treatment is sufficient. It appears that the observed clinical effect is brought about when a volume of about 1.5 litres of blood has been treated with ozone but it reaches a plateau after exposure of about 3 litres of blood. Whether a different schedule (thrice weekly), larger volumes of blood, the use of heparin instead of CPD, a different ozone concentration, or a more prolonged treatment are able to improve the outcome, particularly in NON-responding patients, remains to be investigated. It will not be an easy task because it will involve a great number of patients and time that is beyond my personal possibilities.

It is reasonable to envisage that a beneficial effect at the choroidal-retinal circulatory network and photoreceptor level occurs only when a critical mass of blood, in spite of dilution and erythrocyte turnover, has undergone ozonation and has activated a number of factors. The oral, daily antioxidant supplementation assumed by our patients during the cycle, although in itself is unable to improve ARMD, ought to be continued lifetime, because in elderly people common micronutrient deficiencies cause mitochondrial decay with oxidant leakage leading to accelerated ageing and cell death (Ames, 2004). This interpretation is supported also by the lack of significant increase of plasma levels of peroxidation products that indicates lack of ozone toxicity. Although the majority of ARMD patients gain an improvement of 0.5-2 lines on the visual acuity chart, there is a considerable variability because ozonetherapy may activate the functional retinal reserve with different results depending on the individual morphological and functional changes undergone by the retina. **Nonetheless in almost blind patients, even a small gain cannot be disregarded because they report an improved quality of life and are less depressed.**

While we are also well aware of the possibility of a placebo effect, that Zajicek (1995) considered as the healing force of Nature, we feel that it cannot be important because some patients do not have any improvement and at the best, its duration is not consistent with the different slope of the curves of visual acuity changes during time in the treatment group compared to the control group (Figure 16).

Finally the total lack of side effects and the excellent compliance of all the patients need a comment. There is no doubt that ozone has a potential toxicity, but this theoretical drawback does not represent a problem as we are using well determined, precise and low ozone concentrations in relation to blood volume or, in other words, we are actuating a calculated and very transient oxidative stress not to be confused with the endogenous and chronic oxidative stress. The unusual feeling of well-being following AHT in the majority of patients is not necessarily associated with the improvement of visual acuity and it may be due to a release of unidentified hormones and neurotransmitters.

One aspect that must be discussed is which, **among the following three types of treatment for ARMD: ozone therapy, hyperbaric oxygen therapy (HOT) and rheopheresis, is the most effective?** This is a question occasionally posed by well-informed patients. The HOT does indeed increase oxygenation of blood and tissues but only during the usual two hours treatment and this does not procure any substantial advantage. Rheopheresis is used to remove, via the extracorporeal blood purification technique, substances like fibrinogen, cholesterol, alpha$_2$-macroglobulin and so forth that may contribute to the progression of ARMD. Obviously it does not correct the metabolic disorders that can be achieved with oral drugs and does little to increase oxygenation at the retinal level. Moreover it is more invasive because both arms must be cannulated with large needles (G17), each treatment takes almost two hours and is expensive. As a comparison, the classical ozone therapy is conceptually more rational and indeed far more effective, easy to perform, less costly and well accepted by patients.

Some 20% of ARMD (dry form) do not respond to the therapy probably because the retinal degeneration is no longer reversible and that is the reason why some patients have tried acupuncture with minimal or no advantage. Needless to say that the patient is free to further undergo any other therapy but I hope that he will receive objective advices.

For compassionate reasons and pressing request of the patients, we have tried to perform ozone therapy in a few cases of **ARMD (exudative form),** but the advantage has been minimal and restricted only to peripheral vision so that it is imperative to exclude a hope of real improvement. Patients with subfoveal idiopathic choroidal neovascularization can now be treated with photodynamic therapy using verteporfin (Chan et al., 2003) and perhaps thereafter we could possibly implement an advantage with ozone therapy. However I have been unable to evaluate this approach owing to the scepticism of orthodox ophthalmologists. Although further investigations are necessary, **the possibility of blocking an excessive vascular permeability after an intravitrous injection of PEDF is a promising approach** (Rasmussen et al., 2001; Liu et al., 2004).

We have occasionally treated other **disorders of the retina and optic nerve**, for which there are no other options, with ozonetherapy. We have noted some unexpected improvement in patients with degenerative myopia and ischaemic optic neuropathies and therefore it seems correct to leave no stone unturned.

Retinitis pigmentosa is another dramatic disease due to multigenic and progressive disorders affecting men from a young age. In Italy there are 30-40.000 patients and because mutations are located in the X chromosome, it may be possible with careful analyses to reduce the number of defective newborns in the near future. In this pathology there is no circulatory defect and therefore ozone therapy cannot help these patients and it would be dishonest to elicit even the faintest hope. Nonetheless, owing to parents' pressure, in a few young patients we have gratuitously performed an ozone therapy cycle. They reported a tenuous and transitory improvement and absolutely no side effects. At this stage the prospect of an effective therapy remains dim but, at least theoretically, gene therapy or the possibility of implanting normal staminal (embryonal) cells or a semiconductor microchip in the retina (Humayan, 2001) may offer a possible improvement. However all of these approaches, although fashionable in these days, will take some time to be accomplished. Unfortunately there are always disgraceful quacks around ready to exploit the good faith of these patients claiming great success with the most ridiculous implants of extravagant materials.

Although an evaluation of 10 patients with retinitis pigmentosa performed in Cuba may have been done with good intentions, it appears complicate and eventually useless. A multi-tecnique approach consisting of: a regimen of electrical stimulation, AHT and ocular surgery had not been validated by a distinguished ophthalmologist in Boston, when the American patients went back home. Actually it was suggested that, in comparison to an excessive vitamin A supplementation that, in my opinion can be toxic, this complex intervention may worsen the course of the disease (Berson et al., 1996; Weleber, 1996). Thus, the problem of an effective therapy of retinitis pigmentosa remains open and I would like to make a plea for avoiding useless therapy and disappointment for the patients

I cannot omit to mention that Radu et al., (2003) have suggested the use of isotretinoin (13-cis-retinoic acid or Accutane), a drug in common use in acne (however known to cause birth defects), as a possible treatment for retinal or macular degeneration associated with lipofucsin accumulation. This therapy may be particularly useful in children with **recessive Stargardt's disease**, which is an inherited form of macular degeneration associated with an early accumulation of fluorescent lipofucsin pigments in the RPE. In this disease, the lipofucsin accumulates A2E, a conjugate of vitamin A aldehyde that cannot be degraded and causes a detergent-like effect on cell membranes with deadly results. There is no other treatment for delaying a rapid death of the RPE cells, hence of photoreceptors.

Finally **diabetic retinopathy** is one of the several ocular complications of both type 1 and type 2 diabetes and is a common cause of visual loss in the working age population (see also the dysmetabolic syndrome, Section VII). It is characterized by varying degrees of microaneurysms, haemorrhages, exudates, neoangiogenesis and retinal thickening involving the macula or the peripheral retina or both (Frank, 2004). The earlier the treatment, the better is the outcome (Kohner, 2003). A strict control of diabetes and blood pressure can significantly reduce the progression of retinopathy. **Current and efficacious treatments are carried out with retinal laser photocoagulation and vitrectomy.** On the other hand, clinical trials testing the potential efficacy of aldose reductase inhibitors, aspirin (Kohner, 2003b), aminoguanidine (for blocking the formation of advanced glycation end products, AGEs), and VEGF inhibitors have been disappointing (Kohner, 2003a). A trial evaluating PEDF gene therapy is in progress. A clinical study with ozone therapy has yet to be envisaged but there is a rational basis for using ozone therapy as a supportive treatment either with AHT or with self-administration of ozone via RI.

THE FOLLOWING ANNOTATION CAN BE USEFUL:

In September 2003, I faced the decision to either perform the infusion of the "gluco-peroxide" solution in ARMD (dry form) women with extremely poor venous access or leave them untreated. As it has been clarified in Chapter 6, this solution can be easily infused via a small needle (butterfly G25) in a small vein of the back of the hand. Always applying the concept of inducing tolerance to an acute and calculated oxidative stress, I followed the strategy of the "start low, go slow". Thus, I begin with a solution with a final hydrogen peroxide concentration of 0.03% (8.8 mM) that is slowly raised, by the 7th treatment up to 0.12% (35.2 mM). The 250 ml volume is infused in about 20 to 30 minutes with neither problems, nor side effects. In line with the theory that hydrogen peroxide is one of the most important ROS messengers, the therapeutic effect checked by the ophthalmologist in these patients by the end of the treatment is practically comparable to the one achieved by the AHT. This result will be reported in details in the near future (Bocci et al., manuscript in preparation). ***One limitation is that the "glucoperoxide" solution CANNOT be used in diabetics but nevertheless other patients with a difficult venous access can be helped.***

CONCLUSIONS: it seems to us that, although ozone therapy is a fairly unknown and boycotted (by orthodox ophtalmologists) complementary medical approach, it should not be viewed with scepticism and, with the limitations objectively discussed above, deserves to be applied in suitable patients, Even though they regain only a fraction of their original visual acuity, when there is NO OTHER USEFUL TREATMENT, patients are greatly appreciative as demonstrated by an excellent compliance along the years.

4. NEURODEGENERATIVE DISEASES

The hypotetical aetiology and pathogenesis of neurodegenerative disorders such as Parkinson', Menkes', Alzheimer' and Wilson's disease, senile and vascular dementias, amyotrophic lateral sclerosis, optic nerve dysfunction, primary open angle glaucoma, neurosensorial bilateral hypoacusia and maculopathies have been extensively discussed and, although they have distinct characteristics, **have in common the feature of**

chronic oxidative stress (Ames et al., 1993; Yu, 1994, Cohen et al., 1994; Jenner, 1994; Bondy, 1995; Carlsson et al., 1995; Jaeschke, 1995; Pardo et al., 1995; Yoritaka et al., 1996; Simonian and Coyle, 1996; Back, 1998; Halliwell, 2001; Rowland and Schneider,2001; Perry et al., 2002; Steece-Collier et al., 2002; Butterfield and Lauderback, 2002). These are distressing diseases whether they are affecting young people at the height of their physical performance or great minds that, in a few years, sink into oblivion.

The physiological process of ageing is endowed, luckily to a lesser extent, with similar biochemical anormalities and this fact compels us to understand the mechanisms and put into action innovative ideas to delay both ageing and neurodegeneration. Indeed the progressive prolongation of the human life-span is accompanied by an increase of neurodegenerative diseases: the lifetime risk of Alzheimer's disease has been estimated to be about 13% among the Europeans so that, with some approximation, one in ten women, who live to 80 and one in seven men, who live to 76 will develop the disease. There is good evidence that **the combination of genetic predisposition, familiarity, life-long oxidative damage, an excessive or poorly balanced diet, exposure to transition metal ions, alcohol and tobacco smoke intoxication, lack of physical exercise and diabetes plays a role in accelerating cell degeneration.** Thus, although the primum movens remains unknown, **once it is switched on, it is perpetuated or enhanced by a deranged reduction-oxidation homeostasis.**

The pathophysiology is quite variable: in some cases, there is a chronic inflammation, possibly started by the deposition of advanced glycation end products (AGE) with the release of ROS, LOPs (4-hydroxy-2,3-trans-nonenal) and pro-inflammatory cytokines; in other cases, we can observe a biochemical defect such as low GSH content (Ault and Lawrence, 2003), or a decrease of antioxidant enzymes (GSH-Pxs, SOD, catalase) associated with improper metal binding; in other cases, there is an excessive release of anion superoxide and NO, hence of cytotoxic $ONOO^-$ and nitrotyrosine (Dedon and Tannenbaum, 2004), or of noradrenaline from presynaptic terminals or of glutamate with Ca^{2+} influx and activation of protein kinases, phospholipases, etc. (Pardo et al., 1995; Nakao et al., 1995; Ceballos-Picot et al., 1996a; Markesbery 1997; Aejmelaeus et al., 1997; Sagara et al., 1998; Floyd 1999; Li et al., 1999; Perry et al., 2002; Rotilio et., 2000 ; Rotilio, 2001; Reisberg et al., 2003).

Ozonetherapists must be aware of the intense research activity trying to find drugs able to delay or block the neuronal degeneration and death: the usual hydrophilic and lipophilic antioxidants taken in appropriate amounts via os are not harmful but are modestly effective (McCall and Frei, 1999; Engelhart et al., 2002), because only a small percentage reaches the CNS. Metal chelators may help by reducing free transition metals and $OH^\bullet$ formation, but one must pay attention not to exceed with chelation therapy. Moreover, several inhibitors of the reuptake of dopamine, of $NO^\bullet$ synthesis and of ionotropic receptors to block glutamate neurotoxicity are being tested (Reisberg et al., 2003).

The more biologically oriented approaches are attempting to use neurotrophic factors or to transplant dopaminergic foetal cells or stem cells into selected areas (Weber and Butcher, 2001). At least in theory, embryonic stem cells, if compatible with the recipient, in the presence of appropriate growth factors (?) could be coaxed into producing a line of cells needed to repair or substitute dying dopamine-rich neurons. Among neurodegenerative diseases, Parkinson's disease is the ideal one, because the degeneration is fairly restricted to particular areas of dopaminergic neurons (Lang and Lozano, 1998 a, b). If the ethical problem will be overcome, it will take considerable time to practically achieve therapeutic cloning because, in order to avoid rejection, we must transfer the nucleus of one of the patient's own epithelial cell into a human egg, whose nucleus has been removed and then, after idoneous signals, revert the patient's genome to its embryonic state. A simpler solution has been proposed by Mezey et al., (2003). They have demonstrated that a few SC present in bone marrow transplants from human male donors into cancer-irradiated women could be detected, *post-mortem*, in the hippocampus and cerebral cortex of the recipients. Although this result confirms a previous one in mice, the already small number of "new" neurons in humans is 10-25 fold less than that observed in rodents. Moreover we don't know if irradiation of the CNS may have facilitated the migration and homing of BMSC into the brain. In spite of these caveats, this interesting result encourages pursuing this avenue of research that will avoid ethical and rejection problems and could acquire an enormous practical importance. Obviously **the critical problem is how we can activate a large migration of BMSC towards the CNS for substituting dead or dying neurons** and I am more concerned in accomplishing this first step rather than the successive one of differentiating SC into efficient neurons. I like to hypothesize that repeated "therapeutic shocks" induced by ozonated HAT (via LOPs) are able to stimulate the BMSC release because the transitory and acute oxidative stresses disturb the homeostasis in the bone marrow environment **POSSIBLY ANERGIC** in patients with neurodegeneration. Has orthodox medicine a better option? As I am not aware of any other possibility, I would insist in performing ozone therapy in patients with neurodegenerative diseases and recent strokes. The chance of achieving some benefit has been evaluated in Chapter 8.

The pharmacological therapy is certainly useful (levodopa is still an effective therapy after three decades!) but only for a limited time and it does not arrest progression of the disease. The combination of several experimental therapies promises to improve the present limitations, but we are still fighting a virtually lost war because neurodegenerative diseases are projected to surpass even cancer as the second cause of death by the year 2040 (Lilienfeld and Perl, 1993).

At first glance, it seems irrational to propose a treatment envisaged as a "therapeutic shock" for neurodegenerative diseases, based on a series of brief and calculated oxidative stress. However, **this approach, in combination with pharmacological therapy, may exert a paradoxical effect and reverse or stabilize an otherwise irreversible situation**. The idea is that a gradual escalation of the ozone dose (from 10 to 30-40 mcg/ml per ml of blood) may be able to enhance the cerebral blood flow (Clavo et al., 2004), improve the metabolism and correct the chronic oxidative stress. In practical terms, by gradually receiving trace amounts of LOPs, neuronal cells under constant oxidative stress may reactivate the depressed synthesis of antioxidant enzymes, which is the crucial key to normalize the redox state and avoid cell death. Moreover the local induction of haeme oxygenase-1 would play a critical role in further reducing oxidative damage. It is perhaps useful to remember that this enzyme would cause the local release of CO and bilirubin that acts as a potent antioxidant of peroxynitrite (Minetti et al., 1998).

Today there is no other pharmacological approach able to achieve this objective, which instead can be realized, without any biotechnological complexity, simply by ozonating blood for a few minutes. Obviously the sooner we start the ozone therapy the better, because there is no hope of reviving dead neurons. **It remains hypothetical if some particular LOPs, by reaching in trace amounts the substantia nigra region of the brain, are able to stimulate some dormant staminal cells and induce their differentiation.** Besides Mezey's results, this is another possibility even simpler than the one involving BMSC. If this would be the case, ozone therapy will simply realize the modern dream and avoid cell cloning. More than ever, I persist in my opinion that, if neurodegeneration is not due to an irreversible genetic defect (like amyotrophic lateral sclerosis, ALS, or Lou Gehrig's disease, for example), judicious administration of ozone can be helpful. While I am aware, and I repeat to everyone, that ozone is intrinsically toxic and must be used with care, I do not see any risk in evaluating this problem with either AHT or the "gluco-peroxide" infusion, or BOEX, or simply, at home, with daily RI. At the worst, even if we will not obtain a positive result, patients will not be harmed and probably they will feel better.

If, in our countries, the dogma on ozone toxicity will persist and Health Authorities will continue to neglect this problem, it will be difficult to make any progress. Fortunately Cuban physicians have performed one study: it was a double blind RCT on 60 patients affected by senile dementias: group A (30 patients) was treated with O_2-O_3 by daily RI (50 mcg/ml) for 21 days and group B with oxygen only. Although I am not enthusiastic of the administration route (RI), this is a pioneering study to be taken into account. Using several psychometric tests (mental condition, capacity for self-administered medication and evaluation of daily activities), it demonstrated that 73-90% of ozone-treated patients showed marked improvement without any adverse effect (Rodriguez et al., 1993). Therefore, if ozonetherapy is really useful, we continue to deny a possibility to many patients.

If we can perform a study, it will be important not only to evaluate the therapeutic effects but also to clarify the mechanisms of action. Rodriguez et al., (1993) and Clavo er al., (2004) have already demonstrated that ozonetherapy can simultaneously improve blood flow and oxygen supply to hypoxic tissue. It is then possible to envisage an increase of the aerobic glycolysis in hypofunctional cells, which by resuming normal metabolism might restore the normal ATP content and GSH/GSSG ratio. LOPs generated during lipoperoxidation of plasma or absorbed from the rectal mucosa (RI) or the skin (BOEX) will be diluted in the plasma pool and trace amounts can pass through the blood-brain barrier to reach the sites of neurodegeneration and upregulate the cellular synthesis of antioxidant enzymes, which is the crucial step to readjust the impaired cell redox system. An increased release of either dopamine or/and neuronal growth hormones and the activation of resident stem cells remain speculative, but they are not too far-fetched ideas.

The possibility that Alzheimer's disease, associated with a deposition of insoluble β-amyloid aggregates, reflects an $NO^{\bullet}$/superoxide imbalance has been entertained by Thomas et al. (1996). The therapeutic implication is that a prevalence of $NO^{\bullet}$ over superoxide is advantageous and may inhibit aggregation. This may be achieved by the administration of exogenous SOD mimetics and/or antioxidants but, interestingly, ozonated AHT could correct the imbalance by inducing SOD and the production of $NO^{\bullet}$ at the same time. Two cautionary annotations appear to be in order: the first is that functional recovery may be achieved only in initial or not too advanced patients, and secondly an optimal AHT schedule has not yet been worked out, although it appears reasonable to start with a low ozone concentration (10-20 mcg/ml) and slowly raise it (in 3-4 weeks) to 35-40 mcg/ml per ml of blood, two-three times weekly. For RI, I would suggest beginning with a dose as low as 5 mcg/ml and slowly upgrade to a maximum of 25 mcg/ml and a volume of 600 ml gas, five times weekly. In this case, I think that the concentration (50 mcg/ml) used constantly by Rodriguez et al. (1993) is excessive and frequently causes intestinal cramps. If an improvement really occurs, it may be necessary to continue the treatment at home biweekly for life.

It must be explained and understood that **one cycle of ozonetherapy cannot solve the problem: all cells have a more or less long biochemical memory and must be stimulated by LOPs at short intervals.** Our study on ARMD has been very instructive in this sense and WE MUST BE HONEST WITH PATIENTS CLARIFYING THAT OZONE THERAPY CAN BE REALLY HELPFUL, IN THE SENSE THAT IT MAY BE ABLE TO REACTIVATE MANY BIOLOGICAL FUNCTIONS GONE ASTRAY,

BUT IT CANNOT "CURE" THE DISEASE AND, AT THE BEST, CAN BLOCK ITS PROGRESSION. **It is therefore essential to undergo a MAINTENANCE THERAPY.** Patients with neurodegenerative diseases undergoing ozonetherapy must receive oral antioxidant supplementation because they are frequently undernourished and may have a low antioxidant capacity. Although there is a general consensus regarding the administration of antioxidants, daily doses vary among Authors (Morena et al., 2000; Peng et al., 2000; Halliwell, 2001; Engelhart et al., 2002; Polidori et al., 2004) but the suggested dosage (Chapter 8) is believed to be quite sufficient. The therapeutic value of Ginkgo biloba in reducing symptoms of decline in mental function has been evaluated by Curtis-Prior et al., (1999) but this topic remain controversial.

CONCLUSIONS: Neurodegenerative disorders affect about 50 million people in the world and have a terrific and increasingly negative social-economic impact on families and society. While a better understanding of degenerative events may allow devising medical therapies able to slow down the demise of critical populations of neurons, we should not disregard the corroborant effect of ozone therapy particularly in the early stage of the disease. If ozone therapy is endowed with the capacity of mobilizing BMSC or activating dormant SC in the brain, we may be able to drastically change a gloomy prognosis. At the least patients have only to gain a better quality of life by associating useful medical therapies to ozone therapy.

5. AUTOIMMUNE DISEASES. CAN OZONE THERAPY DO BETTER THAN ANTIBODIES TO TNF ALPHA?

The most relevant autoimmune diseases such as rheumatoid arthritis (RA), Sjogren's syndrome, vasculitis, multiple sclerosis (MS), Crohn's disease, psoriasis, systemic lupus erythematosus and type 1 diabetes affect about 5% of the population in Western countries.

The aetiology of these diseases remains uncertain but genetic susceptibility, unclarified viral or/and bacterial infections, age and sex are playing a role. On the other hand, during the last 25 years, considerable progresses have been made on the pathogenetic mechanisms, which, with a strong prevalence in women and with different locations, present a remarkable similarity suggesting that the primary cause switches on a number of identical offensive mechanisms. Different tissues (articular, gut mucosa, myelin, skin, etc.,) become infiltrated by macrophages, neutrophils and cytotoxic T lymphocytes (CTL), which are responsible for an abnormal release of ROS and proinflammatory cytokines (IL-1β, IL-2, IL-8, IL-12, IL-

15, IL-18, TNFα, IFNγ), while inhibitory cytokines (IL-10, IL-11, TGFβ1) are largely suppressed (Kuruvilla et al., 1991; Brandes et al., 1991; Taga et al., 1993; Akdis et al., 1998; Letterio and Roberts, 1998; McInnes and Liew, 1998; Pizarro et al., 1999; Perdue, 1999; Dinarello, 1999; Herrmann et al., 2000). This is a most interesting aspect to keep in mind for developing a therapeutic approach because the basic concept is to deplete or eliminate offensive cells and re-establish the equilibrium. I must inform the reader that between those immune cells producing either pro-inflammatory cytokines or inhibitory (or immunosuppressive) cytokines, a competition may arise from time to time although normally there is a physiological balance that aims to maintain a healthy condition. The Chinese concept of the yin-yang, or of the darkness opposing the light, is well suited here to explain that the immune system, throughout life, must be ready to respond more positively or more negatively in such a way to neutralize noxious stimuli and ripristinate homeostasis, i.e, equilibrium, as soon as possible. Unfortunately, **in autoimmune diseases, the generation of autoreactive cells and the release of pro-inflammatory mediators will cause tissue injury, swelling and pain.**

Mossman and Sad (1996) have been the first to show that $CD4^+$ lymphocytes (helper T cells), depending upon the type of a stimulus can undergo a profound shift towards either the pro-inflammatory Th1-phenotype (generally producing IL-1, IL-2, IL-18, IFNγ, TNFα) or the immunosuppressive Th2-phenotype (producing IL-3, IL-4, IL-5, IL-10 and TGFβ1).

A schematic representation is shown in Figure 17, although Nature is far more complex than our mind and often some $CD4^+$ T cells cannot be grouped into either a Th1 or Th2 subset (Th3?) because they exhibit a heterogeneous pattern of cytokines. Nevertheless, pathological immune responses at least partly support the pattern of cytokine production linked with the Th1 or Th2 predominant immune state. Th1-type responses are associated with inflammation and defense reactions, including cytolytic reactions, while Th2-type responses are characterized by antibody-mediated immunity. It must be kept in mind that the interaction between the two types of responses is reciprocal and thus Th1-type cytokines are inhibitory to Th2-type responses and vice versa. As an example, IL-4 can inhibit IL-12 production, while IL-4, IL-10 and IL-13 antagonise the macrophage-activating properties of IFNγ and IL-2.

Thus the main therapeutic aim is to reverse and normalise the T-helper type1/T-helper type2 imbalance.

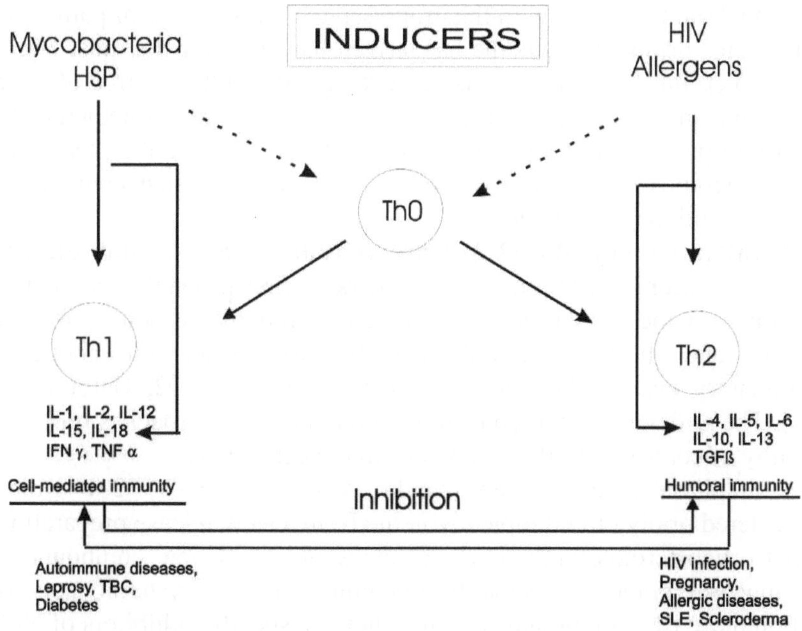

Figure 17. A schematic representation of the immunological equilibrium between $CD4^+$ T lymphocytes with a Th1 or Th2 phenotype. The former release proinflammatory cytokines while the latter release immunosuppressive cytokines: There is a reciprocal inhibition and it would be interesting to investigate if ozone therapy can re-equilibrate a dysimmune state.

Besides cytokines, the activation of enzymes, such as phospholipase A_2 (PLA_2), metalloproteinases (collagenases, elastases), cathepsins B and D, and plasminogen activators causes the release of compounds leading to cell death and disintegration of the intercellular matrix and/or myelin, thus perpetuating and aggravating a negative involution. Local release of substance P, calcitonin grp (gene related peptide), bradikynin, leukotrienes, LTB_4 (a potent chemotactic and hyperpermeabilizing factor), PGE_2, PGD_2, PGI_2 (vasodilatators), TxA_2, and F_2-isoprostanes (both vasoconstrictors) wreak further havoc and elicit oedema and pain (Cracowski et al., 2000). Interestingly, these eicosanoids (2-series PGs and 4series LTs) derive from arachidonic acid, (AA, 20:4n-6), while **5-series LTs and 3-series PGs, deriving from 5, 8, 11, 14, 17 eicosapentaenoic acid (EPA, 20:5n-3) and from 4, 7, 10, 13, 16, 19 docosahexaenoic acid (DHA, 22:6n-3), are far less phlogogenic but are practically absent** (Purasiri et al., 1997). **EPA and DHA well known as fish oils, competitively inhibit the conversion of AA to PGs, thus exerting useful inhibitory effects on inflammation and inappropriate immune responses** (Calder, 1998; Mori et al., 2003). That is why **a diet rich in n-3 PUFAs has been advocated for the treatment of various chronic inflammatory conditions typical of autoimmune diseases** (Belluzzi et al., 1996).

Throughout the years, with the progressive understanding of pathogenetic mechanisms, orthodox medicine has striven hard to offer the most effective therapy. Yet, only recently, it has achieved good results not free of adverse effects and unforeseen complications. Nonetheless, the ozonetherapist has the duty to present the following options extensively described by Hanauer and Dassopoulos (2001). For a didactic purpose I will first enumerate the basic, conventional treatments:

A) Inflammatory Bowel Disease (Crohn's disease and ulcerative colitis) are chronic inflammatory disorders at first probably initiated by a breakdown in the regulation of the mucosal immune responses to enteric antigens and bacteria complicated by ischaemic, thrombotic and inflammatory events (Ardizzone and Bianchi Porro, 2002; Hatoum et al., 2003). The wide number of conventional therapeutic approaches reflects our difficulty of controlling different pathogenic mechanisms:

1) Sulphasalazine (5-aminosalcylic acid or 5-ASA) 2-4 g/die, is administered orally or/and topically in the form a slow release preparation.

2) Corticosteroids, among which budesonide is a new compound with high mucosal potency (enema formulation) and low systemic activity. I mention these two compounds because they are specific inhibitors of NFKB, which allows the synthesis of IL-1β and TNFα (Auphan et al., 1995; Wahl et al., 1998).

3) Antibiotics, such as metronidazole and ciprofloxacin, used alone or in combination. In comparison with placebo, rifamixin did not show any benefit (Gionchetti et al., 1999).

4) Immunomodulatory drugs: azathioprine, 6-mercaptopurine, methotrexate, mycophenolate mofetil, cyclosporine, tacrolimus (FK 506), thalidomide. They have different mechanisms of action, but substantially inhibit the production of pro-inflammatory cytokines (IL-1, Il-2, Il-8, IL-12, TNF α). Probably statins will prove to be useful!

5) Imunosuppressive cytokines to inhibit the Th1-type >> Th2-type excessive response. Il-10 and Il-11 seem to suppress effector functions and Th1-type cytokine production (Taga et al., 1993; Akdis and Blaser, 2001). A few trials have shown the safety and tolerance, but the ultimate efficacy remains unknown. An interesting possibility, so far evaluated in mouse colitis, is the increased release of IL-10 into the gut lumen by genetically engineered bacteria. IL-10 may be absorbed via a paracellular route and may downregulate T cell activation in the submucosa (Steidler et al., 2000). In mice, Lee and Chau (2002) have demonstrated that IL-10 induces the expression of the wonderful enzyme HO-1, well known for reducing the chronic oxidative stress. TGFβ1 may also be efficacious but has not yet been tested. The usefulness of IFNα remains equivocal.

6) Manipulation of the normal gut flora for achieving oral tolerance. If the responsible autoantigens can be identified, their oral administration could

induce an immune tolerance and represent a rational treatment. It seems important to readjust the gut microflora because it continuously interacts with enterocytes and with the mucosal-associated lymphoid system. There are promising, yet unsubstantiated, results after administration of competitive bacteria such as: *Lactobacillus acidophilus, Bifidobacterium bifidum and Streptococcus thermophilus* against pathogenic bacteria. This complementary approach should not be disregarded, since it is non-toxic and may become even more useful by modifying the luminal environment by intermittent hydrocolon therapy. A correct ecological environment may also be restored by microflora administered via enema.

7) Dietetic support. As previously mentionerd, a diet enriched with n-3 PUFAs present in fish oil generates (via cyclooxygenase and lipoxygenase) 3-series PGs and 5-series LTs, which are anti-inflammatory and may re-equilibrate the Th1-Th2 pattern (Hodgson, 1996; Mori etal., 2003). N-3 PUFAs can easily be taken in capsules (Belluzzi et al., 1996) or emulsionated with milk. Although this approach is probably not sufficient to solve the problem, it is recommended because it is also useful for preventing atherosclerosis and can be continued for life.

8) Administration of growth hormone (SC route) for four months (Slonim et al., 2000). The optimal dose, schedule and duration of response remain to be defined.

9) The seemingly useful effect of cigarette smoking in ulcerative cholitis (but not in Crohn's disease) is very intriguing and, despite some encouraging results with transdermal or rectal nicotine administration, its effective therapeutic role remains uncertain.

10) Modulation of vagus nerve activity: the activation of the cholinergic anti-inflammatory pathway may provide a therapeutic advantage for inflammatory diseases. This interesting idea, discussed by Tracey (2002), possibly assessable by implanting a pacemaker-like device, is supported by the clinical finding that nicotine administration reduces the severity of ulcerative colitis.

11) Hyperbaric oxygen therapy has been described as useful in severe or refractory perineal Crohn's disease. Noyer and Brandt, (1999) reported 16 complete responses over 22 patients.

B) The following conventional therapies of **rheumatoid arthritis (RA)** aims to relieve pain, reduce and possibly resolve the chronic inflammation causing degeneration of cartilage and the erosion of juxta-articular bone.The excessive local release of TNF alpha seems the major culprit (Feldmann and Maini, 2001):

1) Non-steroidal anti-inflammatory drugs (NSAID). Besides the old aspirin, they include ibuprofen, indomethacin, naproxen, sulindac, etc. They are all associated with at least gastric irritation. Coated tablets or other administration routes can limit a potential damage. The latest-generation

cyclooxygenase II inhibitors seem to be fairly effective and have less adverse effects (Fitzgerald and Patrono, 2001).

2) Glucocorticoid therapy. It has been widely used and is effective but its prolonged use involves serious side effects.

3) Immunosuppressive therapy. Azathioprine, cyclophosphamide, methotrexate at an intermittent low-dose may be useful, but attention must be given to adverse effects.

4) Disease-modifying drugs, such as D-penicillamine, sulfasalazine, gold compounds, are partly useful, but there is minimal evidence that they delay bone erosion or allow real healing.

5) A statin can mediate modest but beneficial anti-inflammatory effects and reduce cardiovascular morbidity (McCarey et al., 2004).

C) Psoriasis is a chronic inflammatory skin disorders that affects 1 to 2% of people. Although the disease may not be as crippling as Crohn's disease and rheumatoid arthritis, it often causes physical and mental disabilities. Psoriasis is characterized by the infiltration of the skin by activated T cells and an exceptional proliferation of keratinocytes. Here it is discussed with the previous diseases because the concentration of TNF alpha, very high in psoriatic lesions, suggests an important pathogenic role (Bonifati and Ameglio, 1999).

Conventional treatments depend on the type, location and extent of the lesions:

1) Topical glucocorticoids are more effective when used in conjunction with a keratolytic agent. Corticosteroids and cyclosporine A have been used systemically.

2) Ultraviolet A (UV-A) light combined to either oral or topical psoralens is very effective but the potential toxicity limit its use. A similar attention must be paid when using methotrexate or the synthetic retinoid etretinate, which is a potent teratogen.

Conventional treatments of the just described diseases have been scarcely effective and accompanied by relevant side effects. **In the '80s, the pathologic role of IL-1 and TNF alpha became evident and, among several ideas, one was to reduce or eliminate T cell-dependent autoimmune diseases by using either monoclonal antibodies to T cell surface molecules, such as CD3, CD4, CD25 and CD52 or IL 2-diphtheria-toxin fusion protein.** Although they dampen the progression of the disease, they were all associated with long-term T cell depletion. Then, at the annual IFN meetings, I remember that Dr. M. Feldmann was the first to pioneer, among several theoretical possibilities, the use of TNF inhibitors. It has taken considerable time to pass from the laboratory bench to the bed side but **now biotechnology has allowed the preparation of several complex proteins, all aiming at reducing inflammations in the different pathological sites.** Today the following products are available:

1) **INFLIXIMAB** is a humanized, mouse-derived monoclonal IgG1 antibody against TNF (Maini et al., 1999; Present et al., 1999; Feldmann and Maini, 2001; Hanauer et al., 2002; Baert et al., 2003; Sands et al., 2004). It has been used in patients with Crohn's disease and RA. The antibody, administered during a 2-hours infusion, remains in the intravascular pool with an average half-life of 9.5 days. Therapy is repeated usually every two months and it is frequently combined with methotrexate for improving the response and reducing formation of autoantibodies to Infliximab.

2) **ETANERCEPT** is a recombinant human soluble fusion protein of TNF-Type II receptor with human IgG1. It antagonizes the effects of endogenous TNF by competitively inhibiting its interaction with cell-surface receptors. (Lovell et al., 2000; Leonardi et al., 2003). It has been used in patients with psoriasis and RA. This biological response modifier (BRM) is self-administered via SC twice weekly but some patients do not like this route because of pain and irritation at the site of injection.

On the basis that not only T cells but also macrophages, antigen-presenting cells (dendritic cells) and B cells have a role in the disease process, complex molecules have been constructed which can either block intercellular adhesion or prevent the delivery of the second costimulatory signal required for optimal activation of T cells:

1) NATULIZUMAB is a humanized monoclonal antibody against alpha4 –beta1 integrin. In other words, it is an alpha-4 integrin antagonist able to inhibit leukocyte adhesion (Ghosh et al., 2003; Miller et al., 2003). It has been used with patients with multiple sclerosis and Crohn's disease.

2) EFALIZUMAB is again a T cell modulator. The humanized monoclonal IgG1 antibody against leukocyte-function-associated antigen type 1 (LFA-1) is able to inhibit the binding of T lymphocytes to the adhesion molecule 1 (ICAM-1) present on the surface of endothelial cells (Lebwohl et al., 2003). It has been used in patients with psoriasis.

3) FUSION PROTEIN CTLA4Ig. It is constructed by genetically fusing the external domain of human CTLA4 to the heavy-chain constant region of human IgG1. By blocking the second costimulatory CD28 signal on T cells, CTLA4Ig prevents the binding of CD80 and CD86 molecules present on antigen-presenting cells so that T cells become poorly responsive or undergo apoptosis (Kremer et al., 2003). Interestingly, the above binding on APC appears to lead to the production of indoleamine-2,3-dioxygenase, which is associated with down-regulation of the inflammatory responses of T cells, macrophages and dendritic cells (Mellor and Munn, 1999). It has been used for the treatment of RA.

4) ANAKINRA. It is the human recombinant antagonist of the IL-1 receptor and neutralizes the biological activity of both IL-1 α and beta. It is administered daily via SC in a dose of 100 mg associated to methotrexate in patients non-responding to Infliximab. The combined administration of

Etanercept and Anakinra yields neutropenia with a high risk of infections and no improvement.

Very impressive, double-blind clinical trials have been performed in thousands of patients involving huge numbers of scientists, clinicians, analysts, statisticians etc. Biotechnological firms, during the last 10-15 years, must have invested billions of dollars before these products were approved by the FDA and could be sold on the market. The actual cost of a treatment per patient is about $12,000. The diversity of the therapeutic approaches does not necessarily mean that one antibody works better than another. **All of these biologic agents are either meant to block noxious molecules or/and stop the signals starting a chronic inflammation.** Besides reducing the influx of cells into the inflamed tissues, they appear to down-regulate the successive production of TNF alpha, IL-1, IL-6, IL-8, MCP-1, VEGF, ROS, phlogogenic prostaglandins and to reduce the blood levels of matrix metalloproteinases and C-reactive protein. The frequency and route of administration vary between intermittent intravenous infusions or weekly SC injections. Some antibodies need to be combined with a concomitant corticosteroids therapy for reducing formation of neutralizing autoantibodies. With some variations, **these therapies have yielded remarkable improvements in 60-70% of seriously-ill patients and official medicine considers these results as a breakthrough.** Why some patients do not respond remains unclear but it is possible that using a combined therapy using simultaneously an IL-1beta receptor antagonist may improve the outcome. Although unlikely, it remains to be seen if a prolonged (2-3 years) treatment with these inhibitors is capable of turning off these diseases all together. If so, it will be a great success. **What about safety?** The rate of adverse events is high but rarely is life-threatening (anaphylactic reaction) and it appears acceptable in comparison to the tangible clinical benefit. Anybody interested in this subject can read the exhaustive review published by Reimold (2003).

An important concern regards the possibility that patients may develop in a few years' time either cancer, or serious infections (tuberculosis), or a lupus-like syndrome (Keane et al.,2001). This has already happened in a few patients and, by causing immunosuppression, it could be expected (Bell and Kamm, 2000; Sartor, 2000; Day, 2002; Emery and Buch, 2002; Fiocchi, 2004). I am personally fairly optimistic because the main idea of these blocking therapies is to reduce the level of noxious molecules locally, where they are released in excessive concentrations. If this reasoning is correct, **we should refrain to increase the inhibitory dosage, thus leaving intact crucial protective functions in other sites.** Only time will give the definitive answer but we must restrain the enthusiasm and exercise the utmost care of not harming our patients. PRIMUM NON NOCERE!

MULTIPLE SCLEROSIS (MS).

I am discussing this disease separately from the previous ones because the treatment of choice is performed with a different compound. This is a tragic disease because it very often disables young adults just when they are about to show their merit. MS is an inflammatory disease of the central nervous system (CNS), probably triggered at first by a viral infection. All physicians know that MS is a T cell-mediated autoimmune disease directed against CNS myelin or oligodendrocytes causing demyelination and axonal damage responsible for later permanent disability: it can either relapse (relapsing-remitting MS) or be very aggressive (progressive MS). Good reviews of the topic are available (Rudick et al., 1997; Karp et al., 2000; Polman and Uitdehaag, 2000; Keegan and Noseworthy, 2002; Revel, 2003).

Orthodox medical therapy is based on:

1) **Corticosteroids** (Milligan et al., 1986).

2) **Immunosuppressive drugs,** namely azathioprine, methotrexate, cyclophosphamide and cyclosporine. All of these drugs can cause immunosuppression to different degrees and may cause serious adverse effects. Quite interestingly, a preliminary report has suggested that simvastatin, acting as an immunosuppressive drug, may have therapeutic activity (Vollmer et al., 2004).

3) **Experimental biologicals**: IV immune globulins are now rarely used. Copolymer 1 (COP) or glatiramer acetate is a mixture of synthetic polypeptides composed of four amino acids (Duda et al., 2000; Kipnis et al., 2000; Neuhaus et al., 2000; Karandikar et al., 2002; Boneschi et al., 2003;) induces a shift from a Th-1 to a Th-2 cytokine profile in COP-treated patients and seems to inhibit antigen-specific T cells.

4) In a placebo-controlled trial, treatment with **Natalizumab** led to fewer inflammatory brain lesions and fewer relapses over a six- month period in patients with r-r MS (Miller et al., 2003).

5) The present treatment of choice (for r-rMS) has been made possible by advances in biotechnology (Revel, 2004), that have allowed the production of **IFNβ-1a** in mammalian CHO. This IFN has a glycosylation similar to the natural fibroblast IFN. The second type, defined **IFNβ-1b**, produced in bacterial cells, is a mutein because it has one cysteine replaced with serine to maintain structural stability. It lacks also the N-terminal methionine, is not glycosylated and is about 10-14 fold less potent than IFNb-1a. For these reasons, a higher mass of IFN protein must be injected, that may be responsible for an increased immunogenicity and possibly reduced therapeutic activity (Antonelli et al., 1998; Sorensen et al., 2003). However, Durelli et al. (2002) have shown that this is not the case and actually the mutein was more effective after two years treatment.

Despite their biochemical difference, both forms of IFNβ (approved by US and European regulatory authorities) have a useful clinical effect, characterized by a 30% reduction of both the frequency and severity of exacerbations (Arnason, 1993; Rudick et al., 1997; Polman and Uitdehaag, 2000; Filippini et al., 2003; Miller, 2003; Revel, 2004). IFNβs are fairly well tolerated. Unfortunately, owing to striking pharmacokinetic and pharmacodynamic differences (Bocci, 1981b; 1987b; 1988a; 1990a), IFNα-2a, which could be therapeutically useful and is inexpensive, causes adverse events that negatively affect the already poor quality of life of these patients (Nortvedt et al., 1999). However, owing to the improved toxic profile of pegylated IFNα, it would be interesting to evaluate its efficacy.

We clarified that IFNsβ are preferentially absorbed via the lymphatic system and, by hardly appearing in the plasma, elicit only minor side effects (Bocci et al., 1988). These IFNs are now in wide use and, in order to minimize long-term disability, the therapy should commence as soon as possible after diagnosis. Oral administration of IFN beta-1a was found ineffective (Polman et al., 2003). The progressive form of MS is far less responsive to this therapy.

Problems such as the optimal dose and schedule, the appearance of neutralising antibodies (mostly to IFNβ-1b) that may jeopardize efficacy, a possible relapse when stopping therapy and the considerable cost, provide a glimmer of hope that a serious RCT based on ozonetherapy may be meaningful. In the case of MS, nothing serious has been done and my attempt to interest three neurologists was in vain because, as expected, they were well sponsored by firms producing IFNβ. Two ozonetherapists (one in Turin and another in Milan) reported to me that they had achieved "good results" treating MS and RA patients with AHT combined with either magnetotherapy or chelation therapy so that cannot be taken into consideration.

There has been one trial performed at Cuba's Institute of Rheumatology in 1988 on 17 RA patients treated with IM injections of oxygen-ozone (total dose of ozone: 700 mcg) for 8 weeks combined with NSAID. Apparently about 25% of ozone-treated patients scored 25% better than controls but this type of study does not clarify if ozonetherapy may be useful.

However two preliminary reports were published by D'Ambrosio (2002) on Crohn's disease and ulcerative colitis treated by rectally administered oxygen-ozone. 24 women and 6 men, with average disease duration of 2.5 years, were enrolled. The standard therapy for both diseases consisted of rectal insufflations of a gas mixture at a dose of 300-400 ml at each session, initially, for reducing diarrheoa and haemostatic purposes, at high concentration (60 mcg/ml!), subsequently reduced in the course of treatment. Patients underwent a total of 30 sessions with an initial frequency of two treatments weekly followed by one every two weeks. Outcome was decidely positive (stable normalization at endoscopy) in 50% of patients and moderately useful in 37% while 4 patients got worse. It has appeared useful to perform a therapeutic cycle every 6-8 months. Although I do not agree either on the ozone dosages, or the schedule, this study appears encouraging.

This first open study stimulate the question: can ozone therapy be of any use in autoimmune diseases? Should we use it in combination with conventional approaches or can it be efficacious on its own? Among other complementary approaches, I believe that ozone therapy is the only one with meaningful rational bases. If it is true that hyperbaric oxygen therapy improves perineal Crohn's disease (Noyer and Brandt, 1999), OZONE THERAPY COULD BE EVEN MORE USEFUL!

In autoimmune diseases (pattern: Th1 >> Th2), ozonation of blood with low-medium ozone concentrations (20-40 mcg/ml) may upregulate cytokines produced by Th1 cells and accelerate the progression of the disease, while high concentrations (40-80 or more, mcg/ml of gas per ml of blood), by producing a high concentration of toxic LOPs, may kill proliferating autoreactive cells, leading to a quiescent phase. Moreover the decreased production of proinflammatory cytokines may favour the release of Il-10 and TGF beta. In other words, low- medium doses of ozone can enhance the progression of the disease while high doses may down-regulate the inflammatory process. Certainly, even empirical but trustworthy results by private ozonetherapists would have been helpful. Unfortunately, most ozonetherapists neither possess a reliable generator nor precisely check the ozone concentration and the blood/gas volume ratio. Moreover, there is still extreme confusion about the blood volume and the system for ozonation: some ozonetherapists use small glass bottles and only 50-100 ml of blood, others, like me, use 500 ml glass bottles and collect between 150 and 250 ml of blood, while some even use the hyperbaric system, for which we have no laboratory data. Others insist on using toxic PVC bags of different volumes in spite of their toxicity and prohibition by the Italian Ministry of Health. During the last five years, I have tried to no avail to correct this anarchical situation that hinders any progress.

At least, as a working hypothesis, we must try to have a few basic ideas and standard conditions. Let us first consider the crucial parameters:

1) The target is represented by $CD4^+$ lymphocytes present mainly as actively proliferating Th1 phenotypes. Although it may not be completely true, a fair assumption is that these cells are somewhat sustaining the ongoing disease and a possible approach is to suppress the secretion of Th1-type cytokines (with cytolytic and ROS enhancing activity).

2) **The volume of blood** appears critical for three reasons:

a) The number of present and active lymphocytes during the ozonation process, because they will be directly affected (via H_2O_2 and very short half-life ROS);

b) The volume of plasma, because it contains all the substrates undergoing direct peroxidation that will generate long half-life LOPs. These compounds (4-HNE, MDA, isoprostanes, possibly acrolein, etc.) act immediately on proliferating lymphocytes and will also bind to circulating lymphocytes after blood reinfusion into the donor. Activated cells are more likely to be inhibited than resting cells.

c) The ozone concentration (mcg/ml per ml of blood), which can be divided into:

- low (10-30 mcg/ml)
- medium (30-50 mcg/ml)
- high (50-80 mcg/ml)
- very high (80-120 mcg/ml).

Depending on the capacity of the plasma antioxidant system, the formation of ROS and LOPs, although not proportional to the ozone concentration, increases with the ozone dose. The consequence is that the final amounts of these compounds, which are supposed to act as cytotoxic drugs, depend upon the volume of plasma and ozone dose.

Therefore, a low ozone dose may hardly affect the lymphocytes present in the blood during ozonation and, owing to minimal LOPs formation, also may not affect circulating cells. Conversely, a high ozone dose may prevalently deplete Th1-type lymphocytes (via reinfused LOPs), thus slowing down the disease. While it would be naive to think that LOPs will selectively inhibit Th1-type lymphocytes, they might preferentially bind and inhibit these cells because they are in an activated state. Needless to say, the same reasoning can be used for allergic diseases with a pattern Th2 >> Th1.

It must be emphasized that this is only a working hypothesis and much remains to be learned before making definitive recommendations. Moreover, as it has been discussed in Chapters 4 and 8, we strongly advise the ozonetherapist to apply the "up-dosing" system.

In other words, in order to induce ozone tolerance, the "start low, go slow" strategy appears most reasonable. The following is a schematic example of a possible schedule for ozonetherapists performing ozonated AHT in 500 ml glass bottles:

Time (weeks)	Treatment Number	Blood volume (ml)	Ozone concentration (μg/ml)	Total ozone Dose (mg)
	1	270	50	13.5
1st	2	270	50	13.5
	3	270	50	13.5
	4	270	60	16.2
2nd	5	270	60	16.2
	6	225	70	15.7
	7	225	80	18.0
3rd	8	225	80	18.0
	9	225	80	18.0

and so on for at least 26 treatments (8 weeks), unless unforeseeable side effects appear. During the next four months ozone therapy can be continued at high ozone concentration, at least twice weekly (32 sessions).

If the patient has a difficult venous access and is not a diabetic, we can infuse the "gluco-peroxide" solution starting with an hydrogen peroxide concentration of 0.03%, three times weekly, slowly (within three weeks) upgrading the concentration up to the maximum of 0.15%. If the patient improves and does not report adverse effects, we can continue the treatment for six months.

In order to maintain a sufficient antioxidant capacity, the patient must take a daily dose of antioxidants (Chapter 8) and it is then possible to increase the ozone concentration to 90 mcg/ml. Using either AHT or the "gluco-peroxide" solution or BOEX, we may be able to dump autoreactive cells. A simultaneous immunological investigation in the treated patients should aim at clarifying if ozonetherapy induces anergy of the cytotoxic T lymphocytes.

The RI approach, ASSOCIATED with the parenteral ones (AHT and BOEX), may be helpful in inducing immunosuppression in the gut (see Chapter 6). In one patient with Crohn's disease we have successfully administered medium-strength ozonated olive oil (2 g daily in the morning, before breakfast) enclosed in four gastroresistant capsules. Beside gas insufflation, it is also advisable to make a small clisma (50 ml) of mildly ozonated olive oil once a week. RI may also inhibit the bacterial flora that is partly responsible for Crohn's disease. Ozonetherapy can be potentiated by a simultaneous administration of probiotics and fish oil (2 g daily) easily ingested when enclosed in gastroresistant capsules. Fistulae and abscesses in Crohn's diseases can be dealt with by insufflation of ozone or ozonated oil.

How and why could ozonetherapy be beneficial?

a) We know that prolonged ozone therapy induces a generalized induction of HO-1 and antioxidant enzymes, which is extremely important for correcting the chronic oxidative stress. Today, paradoxically, only ozone therapy can strengthen the adaptation to continuous stress. If it succeeds in inhibiting the clone of cytotoxic lymphocytes, the reduced production of pro-inflammatory cytokines may facilitate the production of IL-10, IL-11, TGFβ and perhaps IL-1 Receptor antagonist (IL-1 Ra), which will be a prodigious result.

b) Moreover ozone therapy can progressively inhibit the release of inflammatory enzymes, metalloproteinases etc., with a progressive decrease of plasma levels of PAF, LTB_4, PGE_2, TxA_2 and isoprostanes. The chronic inflammatory process can be slowly turned off only if we can perform 6 months of therapy.

c) The "therapeutic shock" induced by the withdrawl and reinfusion of ozonated blood (AHT) or the "gluco-peroxide" solution or by BOEX or RI induces a transitory homeostatic change that, particularly in severely-ill patients, results in a sudden hormonal release (possibly including cortisol) that explains the feeling of wellness. **This positive response has never been accompanied by any of the adverse effects noted in about 90% of patients treated with inhibitory antibodies.**

Finally I will pose the readers the most relevant question.

Who will support these researches? Who will pay the medical personnel and the huge cost of endoscopic, radiological, histological, biochemical, immunological and clinical exams? On average, a trial enrolling 100 patients may cost about $ 600,000! (Emanuel et al., 2003). We are not backed by any pharmaceutical and/or biotechnological firm because ozonetherapy does not produce profits. However, if with our very good will we can prove the validity of ozone therapy, the National Health Services of all countries, particularly those with few resources may become interested. So far, based on my personal experience, both the Italian National Health service as well as the World Health Organization in Geneve, have proved to be biased and will not support this research.

6. OZONE THERAPY IN CANCER

Although some haematological cancers are now being treated successfully, the common solid cancers, which are the great majority, continue to be a problem for mankind (Bailar III and Gornik, 1997). Owing to earlier diagnoses and some therapeutic advances, for the first time in Western European countries, the total cancer mortality was moderately reduced for both sexes in the period 1990-1994 (Levi et al., 1999). However, due to prolongation of the life-span, the figures for overall mortality from cancer (in Italy about 160,000 and in the USA about 520,000 in 1993) are still dramatic. Moreover, in the same period, cancer mortality was still increasing in eastern European countries. This is not likely to change soon because a highly desirable improvement of chemotherapeutic compounds, so far rather unspecific and toxic, may come too slowly. The search for highly selective drugs is relentless and a few new drugs like imatinib mesylate (a selective tyrosine kinase inhibitor), a monoclonal antibody (trastuzumab) against the epidermal growth factor receptor (EGFr) and another (bevacizumab) for metastatic colorectal cancer (Mayer, 2004) appear as a breakthrough until cancer cells mutate and become resistant (Gorre et al.,2001). An appropriate cancer prevention campaign, aimed at early detection and the use of an appropriate diet rich in fibre and antioxidants (Dreher and Junod, 1996; Bailar III and Garnick, 1997; Kramer and Klausner, 1997), may help up to a point. Yet, on the whole, smoking is not decreasing and has partly shifted from men to women and to Third World countries. A report by the WHO foresees that worldwide cancer rates may double by 2020, unless we take stringent measures for promoting a healthy diet, smoking cessation and improved access to viral immunisation (Eaton, 2003).

The pillars of therapy are surgery first and then radiotherapy and chemotherapy. Hormonal therapy has some more specific applications and since 1891, Paul Ehrlich's dream (the famous magic bullet!) was to make immunotherapy effective. At least theoretically, immunotherapy aims specifically at destroying only neoplastic cells, but unfortunately these cells are poorly immunogenic and diabolically equipped to evade or suppress the immune system. In spite of numerous and theoretically brilliant approaches, none has achieved tangible results (Rosenberg et al., 1987; Rosenberg, 2001; Bocci, 1985a, b; 1987b, 1990, a, b; 1991a, b; Reddy et al., 1997; Ernst, 1997; Motzer et al., 2001).

Immunological gene therapy works well in experimental murine tumours, but so far has been disappointing in patients (Anderson, 1992; Bubenick, 1996; Roth and Cristiano, 1997; Parmiani et al., 2000). **The greatest hurdle for successful cancer therapy is a thorough understanding of the several mechanisms used by tumour cells to evade the immune attack.** In spite of a meaningful rationale, the latest disappointment has been the anti-angiogenic therapy (Carmeliet and Jain, 2000): it works very well in mice (O'Reilly et al., 1997; Boehm et al., 1997; Perletti et al., 2000) but not, as we hoped, in human tumours, even though **angiogenic inhibitors** (Oehler and Bicknell, 2000; Daly et al., 2003; Yang et al., 2003; Eskens, 2004;) **COMBINED with other drugs may still play an important role.** Thus, after all the untimely and deleterious propaganda of the mass media, it is not surprising that desperate patients are always looking for other possibilities, particularly in the vast field of complementary medical practices (Cassileth and Chapman, 1996; Burstein et al., 1999) such as diet, nutrition and lifestyle changes, therapeutic touch (Rosa et al., 1998), mind-body control (Flach and Seachrist, 1994; Sheldon,2004) and anthroposophic medicine based on the use of mistletoe lectins (Bocci, 1993b; Ernst, 2001; Steuer-Vogt et al., 2001).

In June 1995, the National Institutes of Health (NIH, Bethesda, MD, USA) included the use of oxidizing agents (ozone, hydrogen peroxide) in class 5, among chelation and metabolic therapies, cell treatment and anti-oxidizing agents. It is noteworthy that hydrogen peroxide has been evaluated as an anti-neoplastic agent by Zanvil Cohn at the Rockefeller University (Nathan et al., 1979,a,b; Nathan and Cohn, 1981). Another study has been performed by Sasaki et al. (1967). As reported in Chapter 6, Section II, the infusion of the "gluco-peroxide" solution is becoming useful and practical and it will be included in the suggested protocol for cancer treatment.

At an earlier stage, ozone was tested in cancer by Varro (1966, 1974, and 1983) and Zabel (1960). Thus, although ozonetherapy is more than 40 years old, it has been carried out in a few private clinics in central Europe but for several reasons, not totally right, it has never been accepted by official Medicine and is currently despised in France, England, USA and barely tolerated in Italy.

Is ozonetherapy useful in cancer? Varro (1983) claimed that, after undergoing surgery, chemotherapy and radiotherapy, most of his private cancer patients benefited from ozonetherapy, as their quality of life improved and they survived for a long period. However, these statements were not validated by statistical data and have no scientific value. There are other anecdotal reports of major or minor autohaemotherapy having beneficial effects: for example, Beyerle (1996) treated prostate cancer with "phenomenal" (?) results. For other types of cancer (throat, ovarian, colon and breast), he comments:

"We are seeing patients who were bedridden two years ago and sent home to die. They are becoming ambulatory. Their energy level is coming up. They are gaining weight. And we see these spontaneous fractures in the spine are gradually disappearing. Strength is returning to the musculature. There is no spinal pain".

It is unclear why Dr. Beyerle has not reported the data in a peer-reviewed medical journal, because as presented they are worthless. His comments were actually recorded by a journalist (Null, 1996) during an interview published in Penthouse, which certainly is not a scientific journal. Kief (1993a), at his clinic at Ludwigshafen (Germany), has used Auto-homologous Immunetherapy (AHIT) to treat a variety of malignancies. AHIT was administered daily for a period of four months and he claimed that it is:

"cost-effective, individually-oriented, has no-side effects, decreases pain in 70% of all cases and increases the life-quality and vitality in approximately 90% of the cancer patients".

What AHIT really was remains a mystery (apparently a mixture of the patient's blood and urine treated with ozone!) and, to the best of my knowledge, the German Health Authorities have now prohibited its use.

In conclusion, today there is no serious evidence that ozonetherapy can be beneficial to cancer patients because:

- **Randomized, double-blind clinical trials have not been performed** as they should have been done (Ernst and Resch, 1996).
- **It is unclear whether biological and/or clinical effects, if any, are due to either oxygen or ozone or to both, or simply to blood transfusion.**
- **The relevance of the placebo effect is unknown.**
- **Too often ozonetherapy is carried out together with other conventional or natural therapies**, so that any result remains questionable.

In spite of these negative conclusions, it is worth while to discuss the peculiar biological characteristic of the tumour environment in relation to the effects of ozone therapy because we can try this approach only if there is a solid rationale.

Tumour hypoxia is a well recognized mechanism for resistance of neoplastic cells to anticancer drugs and radiotherapy. Warburg's work in the 1920s demonstrated that, even in hypoxia, cancer cells intensely convert glucose to lactic acid, but unless they are in anoxia their intracellular pH remains neutral (pH 7.0-7.2) while the pH is slightly acidic (6.8) in the interstitial fluid. **Tumour hypoxia is also a relevant factor enhancing neoangiogenesis, dedifferentiation and metastasis** (Bush et al., 1978; Coleman, 1988; Gatenby et al., 1988; Vaupel and Hockel; 2000; Hockel and Vaupel, 2001; Brahimi-Horn et al., 2001; Harris, 2002; Fyles et al., 2002; Subarsky and Hill, 2003). **Both primary and metastatic tumors thrive in areas where the average pO_2 is lower than normal tissues and the host appears unable to mount a reaction for reestablishing physiological levels.** An anarchic vascularization usually implies anomalous vessels with variable blood flow, increased permeability, oedema, hypercoagulability, metastatic progression and therefore poor prognosis (Brizel et al., 1996; Hockel et al., 1996; Young et al., 1988; Denko and Giaccia, 2001; Subarsky and Hill, 2003; Helczynska et al., 2003; Denko and Giaccia, 2001).

In physiological conditions, at sea level, the pO_2 in the alveolar space (O_2 equal to 14%) is equivalent to 100 mm Hg (1 atmosphere = 760 mm Hg = 101.3 Pa) and the pO_2 of arterial blood is about 98 mm Hg, haemoglobin is fully saturated to Hb_4O_8 and there is about 0.3 ml/dL of oxygen solubilized in the plasma. Depending upon their metabolism, tissues (retina>kidney>liver>heart>brain, etc,) extract from blood variable amounts of oxygen (on average about 25%, i.e. 5 ml of oxygen/dL blood) so that venous blood has a pO_2 of about 40 mm Hg, with oxyhaemoglobin depleted on average of only one molecule of oxygen. Thus the amount of oxygen physically dissolved in the plasma is grossly insufficient for the requirements of the tissues and the normally necessary 5 ml of oxygen/dL blood derive from deoxygenation of oxyhaemoglobin. The crucial point is that, for reasons mentioned below, erythrocytes of the neoplastic patients are unable to deliver more oxygen to the hypoxic tumor tissue.

Although among different tumors and actually within the same tumor, there is a marked heterogeneity in terms of oxygen supply (Coleman, 1988; Gatenby et al., 1988; Young et al., 1988; Brizel et al., 1996; Hockel et al., 1996; Vaupel and Hockel, 2000; Helczynska et al., 2003; Brizel et al., 1996; Denko and Giaccia, 2001), **there is a general consensus that neoplastic tissues prefer a hypoxic and acid micro-environment.** This seems due to a combination of an aberrant vascular bed, leaky microvessels, elevated interstitial fluid pressure, lack of lymphatics and reduced blood flow. **In comparison to normal tissues, the average pO_2 in tumors is less than 1/4 (40-45 versus 2-10 mm Hg).** For normal tissues, hypoxemia represents a consistent metabolic disadvantage whereas experimental observations led to the conclusion that hypoxia is advantageous for growth and expansion of neoplastic cells (Gatenby et al., 1988; Young et al., 1988; Brizel et al., 1996; Vaupel and Hockel; 2000; Harris, 2002; Helczynska et al., 2003). Overexpression of hypoxia-inducible factor, HIF-1-α was detected in the majority of tumor types in comparison with the respective normal tissues (Ryan and Johnson, 1998; Carmeliet et al., 1998; Zhong et al., 1999 Semenza, 2001, 2003).

HIF-1 is a heterodimer consisting of the hypoxic response factor HIF-1-α and the stably expressed arylhydrocarbon receptor nuclear translocator (ARNT) or HIF-1-β (Semenza, 2001; 2003; Huang and Bunn, 2003). The availability of HIF-1 is determined by HIF-1-α, which is regulated at the protein level in an oxygen-sensitive manner: **under hypoxia, HIF-1-α protein is stable, translocates to the nucleus and, after binding to HIF1-β, activates gene transcription of VEGF, erythropoietin and glycolytic enzymes that allow neoplastic cells adaptation to hypoxia.** ***In contrast, during normoxia, HIF-1-α binds to the Von Hippel-Lindau tumor suppressor protein, that being one of the components of the multiprotein biquitin-E3-ligase complex, targets HIF-1-α for proteosomal proteolysis.*** Thus, the establishment of normoxia in human tumors ought to inhibit overexpression of HIF-1-α, enhance its degradation and may limit tumor progression and metastasis.

As it was mentioned, **in order to block the malignant evolution of tumors, one of the most studied approaches is to inhibit angiogenesis** (Tosetti et al., 2002). This process is clearly stimulated by hypoxia (Carmeliet et al., 1998; Ryan and Johnson, 1998; Zhong et al., 1999; Brahimi-Horn et al., 2001; Denko and Giaccia, 2001; Harris, 2002; Subarsky and Hill, 2003; Semenza, 2003; Huang and Bunn, 2003; Falm, 2004), **but a**

direct correction of the hypoxic state seems a more straightforward method to block cancer progression. ***If this postulation is correct, a novel approach for constantly restoring normoxia in hypoxic tissues can be proposed.***

But will it be feasible to constantly correct hypoxia in cancer patients?

Will it be possible to induce a constant restoration of normoxia in hypoxic tumours?

During the last century **several strategies have been proposed** for enhancing oxygenation of tumors. The most obvious was **breathing pure oxygen** but because of its toxicity, this can only be done for short periods with only a transitory increase of arterial pO_2 (Thomson et al., 2002). **Carbogen breathing** on its own or in combination with other therapies is practical and useful at high altitudes, but it is not resolutive for neoplastic patients (Inch et al., 1970; Siemann et al., 1977; Rubin et al., 1979; Song et al., 1987; Falk et al., 1992; Griffin et al., 1996; Bernier et al., 2000; Imray et al., 2003;). **Hyperbaric oxygen therapy** is a procedure by which almost pure medical oxygen is inspired in an air-tight chamber at about 2.6 atmospheres for two hours (Dische et al., 1983; Bergo and Tyssebotn, 1999; Cianci, 2004). During this period the oxygen solubilized in plasma increases up to 5 ml/dL and it becomes practically sufficient for satisfying tissue requirements so that practically no oxygen molecule is released by oxyhaemoglobin. In this situation neoplastic tissues may temporarily become normoxic but only if organ vasoconstriction does not occur (Bergofsky and Bertun, 1966).

Cancer patients are often anemic and recently, in order to improve therapeutic effectiveness as well as fatigue, **recombinant erythropoietin** is used (Marrades et al., 1996; Littlewood et al., 2001), although, more recently, Henke et al., (2003) have warned that it does not improve cancer control or survival. Obviously **blood transfusion or artificial oxygen carriers** can be used (Song et al., 1987; Teicher and Rose, 1984) provided that they do not excessively increase blood viscosity and, once again, they correct hypoxic microenvironments temporarily. **Vasoactive drugs** (Horsman et al., 1989; Song et al., 1992; Siemann et al, 1994; Honess et al., 1995; Bernier et al., 2000) and **mild hyperthermia** (Dewey et al., 1977; Valdagni and Amichetti, 1994; Overgaard et al., 1995; Griffin et al., 1996; Song et al., 1996; 1997) may also be of some help. Although all of these approaches have some merit, they do not solve the problem of constantly correcting tumor hypoxia.

Is it then possible to constantly improve oxygen delivery into the tumour environment by ozonetheray? Let us see what ozone is able to do!

As any other gas, ozone dissolves in the water of plasma and immediately disappears by reacting with organic compounds (hydrosoluble and lipophylic antioxidants, unsaturated fatty acids, etc) generating a number of messengers acting on various blood components and procuring early (by

ROS) and late (by LOPs) biological effects. While we were assessing the range of the therapeutic window, we found that the ozone concentration must reach a critical threshold to be effective as otherwise it results only in a placebo effect. An early effect is due to a sudden increase of hydrogen peroxide that switches on a number of biochemical pathways in erythrocytes, leukocytes, platelets and endothelial cells (Bocci, 2002; Stone and Collins, 2002). The late effects are due to a number of LOPs with a half-life far longer than ROS. Upon blood reinfusion in the donor that begins 5-10 min after blood ozonation, LOPs will undergo extensive dilution, catabolism and excretion. At the same time some of the LOPs will activate endothelial and parenchymal cells of several organs among which bone marrow is particularly relevant (Figure 2). LOPs may also bind to neoplastic cells.

It is well known that every day about 0.8% of the erythrocyte pool, a fraction corresponding to about 40 ml of blood including $2x10^{11}$ (Bocci, 1981a) four-month's old erythrocytes, is taken up by erythrocatheretic organs. An intensive schedule envisaged for cancer patients includes three major AHTs sessions weekly (including 810 ml of blood) for six months allowing the ozonation of about 20 l of blood, a volume most likely sufficient for correcting the hypoxic state. Thus, **since the first session, ozone causes two important modifications, of which the first happens *ex vivo* and the second *in vivo.***

The first occurs in the glass bottle while ozone dissolves in the water of plasma and generates hydrogen peroxide and lipoperoxides which behave as second messengers: almost instantaneously, they enter into the erythrocytes and activate a number of biochemical pathways. These ROS are almost immediately reduced (H_2O_2 to water and $ROO^{\cdot}$ to hydroperoxide) mostly at the expense of GSH.

While GSH-Rd utilizes the coenzyme NADPH to recycle GSSG to the original level of GSH, the oxidized NADP is reduced after the activation of the pentose phosphate pathway, of which G6PD is the key enzyme. Thus glycolysis is accelerated with a consequent increase of ATP levels. Moreover the reinfused erythrocytes, for a brief period, enhance the delivery of oxygen into ischemic tissues because of a shift to the right of the oxygen-haemoglobin dissociation curve due either to a slight decrease of intracellular pH (Bohr effect) or/and an increase of 2,3-DPG levels.

The second and I believe, **more important modification occurs in the bone marrow** when submicromolar amounts of LOPs present in the reinfused blood reach various organs, among which the bone marrow, where they can influence the differentiation of the erythroblastic lineage. It is emphasized that each AHT represents a calculated, very transitory oxidative stress that, by activating the adaptive mechanism, results in the generation of erythrocytes with improved biochemical characteristics. These **"supergifted erythrocytes"**, as we called them, due to a higher content of 2,3-DPG and antioxidant enzymes, during their life-time, become able to deliver more oxygen into ischemic tissues (Bocci, 2002; Rokitansky et al., 1981; Mattassi et al., 1987; Romero Valdes et al., 1993; Tylicki et al., 2001; 2003; 2004; Giunta et al., 2001; Clavo et al., 2003). The consequence of repeated treatments, obviously depending upon the volume of ozonated blood, the ozone concentration and the schedule is that, after a few initial treatments, a cohort of "supergifted erythrocytes" will enter the circulation every day and relentlessly will substitute old erythrocytes generated before the therapy. This means that, during prolonged ozonetherapy, the erythrocyte population will include not only cells with different ages but, most importantly, erythrocytes with different biochemical and functional capabilities. In four patients with ARMD, after a short cycle of fourteen AHT treatments (about 3.8 l of blood was ozonated during seven weeks), density-gradient separation of old and young erythrocytes (Micheli et al., 1985) has shown a marked increase of G6PD in the young erythrocytic fraction generated during the course of ozonetherapy (Micheli et al., in preparation). Other relevant biochemical changes such as glycolysis activation with increased ATP and 2,3-DPG levels, particularly in patients with basal low levels, have been measured in erythrocytes at the end of the cycle. Moreover, while the enzymatic activity is not modified by the ozonation procedure, it does significantly increase in vivo after a therapeutic cycle: we have found that GSH-Px, GSH-Rd, GSH-Tr and SOD increase 210, 147, 164 and 141%, respectively, amply confirming previous data reported by Hernandez et al. (1995).

That ozone can induce the release of erythrocytes with improved functional activities is not surprising because the phenomenon of adaptation to chronic oxidative stress (De Maio, 1999; Jolly and Morimoto, 2000) defined also as "oxidative preconditioning" (Kume et al., 1996; Bocci, 1996a; 1996b; León et al., 1998; Barber et al., 1999; Borrego et al., 2004) or "hormesis" (Goldman, 1996; Calabrese, 2002), implies that the repeated treatments induce the synthesis of several oxidative stress proteins among which HO-1 (or heat stress protein-32), one of the most protective enzymes, is a prototypic example (Zuckerbraun and Billiar, 2003). Interestingly this happens in all organisms from plants to humans, and has also been simply termed "ozone tolerance" (Sharma and Davis, 1997; Burkey and Eason, 2002; Bocci, 1999a). **Our calculated therapeutic stress on blood ex vivo must be clearly distinguished from the life-long, endogenous, oxidative stress due to oxygen because, although it seems a paradox, ozonetherapy can upregulate the antioxidant defenses.**

On the basis of the clinical improvement of ARMD and chronic limb ischemia patients (Mattassi et al., 1987; Romero Valdes et al., 1993; Tylicki et al., 2001; Giunta et al., 2001; Clavo et al., 2003) after only two months therapy, it is likely that three-four months therapy may bring about a normal oxygenation of the neoplastic tissues. This possibility is supported by very recent experimental findings that have indicated that, after ozonetherapy, oxygenation increases particularly in the most hypoxic tumors (Clavo et al., 2004a, b).

The treatments need to be continuously maintained but this is not a problem given the excellent patient's compliance shown in other diseases (Bocci, 2002). ROS and LOPs not only increase erythrocytic functions (Bocci et al., 1998a), but activate leukocytes (Paulesu et al., 1991; Bocci et al., 1993; 1994; 1998b), platelets (Bocci et al., 1999; Valacchi and Bocci, 1999) and endothelial cells (Valacchi and Bocci, 2000). This multidirectional and simultaneous activation leads to an increased release of NO, adenosine (Riksen et al., 2003), autacoids and contribute to improve tissue vascularization (Jia et al., 1996). Indeed LOPs, by interacting with the endothelium, enhance the formation of NO and NO-thiol, which will further increase the oxygen supply by improving the tumor microcirculation. HO-I will enhance haeme breakdown yielding a higher level of bilirubin, a potent lipophylic antioxidant (Minetti et al., 1998) and CO (Snyder and Baranano, 2001; Dore, 2002; Bak et al., 2002; Lee and Chau, 2002; Zuckerbraun and Billiar, 2003). HO-I indirectly reduces vascular constriction because it suppresses the gene expression of endothelin-I and inhibits the proliferation of smooth muscle cells (Morita and Kourembanas, 1995; Duckers et al., 2001). It is certain that traces of CO cooperate with NO in favoring vascular relaxation (Bak et al., 2002).

Reinfusion of ozonated blood does not mean intravenous infusion of gas that is prohibited since 1984 (Jacobs, 1982), because oxygen can cause a deadly embolism. On the other hand, ozone reacts instantaneously and disappears; nonetheless ozone can be considered a pleiotropic bioregulator because it generates a reaction cascade of several compounds responsible for a variety of biological effects.

The result that ozone could directly and selectively inhibit neoplastic cells growth (Sweet et al., 1980) is absolutely irrelevant in vivo unless ozone is directly injected into a neoplastic nodule, that is a rare event. Hepatic metastasis could be embolised with small volumes of ozone via the hepatic artery. However, besides the normalization of hypoxia, **ozonetherapy can display other interesting biological effects that may enhance the therapeutic result.** Firstly reinfused LOPs are heterogeneous but they include cytotoxic aldehydes such as malonyldialdehyde and 4-hydroxy-2,3-alkenals (Esterbauer et al., 1991). These compounds undergo extensive dilution and are partly excreted and partly catabolised by enzymes such as GSH-Transferase and aldehyde-dehydrogenases. Moreover LOPs can be taken up by neoplastic cells and may undergo apoptosis. If this happens, ozonetherapy will act as a chemotherapeutic adjuvant, although it has been shown that poorly differentiated and rapidly proliferating tumour cells, on one hand produce large amounts of hydrogen peroxide (Szatrowski and Nathan, 1991) and, on the other hand, have a high level of antioxidants, particularly ascorbic acid (Agus et al., 1999), and antioxidant enzymes, particularly SOD and GSH-Px (Kumaraguruparan et al., 2002; Kinnula and Crapo, 2004) probably because they seems to be in a state of enhanced oxidation (Kondo et al., 1999). These new results are difficult to reconcile with hypoxia and indicate the level of complexity and disguising ability of malignant cells!

In a series of old papers (Bocci et al., 1993a; 1993b; 1994; 1998b; Paulesu et al., 1991), we showed that ozone, via the transitory action of hydrogen peroxide, acts as a mild inducer of cytokines in leukocytes and therefore primed lymphocytes and monocytes, by releasing cytokines in lymphoid microenvironments, may slowly bring about a concerted activation of the immune system usually suppressed by tumor growth. This is an interesting possibility because an endogenous and balanced cytokine production is conceptually more effective and toxic-free than the exogenous administration of a single cytokine (Bocci, 1988; 1998c).

Finally, after performing millions of AHTs during the last three decades, we can assure that ozonetherapy does not procure any adverse effects but actually improves the quality of life of the majority of patients. The mechanisms producing the state of wellness and euphoria are not yet experimentally clear but a complex hormonal release of CRH, ACTH, cortisol, DHEA, growth hormone, endorphins and neurotonic transmitters modification is likely to occur during the "therapeutic shock" due to the reinfusion of ozonated blood (Bocci, 2002).

In conclusion we have some rational arguments encouraging the use of ozonetherapy in cancer:

a) Possible improvement of blood circulation and oxygen delivery to ischemic and neoplastic tissues.

b) Improvement of the general metabolism.

c) Correct the chronic oxidative stress by upregulating the antioxidant system. Possible improvement of the cellular redox potential.

d) Induce a mild activation of the immune system and

e) Procure a state of well-being in patients by activating the neuro-endocrine system.

There are now ***three questions that need to be answered:***

1) At what stage of the disease, would ozonetherapy be better used?
2) What kind of experience have we got so far?
3) What is the most suitable therapeutic scheme?

There is a total consensus that, whenever possible, the primary tumour must be surgically removed (or irradiated) because large tumour load or /and extensive metastases induce cackexia and an anergic state (Tisdale, 2002; Argiles et al., 2003). However a complete ablation and cure is rare because haematogenous dissemination of breast tumour cells in the bone marrow occurs at an early stage of the malignancy (Riethmuller et al., 1999; Pantel et al., 1999). Thus we can presume that, even after a successful operation (negative lymph nodes), the patient, at worse, may have a dissemination of 1000-10,000 neoplastic cells that, after overcoming the immunedepression of anaesthesia and surgery, may remain dormant or eliminated through the surveillance of the immune system. There are several conventional immunomodulatory compounds but certainly **the application of ozonetherapy appears ideal for patients with the so-called minimal residual disease.**

If metastases are present, the problem is far more complex and chemotherapy is widely used with mixed results: frequently the first-line combinations can be useful and wipe out a good deal of neoplastic cells. **Further cycles, even if intensive, may or may not be useful because of progressive cell resistance to chemotherapy.** Moreover **side effects and diffused toxicity impoverish the quality of life (QoL). Patients become depressed, anaemic, neutropenic, anorectic and almost invariably, they report fatigue.** At long last, this severe complication is receiving due attention (Gutstein, 2001; Servaes et al., 2002; Stasi et al., 2003).

Can ozone therapy be useful on its own or can it be more useful than chemotherapy and radiotherapy in metastatic cancer? Can we combine the treatments? Which is the best time for performing ozone therapy during the course of the disease?

Most of these questions remain without an answer today because ozonetherapy has been totally disregarded by conventional oncology, particularly by chemotherapists. I cannot avoid commenting that chemotherapeutic drugs are economically rewarding for many people while ozone is not. This is very unsatisfactory, mostly because, in spite of a small progress, the death rate remains high and resolutive breakthroughs are not yet in sight. Because I feel that this is one of the most important issues, I have just discussed hypothetical reasons to pursue the evaluation of ozonetherapy, not as a procedure able to cure the neoplasia but rather as a means to slow down or, possibly, stabilize its progression, or at least to improve the QoL particularly in elderly patients more susceptible to the serious side effects of high-dose chemotherapy.

At long last, on October 2003, in a charity clinic we have been able to initiate an open study applying ozonetherapy to chemo-resistant cancer patients and we have made a few observations. Three patients, who had undergone high-dosage and prolonged (1.5-2 years) chemotherapy, with a Karnofsky performance status at 20-30 %, in spite of an excellent compliance, continued to show disease progression and died in 3-4 weeks. Four patients, also with diffused metastasis (usually liver and lungs), initial ascites, oedema, anaemia, hypoalbuminemia and hyperbilirubinaemia with a Karnofsky status at 40-50%, after 48-53 treatments, reported an improvement of their quality of life but the scan showed tumour progression. The experience so far achieved using low-medium ozone concentrations suggests that colorectal cancer patients, at the preterminal phase of the disease, cannot be recovered but it remains unclear if they had already reached the point of no return or, if a more aggressive ozonetherapy may be capable of stopping the progression. **It resulted clear that the palliative chemotherapy carried out for 1-2 years, not only did not prevent a large tumour expansion but markedly depressed vital functions.** *Some oncologists seem more concerned about following a protocol than the patient and forget Hyppocrates's comandament "nihil nocere"*. **Prof Cesare Maltoni used to say that "it is more important to give a good life to the day rather than horrible days to life".**

Thus a preliminary conclusion is that to embark a preterminal, chemo-resistant patient on ozonetherapy seems incorrect because it appears unlikely to modify and reverse a profoundly intoxicated and anergic biological system. This is hard to admit because patients, literally exhausted by prolonged and useless chemotherapy, are depressed and anxious to find a better treatment. This situation is critical because a desperate patient can fall a prey to a charlatan and we should not forget that several complementary approaches are neither efficacious, nor safe (Ernst, 2003).

At the moment, the most suitable time for performing ozonetherapy appears to be:

a) **After successful surgery in patients with *the minimal residual disease.***

b) **In combination to either first-line chemotherapy or radiotherapy in both inoperable and surgically-treated patients.** There are no contraindications and actually the improvement of tissue oxygenation can potentiate both chemo- and radiotherapy.

c) Moreover **ozonetherapy may reduce their typical side-effects and lead to a better tolerance and outcome.** An interesting aspect is that Jordan et al., (2002), at the Christie hospital in Manchester (UK), have already used an unsuitable (that is my opinion) method of ozone administration for enhancing healing and relieving pain in severe radiotherapy skin reactions. It would be useful to use our methodology for improving this treatment in the near future.

Obviously a few treatments are practically useless and, **if we want to radically change the erythrocytic population, or, induce a "therapeutic shock", we have programmed an intensive cycle of at least six months followed by a maintenance therapy to preserve the benefit as we have observed in ARMD patients.** The scheme is the following:

1) Depending on body weight, via a butterfly needle G-19, we collect no more than 250 ml of blood in a 500 ml glass bottle,under vacuum, having previously added 28 ml sodium citrate, 3.8%, at the usual ratio 1:9 (citrate:blood). Mix gently during blood collection.

2) We then insufflate into the bottle 250 ml of gas ($O2+O3$). The ozone concentration is progressively increased from 20 up to 90 mcg/ml gas per ml of blood, in steps of 5 mcg/ml for each session. The top concentration

is reached after 15 sessions at the end of the 5th week. (THREE SESSIONS WEEKLY on M., W., and F., or T., Th., and S.).

3) The bottle is gently (to prevent foaming with erythrocyte damage) rotated for about 10 min to ensure complete blood oxygenation and ozonation.

4) During this interval, 250 ml of the "gluco-peroxide solution" are infused. The initial concentration of hydrogen peroxide of 0.03% (8.8 mM) is progressively raised up to 0.15% (44.0 mM) in four steps.

5) By using the idoneous blood infusing set (with filter), prefilled with saline, the ozonated blood is reinfused into the donor within 15-20 min, always using the same venous access.

6) The final 4-5 ml of blood are aspirated in a 10 ml syrynge, just prefilled with 5 ml of gas (ozone concentration: 90-95 mcg/ml). The syringe is vigourously shaken for one min and the foamed blood is injected into the donor alternatively, either in the glutei or in two subcutaneous sites. This procedure defined minor AHT (Chapter 6) is meant to act as an autovaccine and a potent inducer of HO-1.

7) Rectal insufflation of oxygen-ozone could be an adjunctive treatment only if the patient agrees to do it. Only one patient of ours did it but we do not know if it was useful. We hope to evaluate soon the BOEX procedure because there is no need for venous puncture and the simultaneous, albeit transitory, hyperthermia may be beneficial (Alexander, 2003).

8) Patients, particularly those with breathlessness, by using ordinary oxygen equipment at home, are adviced to undergo intermittent (1 hour, three times) oxygen therapy every day. Although most work has been done in patients with chronic obstructive pulmonary disease (COPD), the use of oxygen is certainly useful for breathlessness in advanced cancer (Booth and Wade, 2003).

9) Patients must take every day the suggested dose of antioxidants (Chapter 8).

10) Haematological, scan and clinical controls must be programmed at least every three months during and post-therapy.

The session is completed in one hour and we have already performed almost 300 sessions without any problem with the exception to substitute in two patients the brachial access with a central one. Patients have never reported any adverse effects and the majority noticed less fatigue. In our charity clinic, the patient reimburses only the cost of the disposable material (15 Euro).

CONCLUSIONS: in the last few years, I have made an effort to explain that ozone therapy, by triggering different mechanisms of action, may be able to create an environment hostile to cancer cells (Bocci, 1988c). This is a new line of thought stating that the cell malignancy can be tamed through the use of a multiform biological modifier. The rationale

of the approach, a possible timing of application, either alone in patients with minimal residual disease or in combination with orthodox treatments and the already used therapeutic scheme have been described in details.

7. THE DYSMETABOLIC SYNDROME AND OZONETHERAPY

The dysmetabolic syndrome includes several metabolic abnormalities of which insulin resistance is one of the major characteristics. Chronic renal failure (CRF) will be discussed in Section IX, but the chronic damaging stress of haemodialysis, unavoidably leading to accelerated atherosclerosis can also lead to this syndrome.

If at least three of the following five diagnostic traits are present (Wilson and Grundy, 2003 a, b), we can make the diagnosis of the dysmetabolic syndrome:

1) Abdominal adiposity (waist girth >88 cm in women, >102 cm in men).
2) HDL-Cholesterol: <50mg/dL in women, <40mg/dL in men.
3) Triglycerides, fasting, >150 mg/dL (1.69 mmol/L)
4) Blood pressure: >130/85 mm Hg.
5) Fasting glucose >110 mg/dL (>6.1 mmol/L).

Diabetes is a disease caused by either too little of the hormone insulin (type 1 diabetes, or insulin-dependent diabetes affecting about 10% of children), or poor use of the body's insulin (type 2 diabetes, or non-insulin-dependent diabetes, prevalently affecting middle-aged patients and some obese adolescents). In Western countries this pathology affects almost 6% of the population, half of whom are UNDIAGNOSED and nonetheless the annual cost of care exceeded $ 92 billion in 1999 (American Diabetes Association. Diabetes, 1996 Vital Statistics). It is becoming a sort of epidemic (Rocchini, 2002), and the OMS has projected a number of 350 millions in 2025. It is sad that the number of people starving or undernourished is higher than the overfed one.

This situation makes the disease one of the worst if one considers the human suffering and the socio-economic burden. In the USA, diabetic patients account for 27% of the federal medical budget and what is worse is that there are a million diabetic patients suffering from chronic limb ischemia with diabetic foot ulcers. These ulcers have no tendency to heal and actually can deteriorate so that diabetics account for 50-70% of the annual non-traumatic amputations (US Department of Health and Human Services. National Diabetes Fact Sheet. Centers for Disease Control and Prevention,

National Center for Chronic Disease Prevention and Health Promotion, November 1, 1997).

Hyperglycemia (HG), present in both types of diabetes, causes a variety of biochemical derangements leading to a diffused vascular damage responsible for several pathologic manifestations. There is a fervour of studies aiming first to block or slow down the onset of type 1 diabetes, secondly to identify the environmental and genetic factors causing type 2 diabetes and thirdly to suggest possible ways for the prevention or the postponement of crippling complications (Rosen et al., 2001; Diabetes Prevention Program Research Group, 2002). The crucial problem of diabetes is the hyperglycemia due to the inability of several control systems to maintain a normal glycemic plasma level.

A relevant question is: can diabetic complications be prevented or delayed by normalizing hyperglycemia? This can be achieved at least in part if a meticulous control of hyperglycemia is kept with an appropriate diet, oral antidiabetic drugs (Inzucchi, 2002; Holmboe, 2002; Bell, 2004, a), or insulin administration (Pickup et al., 2002), associated with daily exercise and a correct life-style. However, owing to genetic factors and in spite of a serious control, complications are found even in patients with a transitory and slight hyperglycemia. Throughout the years the following complications may develop with different intensity and localization. Circulatory abnormalities are the common denominator (Resnick and Howard, 2002) and they are present under the form of microvascular diseases:

- **Diabetic retinopathy (with incipient cataracts and glaucoma) is a leading cause of blindness in about 85% of patients.** A very strict control of diabetes can reduce the incidence and the progression of retinopathy (Kohner, 2003, a ; Frank, 2004).
- **Diabetic nephropathy is a leading cause of disability, the need for dialysis and premature death.**
- **Diabetic peripheral polyneuropathy is a major cause of morbidity** (pain and impotence).
- **Accelerated atherosclerosis frequently manifests itself with myocardial infarction, stroke and limb vascular occlusion complicated with necrotic *ulcers (the diabetic foot)* leading to *amputation*** (Jeffcoate and Harding, 2003).
- **Lipodistrophy** seemingly due to ineffective leptin activity or/and fatty acids dysmetabolism (Petersen et al., 2002; Unger, 2002; Minokoshi et al., 2002).

There is a wide consensus that **the common denominator is represented by a chronic oxidative stress due to the prevalence of ROS in opposition to a depletion of antioxidants.** The endogenous oxidative stress is intracellular and relentlessly causes cell degeneration and death. Besides the need for correcting hyperglycemia, *it appears important to* *adjust the constant imbalance between oxidants and antioxidants and although the administration of antioxidants is useful, it is not sufficient to restore cell homeostasis.*

WHAT WE KNOW ABOUT THE MECHANISMS OF HYPERGLYCEMIA-INDUCED DAMAGE?

During the last decade the following molecular mechanisms have been implicated in glucose- mediated vascular damage:

- **Increased advanced glycation end-products (AGEs) formation**. Intracellular hyperglycemia is the initiating event in the formation of intra- and extracellular AGEs: AGEs, taken up by cell receptors, stimulate the synthesis of pro-inflammatory cytokines and matrix proteins.
- **Increased polyol pathway flux.** Activation of aldose reductase leads to increased conversion of glucose to sorbitol.
- **Activation of protein kinase C isoforms.** Intracellular hyperglycemia increases the amount of diacylglycerol in vascular cells of diabetics.
- **Increased hexosamine pathway flux.** Excess of intracellular glucose is shunted into the hexosamine pathway leading to increased production of transforming growth factor β1 and plasminogen activator inhibitor-1.

These four mechanisms have been precisely reviewed by Brownlee (2001). Interestingly, in different ways, they induce overproduction of superoxide anion ($O_2^{\cdot-}$) by the mitochondrial electron-transport chain and it must be said that since 1991, Baynes postulated that the alteration in diabetic patients may depend on an increased oxidative stress. West (2000) has proposed a scheme clearly indicating the interaction between hyperglycemia and the enhanced production of reactive oxygen species such as superoxide, hydrogen peroxide and hydroxyl radicals accompanied by a depletion of antioxidant compounds and enzymes.

An increased level of xantine oxidase at the endothelial level (Parks and Granger, 1983; Houston et al., 1999) **increases the local release of anion superoxide** that, by rapidly reacting with nitric oxide, on one side it decreases vascular relaxation (increases platelet aggregation!) and, on the other allows the formation of peroxynitrite anion (Stamler et al., 1992; Stamler, 2004; Kokura et al., 2002). We know already that the **peroxynitrite anion is highly toxic** and inactivates several enzymes crucial for correct cell signalling (Mallozzi et al., 1997; Evans et al.2002).

The unbalanced equilibrium between reactive oxygen species and antioxidants, particularly a low reduced glutathione/oxidised glutathione ratio, usually precedes hypertension in diabetes. Another vicious circle may start when AGEs bound to the erythrocytes membrane stimulate production of lipoperoxides and adhesion to the endothelium. This process in turn favours transendothelial migration of monocytes with a consequent exacerbation of the oxidative stress (Zoukourian et al., 1996; Rattan et al., 1997). In order to interrupt this involution there is a constant need for devising effective therapeutic interventions.

ORTHODOX THERAPIES FOR THE METABOLIC SYNDROME

First of all patients must undergo routine checks to follow the course of the disease, such as daily tests for blood glucose level (finger prick) and every trimester the measurement of glycated haemoglobin A1 c. This is a good marker for revealing the average blood glucose level over time. Moreover they must check the body weight, blood pressure, cholesterol, triglycerides, typical biomarkers (C-reactive peptide levels, aldose reductase activity and fructolysine), C-reactive protein levels and so forth.

As it is clear that **hyperglycemia represents a continuous risk factor**, modalities for improving glycemic control are necessary and they differ for type 1 diabetes and type 2 diabetes. A low fat and a balanced low diet intake (1200-1500 Kcal) is effective in reducing weight and may reverse insulin resistance in patients with type 2 diabetes. A Spartan lifestyle is useful but long-term compliance is usually poor and consequently the prevalent trend is to adopt a pharmacological approach that may reduce the burden of morbidity and mortality due to hyperglycemia. Bell (2004,a) has suggested a flexible triple oral therapy for type 2 diabetes with the possible addition of insulin, if needed. Unfortunately only a few patients understand the importance of a continuous and strict control of glycaemia.

How to re-equilibrate the redox potential remains an open problem. An obvious approach is the life-long administration of an equilibrated multi-antioxidant diet that has provided controversial results. However, while there is no doubt that this is useful in several oxidative stress-related conditions, there is controversial evidence that it can be a definitive remedy (Bridgeman et al., 1991; Levine et al., 1996; 1998; Ting et al., 1996; Packer et al., 1997; Hack et al., 1998; Halliwell, 1999 a,b ; McCall and Frei, 1999; Polidori et al., 2001; 2004 ; Asplund, 2002; Wiernsperger, 2003). Reasons for explaining this problem have been enumerated and discussed in Chapter 8. Thus an appropriate oral supplementation, while not harmful, may not be sufficient to block the complications of diabetes. **This becomes understandable bearing in mind that most of the cell damage is due to an intracellular excessive production of oxidants remaining unquenched by both an abnormally low GSH content and impaired enzymatic function carried out by several enzymes** (superoxide dismutase; GSH peroxidase; GSH reductase; catalase; glucose-6-phosphate dehydrogenase), usually acting in a cooperative fashion. Moreover, at least two of the main enzymes degrading 4-HNE, namely GSH-S-transferase and aldehyde dehydrogenase have been found reduced in liver microsomes and mitochondria of diabetic rats (Traverso et al., 2002). It must be emphasised that even a normal plasma level of antioxidants is unable to abate the intracellular oxidative stress that is a continuous process leading to a diffused damage and eventually cell death. Moreover iron overload in diabetic patients with haemolytic diseases, although usefully treated with chelation therapy (Olivieri and Brittenham, 1997) can only make oxidative stress worse (Loebstein et al., 1998). Finally, **once started, endogenous oxidative stress is life-time long and cannot be compared to an extremely transitory and calculated oxidative stress occurring during ozone therapy.**

CAN OXYGEN-OZONE THERAPY REBALANCE THE OXIDATIVE STRESS AND STABILIZE THE DYSMETABOLIC SYNDROME?

First of all **I would like to emphasize that the proposed ozone therapy is not intended to substitute the orthodox treatment regimen for diabetes and for minimizing cardiovascular morbidity.** Similarly all the lifestyle changes suggested in the last decade such as a congruous reduction of food intake including saturated fat compsumption, increased uptake of fish oil (Mori et al., 2003), fiber and antioxidants plus at least 30 min of physical activity remain of crucial importance. However, in my opinion, **all of this,** as good as it may be, **is not enough because it is unable to sufficiently abate the chronic oxidative stress and to really stabilize or reverse the disease.**

In Chapter 8, all the possible strategies known today for reducing oxidative stress have been discussed and it has emerged that carefully performed ozone therapy, PARADOXICALLY, induces a unique adaptative response capable of reducing the endogenous oxidative stress.

For the dysmetabolic syndrome, I propose the usual AHT treatment based on briefly exposing a volume of the patient's blood (at most 225 ml plus 25 ml of 3.8% Na citrate) to an equal volume of the gas mixture (98% O_2 and 2% O_3) with a low-medium ozone concentration to be slowly upgraded from 20 to 40 mcg/ml of gas per ml of blood (0.42-0.83 mM) thrice weekly for the first month, twice weekly for the second and third months followed by a maintenance therapy of at least one AHT monthly. **Unfortunately diabetic patients with poor venous access cannot undergo the infusion of the "gluco-peroxide" solution** but could undergo BOEX at bland ozone doses at a moderate temperature. Some patients may prefer to do auto-rectal insufflation at home also with bland ozone dosages on alternate days.

That traces of ROS, particularly hydrogen peroxide can act as physiological messengers is no longer surprising because it is now clear that they are able to activate multiple biochemical and immunological pathways in blood cells (Chapter 4). Moreover when the ozonated blood is reinfused in the donor or LOPs are absorbed from the skin or the colorectal mucosa, the endothelium at first and then several organs interact with these compounds. The interaction leads to a reactivation of a number of biological processes that combine to ameliorate the chronic oxidative stress by inducing an upregulation of antioxidant enzymes such as SOD, GSH peroxidase, GSH reductase, GSH transferase, as well as glucose-6-phosphate dehydrogenase.

Moreover the phenomenon of adaptation to chronic oxidative stress implies that the repeated ozone treatments induce the synthesis of oxidative stress proteins, of which HO-I is a prototypic example. This prodigious enzyme will yield a higher level of bilirubin (an equally potent lipophylic antioxidant as α–tocopherol) and CO. The enzyme indirectly reduces vascular constriction because it suppresses the gene expression of endothelin-I and inhibits the proliferation of smooth muscle cells (Morita and Kourembanas, 1995; Duckers et al., 2001). It is known that nitric oxide, the release of which is enhanced by ozone therapy (Valacchi and Bocci, 2000), is the most important physiological vasodilator and inhibitor of platelet and leukocyte aggregation and adhesion to the endothelium and certainly traces of CO cooperates with NO in enhancing vascular relaxation. Although some of the released NO is immediately scavenged by the Fe^{2+} haeme of haemoglobin, some is converted into more stable compounds such as S-nitrosohaemoglobin and a variety of S-nitrosothiols (Jia et al., 1996; AlSa'doni and Ferro; 2000; Rafikova et al., 2002; Rassaf etal., 2002), which can relax and increase the flow of blood in vessels distant from the site of origin. However in diabetes, the endothelium generates more anion superoxide, which counteracts the functional activities of NO and causes vessel vasoconstriction, as well as platelet activation and therefore is at least in part responsible for the microvascular damage. The excessive production of anion superoxide and the consequent lack of balance of the physiological equilibrium between NO and superoxide is not only due to the dysmetabolic consequences of hyperglycemia (West, 2000) but also to an increased amount of xantine oxidase bound to endothelial cells (Houston et al., 1999). This phenomenon does not occur only in diabetes but in several pathologies such as chronic hepatitis, ischaemia-reperfusion and haemolytic anaemias (Tan et al., 1993; Sarnesto et al., 1996). There is no doubt that a constant increase of superoxide, hence of hydrogen peroxide and possibly hydroxyl radicals, impairs vascular functions and instaurates a chronic oxidative stress. Furthermore the superoxide not only consumes some of the NO but converts it into peroxynitrite responsible for protein and lipid oxidation that well explain the progressive tissue injury (Beal, 2002). Also the activation of granulocytes, via mieloperoxydase, produces hypoclorous acid that is another potent oxidizing compound which introduces carbonyl groups into proteins (Levine, 2002). The damage may be extended to the remaining β cells in the pancreas or it may alter insulin receptors in target tissues. Thus, one of the scopes for the proposed therapy is to interrupt this involutive cycle of events by renormalizing the balance nitric oxide/superoxide ratio at the endothelial level, which eventually should restore a normal blood flow and slowdown the subtle inflammatory state that perpetuates the process. Thus **ozonetherapy not only improves the physiology of circulation, a**

well ascertained fact, but possibly enhances the insulin secretion and/ or may decrease the resistance to insulin action. In other words, ozone therapy can turn a "vicious" into a "virtuous" circle.

EXPERIMENTAL AND CLINICAL EVIDENCE THAT OZONETHERAPY IS USEFUL IN DIABETES

One paper dealing with ozonetherapy has shown that streptozotocin-diabetic rats treated 10 times in two weeks with oxygen-ozone (with an ozone concentration of 50 mcg/ml) via rectal insufflation showed, in comparison to controls, a reduced hyperglycemia and of biomarkers (aldose reductase and fructolysine) related to diabetes (Al-Dalain et al., 2001). Concurrently total hydroperoxides and malondialdehyde levels did not differ from the control group and moreover the adaptation to ozone treatment was shown by a significant increase in the soluble fraction of pancreas homogenates of GSH, superoxide dismutase and catalase. By considering that rectal insufflation of ozone, in comparison to the AHT method, is a fairly empirical approach and was carried out for a short time, these results are almost too good to be true! It is unfortunate that the evaluation of the AHT approach is not technically feasible in the rat but nonetheless the rectal insufflation of gas, if it does really work, has the advantage to be not invasive, simple and inexpensive. At this stage clinical evidence that ozonetherapy is useful has been noted several times by ozonetherapists, including myself, but a controlled clinical trial as yet has to be performed. Chronic limb ischemia is often accompanied by type 2 diabetes and these patients have been advantageously treated with AHT. The need to reduce the insulin dose, suggesting either an improved insulin secretion or/and an increased receptor sensitivity, has become a common observation. However it appears urgent to organize an appropriate clinical trial in order to evaluate whether an initial cycle, including 28 treatments during three months (as previously mentioned) can modify critical parameters including glycemic and C-reactive peptide levels, non-enzymatic glycosilation, aldose reductase activity, AGEs and the antioxidant-prooxidant balance. Owing to the precise stoichiometry of the AHT, we would prefer this approach, rather than the rectal insufflation. The adopted strategy "start low, go slow" appears the most idoneous for inducing ozone tolerance and the rebalance of the redox system. Bearing in mind that ozonetherapy may modify glycemic levels, a strict control of it is imperative and obviously it would be very interesting to follow it for several months.

It is almost needless to say that if ozonetherapy improves the diabetic condition, it must be continued for an undefined period. In ARMD and chronic limb ischaemia we have observed that after the initial cycle, at least one AHT treatments per month appears sufficient to maintain the clinical improvement. Moreover it may be worth while to evaluate in a set of patients if rectal insufflation of ozone merits consideration. This is because interested patients properly instructed and routinely checked, can do self-administration for long periods, as is often performed by HIV-infected patients.

CONCLUSIONS. The dysmetabolic syndrome is recognized as one of the most serious disease in Western countries caused by a number of metabolic alterations such as type 2 diabetes, hypercholesterolaemia, atherosclerosis, renal dysfunction with the common denominator represented by a chronic oxidative stress. Although orthodox medicine has several good drugs for blocking the progression of diabetes and atherosclerosis, it continues to ignore the capacity of ozone therapy which is able to improve: a) blood circulation and oxygen delivery to ischemic tissues; b) corrects the chronic oxidative stress by upregulating the antioxidant system; c) induces, without side effects, a state of wellness and euphoria and d) may improve insulin secretion or its effectiveness. Diabetic patients, particularly those with foot ulcers, are critical and today they still have a gloomy prognosis. This is because they need a multiform therapy aiming to eliminate infection, the peripheral ischemia and the neuropathy. While we are not yet sure about correcting the dysinsulinemia, we have witnessed dramatic improvements in patients ready for amputation by performing AHT and topical, daily application of ozonated oil. While certainly we are not overlooking the importance of antidiabetic drugs, statins, antihypertensive agents and so forth, we judge it deplorable to disregard the benefit of a combined ozone therapy.

8. IS ANY HAEMATOLOGICAL DISEASE TREATABLE WITH OZONE THERAPY?

Haematological malignancies in children are dealt with remarkable success by orthodox medicine and I doubt that ozone therapy would be useful. After the IFN's failure, chronic myeloid leukaemia, in adults, is now treated with a new drug: imatinib mesylate, which is a selective tyrosine kinase inhibitor able to stop tumor cell proliferation. The "molecular therapy" is a new advancement, and it is hoped that the possibility of adding simultaneously similar drugs will counteract the tendency of neoplastic cells to become resistant.

There are two diseases: sickle cell anemia (SCA) and beta thalassaemia major (TM), which are leading to oxygen blood deficiency accompanied by other serious manifestations where the application of ozone therapy could be helpful. Patients with TM syndrome can survive if they receive regular blood transfusion and desferrioxamine infusion or bone-marrow transplant from a suitable donor. **Among haemolytic anemias, SCA and TM stand as the most relevant and common hereditary chronic anemias due to either**

altered or impaired globin chain synthesis. The altered ß-globin biosynthesis leads to a series of problems such as ineffective erythropoiesis, accelerated erythrocyte breakdown, iron overload, tissue hypoxia, impaired growth and a shortened survival. Besides prevention and whenever possible bone marrow transplantation or gene therapy, orthodox therapy is modestly effective. Recently it has become clear that oxygen-free radicals and peroxidative tissue injury accompany the anaemia and represent a unavoidable complication that accelerates the multi-organ abnormalities (Livrea et al., 1996; Angelucci et al., 1997; 2000; Cighetti et al., 2002). Is there any further possibility of correcting the chronic oxidative stress that from day to day establishes a negative involution? Improving chelation therapy and a supplement of antioxidants (Asplund, 2002), can be useful but they are unable to abate the chronic oxidative stress.

I am sure that the reader thinks immediately that ozone, although strongly oxidant, if carefully dosed, can paradoxically induce an adaptative response capable of reducing the excessive oxidation. Although haematologists have repeatedly refused to evaluate this approach either on its own or in combination with conventional therapies, I would like to examine the validity of several mechanisms that represent a rational basis for evaluating the efficacy of ozone therapy.

Sickle cell anaemia (SCA) or drepanocytosis is a common genetic disease among the black population due to an autosomal recessive disorder involving a single amino acid substitution in the beta subunit of a peculiar haemoglobin, referred to as haemoglobin S (HbS) to distinguish it from the normal adult haemoglobin A (HbA). Vernon Ingram in 1954 made the memorable discovery that **HbS contains valine instead of glutamate at position 6 of the ß chain** and Linus Pauling in 1949 had already shown that HbS has an isoelectric point of 7.09 (oxyHb) and 6.91 (deoxyHb) in comparison to normal Hb (6.87 and 6.68, respectively).

Patients with SCA are homozygous for the abnormal gene and up to 35% of erythrocytes are sickled while heterozygous subjects are normally not symptomatic and 1% only of erythrocytes may become sickled. Homozygous SC patients have usually less than 20% HbF, 3% HbA2 and 70-80% HbS. **Sickling occurs when the erythrocytes, passing through the capillary circulation (the pO_2 decreases from 98 to about 40), release oxygen to the tissues.** The process of deoxygenation causes a brisk change of the tertiary structure of HbS with the formation of an intracellular precipitate consisting of fibers 21.5 nm thick. Interestingly, HbF inhibits the polymerization of HbS so that erythrocytes with a high content of HbF are somewhat protected from sickling.

Consequently the sickled erythrocyte becomes rigid and deformed and by obstructing the circulation provokes ischemia and infarction. The vessel occlusive crises due to physical trapping or increased adhesion of the sickled erythrocytes to the vascular endothelium occur in various organs and can be painful, particularly those successive to bone marrow necrosis. The enhanced haemolysis is accompanied by haemochromatosis, anemia and a chronic inflammatory disease. Indeed there is an activation of macrophages, an increase of leukocytes with release of cytokines and consequently an alteration of cell adhesion regarding monocytes and neutrophils (Muller, 2002). Although any organ may be involved, impairment of cardiopulmonary, renal hepatic, skeletal, ocular and neurologic functions is most common (Prengler et al., 2002).

Thus SCA is a serious disease and only 2% of about 120,000 affected babies born in Africa survive to the age of five. Conventional medicine does practically nothing to help patients in poor countries. In theory, African Americans could undergo bone marrow transplantation but this is rarely performed and is accompanied by significant mortality (Hoppe and Walters, 2001). In the future gene therapy may become useful (Pawliuk et al., 2001). Administration of an oral drug could be practical but to date, among potentially ameliorating agents such as hydroxyurea (HU), cyanate, methylprednisolone (Steinberg, 1999) and Polaxamer 188 (Orringer et al., 2001), only the first is widely used.

HU increases the percentage of HbF, reduces HbS and the rate of painful crisis but the drug is somewhat toxic, mutagenic, and possibly immunosuppressive (Steinberg, 1999). Clotrimazole, a specific Ca^{2+}-activated K^{+} channel inhibitor, may reduce the deleterious dehydration of sickled erythrocytes but it remains to be validated (Brugnara et al., 1996). Similarly the use of antibodies against adhesive integrins, although it may work as an anti-occlusive strategy, remains to be tested (Kaul et al, 2000). Painful crises can be treated with an analgesic, hydration and oxygen administration (Steinberg, 1999). A daily oral supplement of folic acid is somewhat helpful and blood transfusions must be used sparingly to avoid isoimmunization, hepatitis and iron overload.

TM is one of the thalassaemias that ranges from small erythrocytes abnormalities to a life-threatening disease due to wide differences in the synthesis of the globin chains. In contrast to the previously discussed SCA, **the ß chains of patients with TM have a normal structure but are often almost undetectable.** The gene frequency for TM is about 0.1 in Sicily and other Mediterranean islands but the disease is also present in Asia and Africa. Two heterozygotes parents (ß-thalassaemia trait) statistically will generate one in four children in the homozygous state with ß thalassemia major (TM) or Cooley's anaemia**. Erythrocytes contain an excess of α-chains and practically little or no ß-chains** (Scott et al., 1993). Owing to decrease solubility, free α-chains form insoluble aggregates within the erythrocyte precursors in the bone marrow. The result is extensive intramedullary erythroid destruction and in any cases a short life span of the circulating erythrocytes. These defects cause severe anaemia, peripheral hemolysis, release of free iron, haemosiderosis, impaired growth, abnormal development and short life expectancy. Hepatic and splenic extramedullary haematopoiesis is to no avail. Patients with TM, who are able to upregulate γ-chain production have a less severe clinical course because γ-chains combine with the free α-chains to form the stable fetal haemoglobin (HbF), which is however unable to perform the oxygen delivery as requested in normal life.

For preventing TM, genetic counselling and antenatal diagnosis are essential but not always sufficient. Patients can be supported with daily supplement of folic acid and, in order to maintain at least a level of 9 g Hb/dL, transfusion therapy from normal donors is necessary but this, in the long run implies alloimmunization, risk of viral infections and unavoidably fatal iron overload. Constant infusions of desferrioxamine as well as phlebotomy are effective (Angelucci et al., 1997), while the value of oral administration of deferiprone remains uncertain (Pippard and Weatherall, 2000). Thus **in some patients an excess of Fe^{2+} enhances the formation of radical species not sufficiently neutralized by the antioxidant system**. Bone marrow transplantation, even though with some risk, is able to modify the prognosis but it cannot be applied on a large scale. Interestingly it has been discovered that alpha haemoglobin stabilizing protein acts as a chaperone and blocks the deleterious effects of free αHb precipitation. If the lack or a mutation of AHSP in TM proves to be really detrimental, gene therapy may help these patients for the future.

Can oxygen-ozone therapy be useful and why ?

We have already the availability of some clinical data provided by National Center for Scientific Research at Havana. Cuban physicians performed a randomized clinical trial in 55 SCA patients (30 experimental and 25 controls). A gas mixture composed of about 97 % oxygen and 3 % ozone was administered daily (5 days per week) for 3 weeks in 30 patients through the rectal route by insufflation. The control group received only analgesics, vasodilators and IV saline infusion. The ozone treated group displayed a rise in arterial pO_2 and a significantly reduced (by about 50%) frequency and severity of painful crises. No side effects were recorded (Gomez et al., 1995).

Recent basic advancements and clinical results achieved in vasculopathies using this therapy appear very encouraging (Chapter 9, Section II) and they entice testing it in haemoglobinopathies.

Let me examine pros and cons of a treatment cycle based on the well standardized procedure of the ozonated authoaemotherapy by using 225 ml blood (+25 ml Na citrate at 3.8%) and 225 ml of gas at low ozone concentration (starting with 10 mcg/ml per ml of blood and slowly escalating up to 30 mcg/ml). The autologous transfusion is quite safe and the use of low ozone concentrations does not cause any damage to either normal or pathologic erythrocytes and infact the increase of haemolysis remains negligible (±0.2%). This is because the oxidizing activity is exhausted when ozone is solubilized in the plasmatic water and instantaneously reacts with a variety of biomolecules, namely PUFA, hydrosoluble antioxidants generating ROS, mainly hydrogen peroxide and a variety of LOPs.

On the basis of our working hypothesis, ROS and LOPs are the ozone messengers able to activate multiple biochemical and immunological pathways in blood cells Moreover upon blood reinfusion in the donor, the endothelium at first and then parenchymal cells interact with LOPs. We now have good evidence that a prolonged course of AHT is able to reactivate a number of biological processes that, either simultaneously or successively, combine to improve the physiology of circulation and to reduce the chronic oxidative stress. Needless to say **ozone therapy cannot modify the genetic irregularities**. However we have shown that owing to the upregulation of antioxidant enzymes coadiuvated by G6PD, newly formed erythrocytes are more resistant to oxidative stress and more or less rapidly depending upon the therapeutic schedule, become a large proportion of circulating cells.

While any AHT represents a small oxidative stress, this is quite transitory, calculated and promptly corrected by the antioxidant system. The treatment is interpreted as a "therapeutic shock" occurring ex vivo during the exposure of blood to ozone and transmitted into the donor during blood reinfusion. **It must be clear that without stress, no biological effect will ensue.** The synthesis of oxidative stress proteine (OSP), particularly of HO-I or HSP-32, is a clear example. HO-I will enhance haeme breakdown, hence will yield a higher level of bilirubin (a powerful lipophylic antioxidant) and CO (Morita and Kourembanas, 1995). It has been shown that HO-I expression reduces vascular constriction because it suppresses the gene expression of endothelin-1 and inhibits the proliferation of smooth muscle cells (Duckers et al., 2001).

We have demonstrated that human endothelial cells coming in contact with ozonated plasma, hence LOPs, enhance the release of NO (Valacchi and Bocci, 2000). This compound, after binding to the receptor on smooth muscle cells activates guanylate cyclase, so that an increased level of cyclic guanosine monophosphate (cGMP) causes relaxation and thus vasodilation. It is well known that NO inhibits platelet and leukocyte aggregation and adhesion and certainly cooperates with CO in enhancing vascular relaxation. Although the intravascular half-life of NO is about 2 msec, important biochemical pathways describing **the formation of S-nitrosohaemoglobin and S-nitrosothiols have been described** (Jia et al., 1996; Rafikova et al., 2002; Rassaf et al., 2002; Zhang and Hogg, 2004; Stamler, 2004; Gladwin et al., 2004) **for relaxing and increasing blood flow in vessels of ischemic tissues distant from the site of origin.** The possibility of an increased vasodilation cannot be underestimated because in SCA, vaso-occlusion is not only caused by sickle erythrocytes but is facilitated by vasoconstriction and obstruction due to adhesion of platelets and leukocytes to the endothelium. A subtle inflammatory state with release of proinflammatory cytokines and platelet activation does further aggravate the process.

An initial report showed that low concentrations of NO would augment HbS oxygen affinity when SCA patients inhaled NO at 80 p.p.m. in air (Head et al., 1997). This would have been a useful therapeutic approach but recent data (Gladwin et al, 1999; Hrinczenko et al., 2000) have clarified that the induced left shift in P50 correlates with an unacceptable increase of methaemoglobin formation. Another mechanism that has been pursued is the possibility that a high plasma level of arginine may increase NO production (Enwonwu, 1989; Morris et al., 2000). Interestingly HU metabolism in rat (Jiang et al., 1997) and in SCA patients enhances the release of NO, and detectable amounts of nitrosyl haemoglobin (Glover et al., 1999). Thus HU efficacy may be due not only to the ability of stimulating the production of HbF but also to induce vasodilation and decrease platelet activation.

However the important role of NO may be jeopardized by an excessive release of anion superoxide: in physiological conditions, the endothelium produces minute amounts of 1-10μM NO and 1 nM superoxide but NO is rapidly scavenged by erythrocytes (actually the iron II haeme of Hb) that explains its short half-life. **Although anion superoxide displays functional activities (vasoconstriction, platelet activation etc) just the opposite of NO, in normal conditions there is a sort of equilibrium.** However, in pathological circumstances such as chronic hepatitis, ischemia-reperfusion and severe haemoglobinopathies, the liver, that is the main repository of xantine dehydrogenase (XDH) allows its conversion to xantine oxidase (XO) and to its release in the circulation (Parks and Granger, 1983; Tan et al., 1993; Sarnesto et al., 1996; Houston et al., 1999). **When an excess of XO binds to endothelial cells, the consequent excessive generation of superoxide and hydrogen peroxide impairs vascular function and instaurates a chronic oxidative stress (Aslan et al., 2001).** Moreover the anion superoxide enhances NO consumption and formation of peroxynitrite ($ONOO^-$), a deadly compound inducing protein and lipid oxidation, thus extending tissue injury. Not to be forgotten that in SCA, sickle erythrocytes are already generating great amounts of ROS and LOPs. **Clearly the unbalanced NO/superoxide production contributes greatly to the diffused vascular damage and to a progressive involution of SCA.**

In spite of chelation therapy and phlebotomy, TM patients present a progressive oxidative stress generated by the imbalance between the α and β chains and worsened by hepatic and cardiac iron overload.

CONCLUSIONS: life-long ozonetherapy is feasible as we have shown in age-related macular degeneration, in chronic limb ischemia and in angina abdominis. After an initial cycle including 24 treatments in three months (twice weekly), the therapeutic effect can be probably maintained with three treatments per month. Upregulation af antioxidant enzymes and 2,3-diphosphoglycerate is likely to occur

during the first two months, while rheological improvement (decrease of arterial pressure is the norm) due to NO·/Superoxide rebalance may take two-to three months.

Ozonation of patient's blood must be carefully performed, firstly evaluating the antioxidant capacity in order to employ the optimal ozone concentration. The usual strategy "starts low, go slow" is the most idoneous for inducing ozone tolerance and the rebalance of the redox system. This approach will likely diminish the frequency of allotransfusion, the severity of painful vaso-occlusive crises in SCA and will improve the metabolism and the quality of life. Chelation therapy with desferrioxamine must be continued regularly and, for potentiating the plasma antioxidant capacity, we must prescribe the usual oral daily antioxidant supplementation one week before starting the therapy. Haemoglobinopathies are often complicated by chronic hepatitis C infection and, although the combination of interferon alpha and ribavirin is effective (Li et al., 2002), it may well be strengthened by the ozonated AHT.

The treatment proposed by Cuban physicians of ozone insufflation via the rectal route has been evaluated in the rabbit (Bocci et al., 2000) but in comparison to the stoichiometry of AHT, it is too approximate. However it is even cheaper and amenable to self-administration. If ozonetherapy can be proven to be useful in haemoglobinopathies, a re-evaluation of the RI route is warranted also because the patient, once properly instructed, can do it at home. One drawback of ozonetherapy is that lack of electricity and medical oxygen may impede ozone production for SCA therapy in remote parts of Africa. The same problem attains for treating malaria and HIV infections.

In order to overcome these difficulties, one promising option is the infusion of the "glucoperoxide-solution" with hydrogen peroxide concentrations in the low-medium range (0.03-0.09%).

9. CAN OZONE THERAPY SLOW DOWN THE PROGRESSION OF OXIDATIVE STRESS IN RENAL DISEASES AND HAEMODIALYSIS?

There is no doubt that either **infective or autoimmune glomerulonephritis as well as end stages of renal failure associated with hemodialysis are characterized, to a different extent, by an imbalance between pro- and antioxidative mechanisms.** Already three decades ago, Lindner et al. (1974) observed a more rapid progression of atherosclerosis in prolonged maintenance haemodialysis. Today there is a cornucopia of reports linking inflammation (hence, chronic oxidative stress) due to renal diseases and haemodialysis (Knudsen et al., 1989; Ceballos-Picot et al., 1996,b., Hasselwander and Young, 1998; Witko-Sarsat et al., 1998; Morena et al., 2000, Rousseau et al., 2000). Nephrologists have several drugs at their disposal but unfortunately, some patients progress towards renal failure, likely because we are unable to correct the vicious circle initiated and perpetuated by a deranged redox system. Moreover the kidney does not have the regenerative ability of liver and this is one of the reasons for explaining why too often "nephropaties lack a specific treatment and progress relentlessly to end-stage renal disease" (Ruggenenti et al., 2001). I believe that another important reason is that **orthodox medicine has not yet a valid strategy to reduce oxidative stress in renal diseases and Nature is not always benevolent.** So far it has not yet been recognized that ozone therapy, not only can correct a chronic oxidative stress, but can stimulate untapped resources able to afford a natural recovery.

I would then like to suggest the combination of conventional treatments with ozone therapy in any initial nephropathy for preventing the risk of progression towards a chronic disease. This can be easily achieved by a bi-weekly, mild ozonated autohaemotherapy or by the daily infusion of the "glucoperoxide" solution or by RI as the last resort. At least in cisplatin-induced nephrotoxicity in rats, it has been shown that intrarectal ozone therapy prevented and corrected the renal antioxidant unbalance caused by this toxic chemotherapeutic agent (Borrego et al., 2004; Gonzalez et al., 2004). Previously Zamora et al. (personal communication) have shown in rats that oxidative preconditioning, achieved by intraperitoneal or rectal administration of ozone, inhibits the release of TNF alpha during endotoxic shock. While I admit that experimental results in rats may not always be duplicated in patients, we must acknowledge **the paradoxical power of ozone in rapidly inducing the defensive upregulation of antioxidant enzymes.** It would be wrong to accept the concept of unavoidable irreversibility of nephropaties mostly because we have no idea of the pharmacological effects of ozonated blood on the renal circulation, metabolism and possible release of nephropoietins. If the proposed therapeutic combination will yield positive results, we will be able to spare later on the misery of many patients and the cost of haemodialysis.

However, if the patient is already undergoing haemodialysis, can a well constructed programme of ozone therapy improve the quality of life, reduce morbidity and possibly delay mortality? For several reasons, haemodialyzed patients undergo a dysmetabolic syndrome (this chapter, section VII) culminating in vascular complications, neuropathies, chronic infections, diabetes and so forth, which are maintained and worsened by the chronic oxidative stress.

During the last couple of years, a Polish group have already presented important results showing that ozonated AHT, in POAD patients on maintenance haemodialysis, normalizes the lipid profile and the vascular metabolism so that walking ability is improved and clinical signs of ischemia are significantly attenuated. Moreover no sign of toxicity, as evaluated by examining several biochemical parameters and even natural killer activity, has become apparent (Tylicki et al., 2003; 2004; Biedunkiewicz et al., 2004). At the Nephrology Unit of Siena Hospital, prof. N. Di Paolo and I also had the opportunity of carrying out ozone therapy, with the EBOO technique, in very critical patients, one of which was affected from necrotizing fasciitis (Figure 12). Results have been good or excellent and again without any adverse effects (Di Paolo et al., 2000; 2003).

Which may be the most suitable, less traumatic and practical ozone therapy procedure to be applied for very prolonged periods in haemodialysis patients? Careful preservation of vascular accesses is a matter of life or death but either AHT, or the "glucoperoxide" infusion, or EBOO are not adding a superfluous stress. We are planning to evaluate a protocol applying in parallel a dialysis filter and a gas exchanger, which is a special oxygenator resistant to ozone and biocompatible. In order to achieve an ideal blood oxygenation and ozonation, we have overcome several problems described in Chapter 6. As today we have a small but highly efficient gas exchanger requiring a minimal priming volume, this can be filled with saline and remain in stand-by during the whole dialysis period. Everyone knows that this is a critical period due to the filter membrane haemoreactivity, presence of trace amounts of contaminants, release of pro-oxidants worsened by important loss of filterable antioxidants (Morena et al., 1998). Thus, my personal opinion is that, ***immediately after dialysis***, we have **first to infuse a bolus of ascorbic acid (1.0 g) to reconstitute a sufficient antioxidant capacity** in a subject that is already taking the daily supplement of antioxidants, particularly including two doses (0.6 g. each) of NAC (Tepel et al., 2003). **Then, after five minutes, we can open the ozonator line and allow a brief blood ozonation.** If the blood flow is about 250 ml, a period of ten min will ozonate about 2.5 L of blood. Ozone concentration will be very low, probably around 0.3-0.5 mcg/ml, well within the therapeutic range that must be accurately determined with several criteria for a number of patients. **The treatment, carried out once a week, may suffice to correct the dysmetabolic syndrome, with no discomfort to the patient.** In comparison with the classical HAT, or the "glucoperoxide" solution, the EBOO is certainly more expensive and we will have to evaluate the cost-benefit and the total lack of side-effects. As I have often mentioned, as an alternative, the patient can do a low-dosage RI at home as an automedication

CONCLUSIONS: The discovery that nephropaties are progressively worsened by a state of oxidative stress not yet controllable by orthodox medicine compels me to strongly advise the application of ozone therapy either in acute, chronic and terminal stage of the disease. I hope that nephrologists will endorse this idea and test this new approach. The real possibility of controlling the hyperoxidative state and inducing a feeling

of wellness are eloquent and encouraging advantages. The study of gene and stem-cell biology is most important and likely will produce amazing therapeutic innovations but, realistically, growing replacement organs is still a long way off (Soares, 2004). As renal transplantation is still unable to satisfy the global need, what is wrong in trying to help patients with ozone therapy? I honestly cannot justify the obstracism of orthox medicine and the the negligence of Health Authorities in disregarding the beneficial help of this approach. I remain faithful to the concept that only the combination of treatments is the best way to correct the multiform derangements typical of chronic diseases.

10. DERMATOLOGICAL DISEASES AND OZONE THERAPY

My first experience with ozonated autohaemotherapy happened in the Dermatology Institute of the University because, in 1988, they wanted to evaluate ozone therapy in psoriasis, on the basis of great successes claimed by a private dermatologist. It was a failure but, in retrospect, was useful because I realized how badly it had been carried out with doubtful and too low ozone concentrations (probably less than 5 mcg/ml). Surprisingly, after about ten treatments, one patient showed extraordinary improvement, another was slightly better and three patients remained the same. They tried to publish a paper but it was rejected because there were no controls with oxygentherapy. Thus, my first clinical experience was disconcerting and, although I have heard several other anecdotes of splendid results, I remain doubtful and I suppose that the placebo effect could be responsible for occasional improvements. Since that time, great strides have been made in understanding the immunologic derangements occurring in **psoriasis** and two studies (reported in Section V, Autoimmune diseases) have shown that administration of biological response modifiers, namely of two antibodies: the first against TNF-alpha (etanercept) and the second against a leukocyte-function-associated antigen1, LFA-1 (efalizumab) have yielded a remarkable improvement (albeit not definitive) in the majority of patients (Kupper, 2003). Previous therapy modalities employing cyclosporine and methotrexate have dose-limiting toxic effects and similarly, we do not yet know if unnatural antibodies, which disturb the immunologic homeostasis, will procure adverse effects with prolonged use. On this basis I still see a valid reason for exploring the effect of ozone therapy, provided that the study is meaningful and seriously performed. I repeat the suggestion given before for autoimmune diseases, where the clone of autoreactive cytotoxic cells should be suppressed. A cycle of six months therapy will include at least two major AHTs per week with ozone concentrations progressively upgrading from 30 mcg/ml up to 80 mcg/ml, in five weeks time (a weekly step of 10 mcg/ml). **It would be very interesting to evaluate the quasi-total body exposure (BOEX), not only in patients with poor venous accesses but in all patients because this method combines the systemic effect to a direct action on the psoriatic skin.** Occasionally a few patients have reported a marked improvement after a casual application of ozonated oil on the psoriatic areas.

Eczema and atopic dermatitis (AD) are the other two distressing diseases, which have been treated by Russian dermatologists and a German ozonetherapist. It has been claimed that a prolonged therapy, using AHT in adults and rectal insufflation of ozone in children, has provided "good"? results but data have not been published. From an immunological point of view, **the interesting hallmark of AD is a Th1/Th2 imbalance** (Campbell et al., 1999) with a reduced production of IFNγ and an elevated release of IL-4 and IL-5, which favours IgE production and eosinophilia, a typical disorder of atopic diathesis (Beltrani, 1999; Leung, 1999). Prophylactic measures such as avoidance of irritants, allergic food (eggs, soy, peanuts, etc.), contact with house-dust mites or other aeroallergens, are helpful but **the mainstays of therapy have been topical corticosteroids, which appear to be still safe and effective in the medium term provided precise guidelines are followed** (Atherton, 2003). **In severe forms, phototherapy, cyclosporin A, and azathioprine appear to be effective but with some side effects** (Rudikoff and Lebwohl, 1998; Hanifin and Tofte, 1999). Recently two new immunosuppressant drugs used in solid organ transplantation: **tacrolimus and pimecrolimus ointments have been used and seem to have the advantage of absence of nephrotoxicity and do not cause skin atrophy, at least in the short period** (Fleischer Jr., 1999; Williams, 2002). Lacking valid data, I can only guess that precisely performed ozone therapy may be useful and we should progressively test the effect of major AHT rising up ozone concentrations from 20 mcg/ml up to 40 mcg/ml to readjust the Th1/Th2 balance. Whenever possible we should use heparin as an anticoagulant because it enhances the release of IFN gamma. Also for AD, the BOEX procedure, combining systemic and cutaneous treatment, may be an ideal approach.

CONCLUSIONS: There are rational bases for entertaining the application of ozone therapy in dermatological diseases such as psoriasis and atopic dermatitis. However, orthodox medicine, thanks to colossal commercial enterprises, has made available new interesting drugs, which are effective but not totally devoid of risks. This is one reason for dermatologists to obstruct the evaluation of ozone therapy with the consequent difficulty of recruiting patients for clinical studies. Moreover patients with these diseases are often very distressed and understandably anxious to receive the most effective treatment immediately. Ozonetherapy may yield some benefit at a slow pace and

patients will accept it only if, at least in the initial period, they are assisted with the proven topical drugs. On the other hand, the use of ozonated water and oil for chronically infected wounds and ulcers yields wonderful results and official medicine will have, sooner or later, to acknowledge the value of ozone in these dermatological affections, which worry so much diabetics and old people.

11. OZONE THERAPY IN PULMONARY DISEASES

This section is dedicated to the memory of Dr. Maria Trusso

Ozonetherapy has not yet been tested in pulmonary diseases, probably because everybody knows that breathing air polluted with ozone is toxic to the respiratory system (Kelly et al., 1995; McConnell et al., 2002). This daily observation has greatly contributed to establish the dogma that "*ozone is always toxic and should not be used in medicine*". However, an almost irrelevant episode that occurred about four years ago suggested to me that this fact has misled us.

Among our numerous ARMD patients treated with ozonated AHT, one, with emphysema, told us that, after about fourteen sessions, his dyspnea was alleviated and he could walk up to the third floor of his apartment with little effort. I sensed that he had given us a good tip and I took him to the Pneumology Unit where the specialist, Dr. Maria Trusso, was bewildered by the result. Actually, at first she imagined that the treatment for ARMD was based on breathing ozone and the proposal to continue this sort of treatment appeared crazy to her. After I explained that we simply ozonated and reinfused the patient's blood, she became interested and correctly asked how ozonated blood could improve lung function and oxygenation. As we were treating many ARMD patients, we then searched for other cases. We found another two patients, a man with chronic obstructive pulmonary disease (COPD) and an emphysematous woman, who after two cycles of therapy had noticed an improvement in their performance of daily activities. This response was subjective and could have been due to a placebo effect, but it encouraged us to make a protocol. Although it elicited a strong scepticism, the protocol was prepared, submitted to the Ethical Committee and, after revision, approved after about seven months. Unfortunately, the health of Dr. Trusso deteriorated (she had a metastatic breast tumour) and she died shortly afterwards, leaving four young children practically alone. We lost a very nice, energetic woman, who after accepting the idea became very enthusiastic to try this unusual therapy. The clinical trial was cancelled but I remained with the idea that, for several good reasons, **ozone therapy could be useful in the following diseases: emphysema, COPD, idiopathic**

pulmonary fibrosis (IPF), acute respiratory distress syndrome (ARDS), and asthma. I will explain why:

First of all, ozonetherapy improves blood oxygenation in ischaemic areas. This is not a direct effect because, although we reinfuse hyperoxygenated blood (pO_2 rises up to 400-500 mmHg), the infusion rate is so small (about 15 ml per minute, compared with a cardiac output of about 5 L) that the pO_2 of venous blood reaching the lungs is hardly modified. However, **ozonetherapy enhances the delivery of oxygen particularly in ischaemic tissues** and therefore metabolic conditions improve, even though findings, such as decreased blood viscosity and increased 2,3-DPG levels in erythrocytes, have not been definitively demonstrated.

The second realistic possibility is that **LOPs,** present in mildly ozonated blood, act on endothelium and **enhance the release of prostacyclin, NO and IL-8,** while release of endothelin-1 is depressed (Bocci et al., 1999c; Valacchi and Bocci, 2000). It is well known that the release of NO and NOthiols represents the physiological mechanism for vasodilation (Barnes and Liew, 1995; Warren and Higenbottam, 1996; Jindal and Dellinger, 2000; Zhang and Hogg, 2004; Gladwin et al., 1999; 2004; Stamler, 2004) and contrasts the release of the anion superoxide, which causes vasoconstriction and deploys negative influences on platelets and endothelial cells.

The third, extremely important and paradoxical advantages is the adaptation to the small, calculated and therapeutic oxidative stresses induced by the ozone treatments. I already mentioned the upregulation of intracellular antioxidant enzymes and the increased expression of HO-1, a highly protective enzyme. I would like to remind that, with a few conventional drugs (Chapter 8)**, ozone therapy is today the unique approach able, when is not too late, to rebalance the altered redox system**

The fourth advantage is **the mild but continuous stimulation of the immune system,** which, by reinforcing the innate and acquired defence system, can contain acute and chronic pulmonary infections.

The fifth advantage is **the frequent improvement of cenesthesis** due to a comprehensive neuro-endocrine stimulation reported by most of the patients.

All together these diseases represent the third leading cause of death that, owing to the morbidity and mortality, is one of the worst socioeconomical problems. This may well be **the sixth good reason** for seriously implementing ozone therapy in combination with effective orthodox therapies.

For the sake of brevity, I cannot discuss the aetiology and pathophysiology of ARDS, COPD, IPF, emphysema and asthma, which, to a different extent, are characterized by inflammation and chronic oxidative stress. These processes are demonstrated by an increase of ROS and isoprostanes (Morrow et al., 1995; Morrow and Roberts, 1997; Basu, 2004), activation of NF-KB with increased synthesis of IL-1,TNF-α, IL-4, IL-6, IL-8, and inactivation (by oxidative damage) of α1-antitrypsin and leukoproteinase inhibitors, unable to counteract elastase, cathepsins and matrix metalloproteinases (Smith et al., 1997; Barnes, 2000; Gross and Hunninghake, 2001; Kamp, 2003; Langen et al., 2003). Interestingly Maestrelli et al., (2003) have demonstrated that in severe COPD patients compared to control smokers the level of HO-1 is decreased in alveolar macrophages. Ozone therapy could correct this deficiency.

IPF and asthma have been also characterized by dysimmunity, with a prevalent Th1>Th2-like cytokine pattern in IPF (Keane and Strieter, 2002) and a Th2>Th1-like pattern in asthma (Robinson et al., 1993; O'Byrne et al. 2004). With regard to asthma, **thanks to biomedical Cuban scientists, we have already the demonstration that ozone therapy is effective in this disease** (Hernandez et al., 2003). They have treated 113 patients with either ozonated AHT or RI. Particularly using AHT (ozone concentration at 40 mcg/ml), after the completion of a cycle of 15 treatments, they measured significant reduction of IgE and HLA-DR levels and a net increase of GSH as well as of GSH-Px and GSH-T. **Rectal insufflation of ozone** (10 mg per session), in one group of patients, **was found less effective,** even though the ozone dose and the number of sessions (20) were higher than the number of AHT treatments. This difference reinforces my conviction that, whenever possible, we must use the classical autohaemotherapy.

Administration of IFN-gamma-1b in IPF patients, unresponsive to corticosteroid therapy, during about 58 weeks was practically ineffective (Raghu et al., 2004). IFN gamma, a Th1, antifibrogenic cytokine, was intended to down-regulate the expression of TGF-beta-1, a cytokine enhancing fibrosis (Roberts et al., 1986) but one wonders if, in order to quench inflammation and oxidative stress, it would not have been better to administer antibodies to TGF-beta1? This clinical trial exemplifies the difficulty of treating a disease with a complex and obscure pathogenesis. Under these circumstances, it is also difficult to envisage the optimal ozone dosages and schedules for performing ozone therapy.

However I will try to give the following cautious guidelines:

ASTHMA: major AHT (from 20 to 40 mcg/ml ozone per ml of blood), whenever possible using heparin. One cycle within two months.

IPF: major AHT (from 20 up to 80 mcg/ml ozone per ml of blood), using Na citrate.

One cycle within six months.

COPD and EMPHYSEMA: major AHT (from 20 to 40 mcg/ml ozone per ml of blood), using Na citrate. Moreover it appears useful to perform minor AHT via IM route for stimulating the expression of HO-1. One cycle within six months. If the patient owns an ozonator, he can do domiciliary mild RI every day.

ARDS: major AHT (from 10 up to 20 mcg/ml ozone per ml of blood), using Na citrate. This is an emergency situation and we can perform up to four AHTs daily, until needed.

Except ARDS, it is practical to perform two AHTs per week (M and Th or T and F). If the first cycle is beneficial, the therapy can be maintained with three treatments monthly, followed by a resting period of one month. If the venous access is difficult, AHT can be substituted with the "gluco-peroxide" infusion (from 0.03 up to a maximum of 0.12% hydrogen peroxide). Alternatively, RI can be performed at least four times weekly for the same period starting with small gas volumes and low ozone concentrations (3-5 mcg/ml), slowly escalating both the volume (450-600 ml) and ozone concentration (15-20 mcg/m). The system of the quasi-total body exposure at physiological temperature is advantageous and appreciated by the patient but, unfortunately, it is not frequently available.

It is almost needless to repeat that patients must undergo orthodox therapy at the same time. Ozonetherapy is not a superfluous treatment and is intended to complement and improve ordinary therapy, which, on its own, is often insufficient. The case of emphysema is typical. In addition to rehabilitation with exercise training, anti-smoking measures and domiciliary oxygen therapy, new bronchodilators and appropriate antibiotics can control acute exacerbations. After a long incubation (1957), surgical removal of the most emphysematous parts of the lung has come of age; when the operation is successful, short-term results are fairly good, with marked improvement of the quality of life (Hillerdal 1997; Barnes, 2000). However, some of the patients do not benefit from surgery and the value and cost-effectiveness of the volume reduction surgery remain uncertain in the long run (Fishman et al., 2003). Moreover, medical expenditures to treat COPD, associated with invalidity, represent a significant economic and social burden for Health Authorities and society in general. I believe that these are sufficiently good reasons to justify serious and wide-ranging experimentation with ozonetherapy.

I ought to spend **a few words for the severe acute respiratory syndrome-associated coronavirus (SARS-CoV)** even though, hopefully, may not reappear. Just in case, besides **the well-timed use of IFN-beta** (Cinatl et al., 2003**), administered via IM and even better via aerosol**, we could take advantage of ozone therapy with the scheme and schedule suggested for ARDS

CONCLUSIONS: I have reviewed a good clinical study, meaningful assumptions and a few anecdotal hints for justifying the use of ozone therapy in asthma, COPD, IPF, ARDS, and emphysema. It is felt that ozone therapy could act as a synergistic adjuvant when combined to orthodox therapy. The acceptance of this proposal will imply reduction of medical and social costs but, above all, a better and longer life for many patients. It is unbelievable and regrettable that the medical

establishment and World Health Authorities remain sceptical and do not help evaluating the application of ozone therapy.

12. THE PROBLEMS OF TINNITUS AND SUDDEN HEARING LOSS (SHL). IS OZONE THERAPY OF ANY HELP?

Tinnitus (phantom auditory perception) is a poorly understood medical problem affecting some 40 million Americans (frequently men, who are 65 to 74 years of age) with 12 million being severely affected. Sounds, described as ringing, buzzing, hissing and humming, have been reported to be unilateral or similar in both ears. Tinnitus may be caused by otologic (cochlear damage), neurologic, vascular (possibly with turbulent blood flow), infectious, drug related (aspirin and citotoxic drugs) causes, and muscular spasms. If the loss of silence is permanent, it is likely to cause severe hearing impairment (Lockwood et al., 2002).

Incidence of SHL is at least 20 cases per 100,000 inhabitants per year and is caused by either a disturbance of cochlear microcirculation, or an array of viral infections, possibly complicated by an autoimmune process. At our University, De Capua et al., (2001) determined that some patients may undergo a sudden idiopathic hearing loss owing to hypoplasia or inactivation of the posterior communicating arteries in the absence of cerebro-vascular pathologies evaluated by transcranial Doppler.

The variety of therapeutic approaches reflects the uncertain and variable aetiology and pathogenesis of SHL and tinnitus. On the postulation of an autoimmune disorder, methotrexate administration was no more effective than placebo and less effective than prednisone therapy (Harris et al., 2003). When a vascular cause was suspected, fibrinogen and LDL apheresis appeared to modestly improve speech perception two days after the treatment (Suckfull, 2002). I have criticized the use of apheresis for simply and modestly improving the haemorheological parameters in ARMD patients because this is a complex, invasive and expensive technique when similar or better results can be achieved by the use of oral drugs (statins, etc). Moreover, as it can be expected, the reduction of fibrinogen level is transitory.

Treatment for tinnitus included tricyclic antidepressants, which improved the symptom in 67% of the patients taking nortriptyline (Dobie et al 1993). Many patients have tried complementary therapies: Ginkgo biloba seems to procure an improvement but a firm conclusion about efficacy was not reached (Soholm et al., 1998). Six randomized clinical trials of acupuncture failed to demonstrate any efficacy.

Why we tried ozone therapy and was it useful?

On the postulation of a vascular defect with consequent ischemia in the inner ear, the otolaryngologist, Dott. De Capua had previously observed the inefficacy of the hyperbaric oxygen therapy. However, on the basis of our positive results in ARMD and terminal POAD patients, he asked Prof. N. Di Paolo and me to evaluate ozone therapy and we performed the EBOO treatment in a patient (a man 28 years old) with SHL. During the previous three weeks, the patient has undergone antibiotic and corticosteroid therapy to no avail. Surprisingly, thc morning after the first EBOO treatment, a clinical and instrumental examination revealed an almost complete recovery! This first result encouraged us to evaluate a further eleven patients, four of which reported tinnitus as well. To our dismay, results were disappointing and we noticed that, only if the SHL was very recent, we could note a slight but erratic improvement, while, if the SHL had become chronic, there was no advantage. These results are at variance with what we continuously observe in ARMD (atrophic form), surely due to ischemia and hypoxia of the visual receptors, with the simple AHT treatment. Moreover one must know that spontaneous remission of hearing is estimated at about 65%! (Mattox and Simmons, 1977). In the charity clinic, where, gratuitously, I perform ozone therapy twice weekly, I had the opportunity to treat three elderly patients with tinnitus. Two received the regular AHT and one preferred the RI. Both approaches have not improved the symptoms either during or after the suspension of therapy.

CONCLUSIONS: Familiarity (hence, genetic factors), age and sex and an extremely variable number of causes are responsible for tinnitus and SHL. If vascular defects were predominant, we ought to have noted an improvement in at least a few patients. It is however possiblc that, once the symptoms appeared, the lesions are either already irreversible or cannot be modified by ozone therapy. To my knowledge, ozone therapy had not been evaluated before in these pathologies and I will be grateful to exchange information with anyone more knowledgeable

13. THE PARADOXICAL EFFECT OF OZONE IN ORTHOPAEDIC DISEASES. THE PROBLEM OF BACK-ACHE.

I believe that in the 70s, Dr. Alexander Balkanyi in Zurich has been the first to have the idea to inject small volumes of ozone in patients affected by tendinitis and myofascial pain. After him, a number of ozonetherapists (Riva Sanseverino, 1989; Verga, 1989; Siemsen, 1995) have begun to treat acute and chronic polyarthritis (osteoarthritis of the hip, knee, interphalangeal joints, sacroiliac joint, etc.), epicondilitis and carpal tunnel syndrome with intra-articular or peri-articular insufflation of small volumes of O_2-O_3 (5-10 ml in one or three sites with ozone concentrations from 5 to 15 mcg/ml) with very encouraging results. In Morton's disease (neuroma), up to six infiltrations of gas (4 ml each at 20 mcg/ml have yielded great pain relief. In a very informative review, Siemsen (1995) reported that application of medical ozone in acute and chronic painful diseases of the joints is a complementary method of treatment to obtain rapid pain relief, decongestion, disappearance of oedema, reduction of local temperature and increased mobility. If performed by an expert orthopaedic surgeon, the treatment is not risky and causes only transitory local pain that disappears in 5-10 min without any other adverse effect.

The pathophysiology of these diseases is complex and characterized by the softening and even distruction of the articular cartilage, with increased matrix degradation due to collagenase and proteoglycanases. The enzymes may be secreted by activated chondrocytes and monocytes, which release IL-1 and TNFα. Synthesis of PGs increases several fold and there is a natural attempt to maintain a biomechanically adequate matrix. In contrast to RA (Section V), pannus does not develop. Joint pain may be aggravated by concomitant synovitis.

Drug therapy is symptomatic, aiming to reduce pain and disability. Inhibitors of cyclooxygenase I are in wide use, with possibly some side effects, and are being substituted, less successfully, with inhibitors of cyclo II. Local injection of glucocorticoids into a given joint can be carried out no more than twice per year.

Because conventional medicine does not provide a "cure", patients search for complementary therapies. **On the basis of the pathophysiology, ozonetherapy should be the last treatment to perform,** because ozone (a potent oxidant) injected into the synovial space should elicit further inflammation or degeneration. Therefore, it is INCREDIBLE that, after an initial but tolerable pain, ozone produces great relief for a long time. **By now, innumerable patients have been treated and, although appropriate controls with oxygen alone have been evaluated only in one trial, we cannot doubt the results.** Obviously ozone is not a "miraculous" medicine and we must try to understand how ozone acts. This is another ozone paradox!

On several occasions, I have asked orthopaedic surgeons to collaborate with us because I think it would be interesting to examine the synovial fluid before and after ozonetherapy. So far this has not been possible, either because most patients are treated privately or because it is difficult to collect samples. Thus I can only advance a few speculations.

Once ozone dissolves in the synovial fluid, as usual, it reacts with biomolecules (antioxidants, PUFA, proteins), generates ROS and LOPs responsible for:

a) Possible inactivation and inhibition of the release of proteolytic enzymes and of proinflammatory cytokines.

b) Stimulation of the proliferation of chondrocytes (probably via H_2O_2) and fibroblasts, with increased synthesis of matrix and possibly of articular cartilage. **Induction of the synthesis of antioxidant enzymes (SOD, GSH-Px and catalase) may be a crucial event as an adaptive response to COS and to ozone**. That is the reason why I would start infiltrating ozone at low doses.

c) Release of bradykinin and synthesis of inflammatory PGs is probably inhibited, with reabsorption of oedema and pain relief.

d) An increased release of IL-1 soluble receptor or of other soluble receptors and antagonists able to neutralize proinflammatory cytokines such as IL-1, IL-8, IL-12, IL-15 and TNF.

e) Conversely **the release of immunosuppressive cytokines, such as TGF-β1 and IL-10, may inhibit inflammation.** Among several growth factors, TGFβ1 is interesting because it modulates the expression of integrins and stimulates the synthesis of matrix proteins such as collagen and glycosaminoglycans (Trippel, 1995; Qi and Scully, 1997; Grimaud et al., 2002). If this is the case, the long period of remission can be explained.

These are just hypothetical ideas, which should be verified by examining the synovial fluid and bioptic fragments to clarify these really paradoxical ozone effects. **Ozone never ceases to surprise us!**

Low back pain is a very disturbing symptom that **can affect**, at least for a while, **up to about 80% of the world's population.** Luckily, in most cases, physical therapies (exercise, manipulation orthodox therapy, etc.) as well as a number of complementary therapies can solve the problem (Cherkin et al., 1998; Samanta and Beardsley, 1999). If a herniated disc (protrusion of the nucleus pulposus through the annulus fibrosus) is present and causes considerable pain, it must be removed with the least invasive procedure. **However inflammation, rather than compression, seems the cause of pain** because, by using Nuclear Magnetic Resonance (NMR), an extensive evaluation has shown that 76% of apparently normal people have hernias without any symptom.

Up to the 1970s, the typical orthopaedic operation removed the compression but often destabilized the mechanical and functional stability of the vertebral column. Thus it has been substituted by mini-invasive interventions. This trend was started by chemonucleolysis, introduced by Smith in 1969. However, the intradiscal injection of chymopapain and collagenase, potent enzymes able to digest the components of the nucleus pulposus, has been abandoned because of occasional risk of allergic reactions and the exorbitant cost of the pure enzymes. Subsequently, Onik et al. (1987) introduced the alternative concept of aspirating the degenerated disc including part of the herniated material, thus reducing the abnormal pressure and relieving the nerve root compression. This technique is still in use with a success rate of about 75% (Bocchi et al., 1998). There are other variants of this type of approach, the latest being nucleoplasty.

In 1988, Verga, a private ozonetherapist, noted pain relief after infiltrating trigger points in myalgias with oxygen-ozone and proposed to use an indirect technique by injecting the gas into the points localizable in the paravertebral muscle (locus dolendi) corresponding to the metamer of the herniated disc. This approach is now widely used by many ozonetherapists in Italy and it can be defined as the indirect approach, or as I call it: "chemical acupuncture" (Bocci, 1998a).

The "chymopapain model" probably inspired a neurosurgeon, Jucopilla et al. (1995), to test whether **intradiscal injection of ozone** would be nucleolytic and beneficial. **This can be defined as the direct intradiscal injection of ozone.** More recently, another indirect variant has been introduced by the epidural injection of ozone in correspondence to the lesion (Figure 18). This is being performed by anaesthesiologists and, unless is carefully performed with small volumes (1-3 ml of gas) can cause side effects, of which the most frequent is headache. The use of ozone to treat back pain syndrome is now widely used in Italy, while it is becoming to be used abroad. As it is a minimally invasive treatment with a negligible cost and rare side effects, it is worth trying before surgical intervention. At our University, on the basis of our protocol, over 100 patients have been treated and about 80% have shown shown marked improvement (Bocchi et al., 2000). Thus there are as many as three technical approaches, which are exemplified in Figure 18.

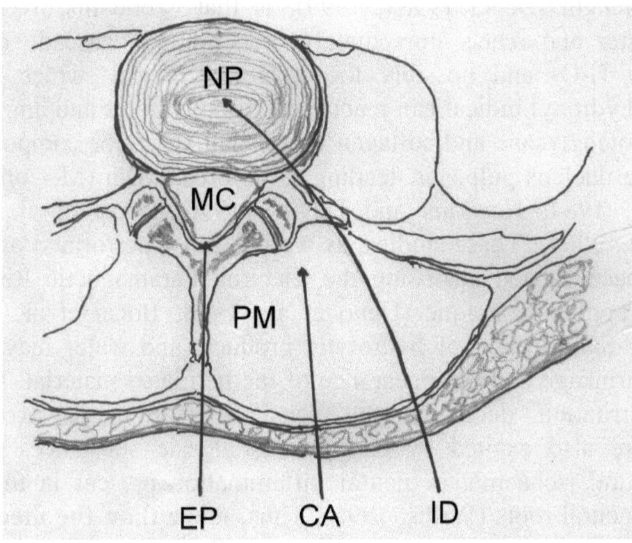

Figure 18. Schematic view of a transverse section of the lumbar region: NP: nucleus pulposus. MC: medullary canal. The arrows indicate the three possible routes of O_2-O_3 administration. ID: intradiscal; CA: ***"chemical acupuncture"*** *in the paravertebral muscle: PM. EP: epidural route.*

The Direct Approach

The direct approach is carried out under radioscopic control: the needle is inserted in the centre of the pathologic intersomatic space just before direct insufflation of the gas mixture (Figure 18). An expert can do it in about 10 min. After a rest of 10-15 min, the patient can get up and often he/she is amazed by the disappearance of the pain, as occurs after nucleoplasty. If necessary, the application can be repeated a second time before changing the approach.

Good results have been obtained after either intradiscal or intraforaminal injection of a variable volume (3-15 ml) of gas at an O_3 concentration of 27-30 mcg/ml. Several thousand patients have been treated, with a success rate of 54-86% (Jucopilla et al., 2000; Bonetti et al., 2001; Fabris et al., 2001; Petralia et al., 2001; Alexandre et al., 2002). An extensive study had been performed in 600 patients, who had failed to respond to conservative management (Andreula et al., 2003): 70.3% of the first half of patients, treated only with ozone, showed a good outcome. This was further improved (78.3%) in the remaining 300 patients, by combining ozone treatment with a periganglionic injection of corticosteroid and anaesthetic. Unfortunately controls (either oxygen or oxygen-corticosteroid-anaesthetic alone) were not evaluated, probably, for ethical reasons. Nonetheless, from a scientic point of view, it will be important to perform a randomized study to evaluate the role of the needle, oxygen and so forth, which are probably relevant.

It remains unclear how ozone acts. One real possibility, previously discussed at length (Bocci 1998a, 1999), is that ozone dissolves in the interstitial water and reacts immediately, generating a cascade of ROS, among which H_2O_2 and possibly the hydroxyl radical, which is most reactive. The hydroxyl radical can react with carbohydrates and amino acids composing proteoglycans and collagen type I and II, major components of the degenerate nucleus pulposus, leading to its breakdown (McCord, 1974; Curran et al., 1984; Hawkins and Davies, 1996; Bocci et al., 2001b; Leonardi et al., 2001). These studies, as well as those performed on human blood, have been carried out using the Electron Paramagnetic Resonance (EPR) spin trapping technique (Ueno et al., 1998; Bocci et al., 2001b). Consequently, reabsorption of hydrolytic products and water may lead to progressive shrinkage and disappearance of the herniated material. Reduced mechanical irritation decreases the sensitivity of nerve axons, but nociceptors are also excited by endogenous algesic substances released during perineural ischaemia or neural inflammation present in the spinal ganglion and neural roots (Willis, 1995). Thus, **more than the mechanical compression as primum movens, it is the inflammatory reaction that sustains chronic pain by releasing PLA2, several proteinases and cytokines.** The continued release of ROS, PGE_2, serotonin, bradykinin, cathepsins, IL-1, IL-6, substance P and TNF alpha causes oedema, possibly demyelination and increased excitability of nociceptors (Fields, 1986).

Indeed, it has been observed that, in absence of inflammation, even a large hernia can be painless. Moreover, the hernia may remain after an operation (as seen radiographically), but the pain disappears once the inflammatory disorder dies down. Interestingly, epidural injections of the anti-inflammatory methylprednisolone transitorily improve leg pain and sensory deficits in patients with sciatica due to a herniated disc (Carette et al., 1997). But even more interesting is the observation that an intravenous infusion of INFLIXIMAB (an antibody against TNF alpha) produced a very rapid and dramatic improvement in leg pain among patients with severe sciatica (Karppinen et al., 2003).

Table 7 intends to summarize the complex reparative process induced by ozone used in substitution or in combination with orthodox remedies.

Table 7. Low back-ache: conventional remedies versus ozone.

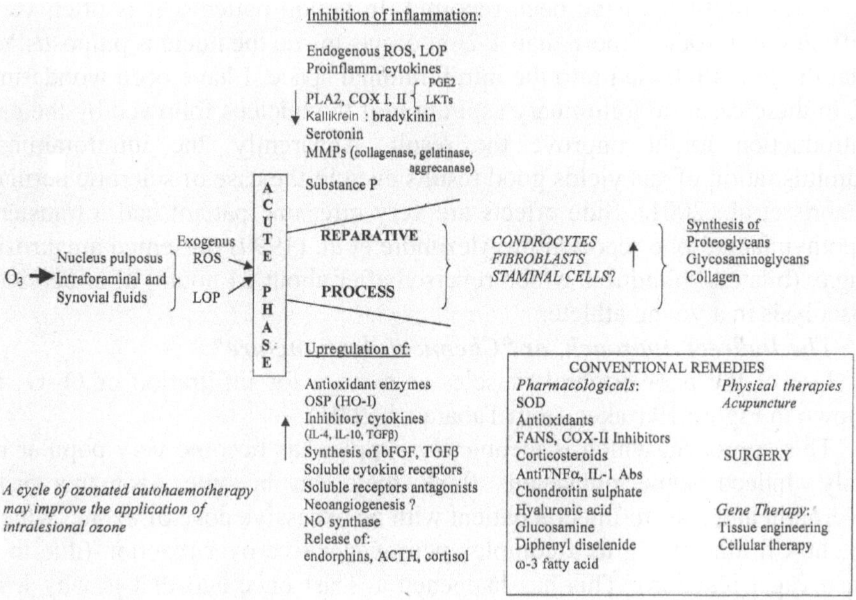

So, **how does ozone act?** We are again facing the ozone paradox: although hydroxyl radicals can degrade the degenerated material and reduce pressure, it often exerts a rapid "anti-inflammatory action", particularly because only a few ml of gas can be introduced inside the nucleus pulposus and most of the gas invades the intraforaminal space. **This may mean that ozone rapidly blocks inflammatory reactants and stimulates the restitutio ad integrum.** What is even more surprising is that this change remains stable (unlike corticosteroids) and it does not necessarily coincide with the disappearance of the herniated material. In fact, CAT or NMR controls in 612 patients, 5 months after treatment, showed that the hernia disappeared in 226 (37%), was reduced in 251 (41%) and was unmodified in 135 (22%). After another 5 months, CAT/NMR controls were performed again in 200 (of 251) patients in whom the hernia was reduced: a further reduction and improvement was noted in 44 patients (22%). In 120 patients (of 135) in whom the hernia was unmodified, there was an improvement in 11.6% (14 of 120) (Alexandre et al., 2000).

Thus the ozone effect is deployed in successive phases: there is an initial rapid change, probably with disappearance of oedema and improvement of circulatory and metabolic conditions, followed by a stasis and then a further improvement possibly due to release of TGFβ1 and bFGF (Silver and Glasgold, 1995; Trippel, 1995), **favouring the reorganization of the residual nucleus pulposus with incipient fibrosis.** So far, attempts to examine the histopathological changes have been inconclusive.

A few problems have been reported. In young patients, it is often very difficult to introduce more than 1-2 ml of gas inside the nucleus pulposus, so that the gas is released into the intraforaminal space. I have been wondering if, in these cases, a preliminary aspiration of the nucleus followed by the gas introduction might improve the result. Apparently, the intraforaminal administration of gas yields good results even in the case of sclerotic hernias (Fabris et al., 2001). Side effects are very rare: one patient had a transient lipothymia and one reported by Alexandre et al. (1999) presented amaurosis fugax (bilateral blindness which reversed after about 24 hours) after cervical discolysis in a young athlete.

The Indirect Approach, or "Chemical Acupuncture"

Use of the paravertebral muscles as a route for infiltration of O_2-O_3 is shown in Figure 19 (taken from Tabaracci, 2001).

This approach, which is technically simple, has become very popular in Italy. Indeed some physicians think they can become ozonetherapists overnight and start to inject a patient with an excessive dose of ozone, which might kill him owing to a complex neurovegetative over-reaction (due to a vaso-vagal reaction). This has happened at least once and that is why it is important to have precise guidelines and rules for the practice of ozonetherapy.

In reality, it is an easy approach consisting in one or several (up to four) injections of 5-10 ml of gas per site (Figure 18). The ozone concentration, normally, must not exceed 20 mcg/ml because it is painful. At first, it is wise to test the patient's reactivity with an injection of sterile saline and then start with 10 mcg/ml ozone. The injection must be done very slowly into the trigger points corresponding to the metamers of the herniated disk. The length of the needle varies (from G22 to G25) depending on the patient's obesity. Usually two symmetrical injections (total dose 10-20 ml gas with at most 200-400 mcg ozone) repeated twice per week for about 5-6 weeks (10-12 sessions) are sufficient; if not, the patient is unresponsive to this approach. This point remains controversial because some ozonetherapists continue treatments for up to 30 sessions. I have noticed that **the pain at first elicited with an ozone concentration of 20 mcg/ml tends to subside because of a progressive elevation of the pain threshold.** In such a case, I slowly increase the ozone concentrationup to 35 mcg/ml. **It appears that the stimulation of nociceptors, hence of a tolerable and transitory pain is an essential requirement for achieving the final therapeutic effect.** Indeed I often remind the patient: **"no pain, no gain".**

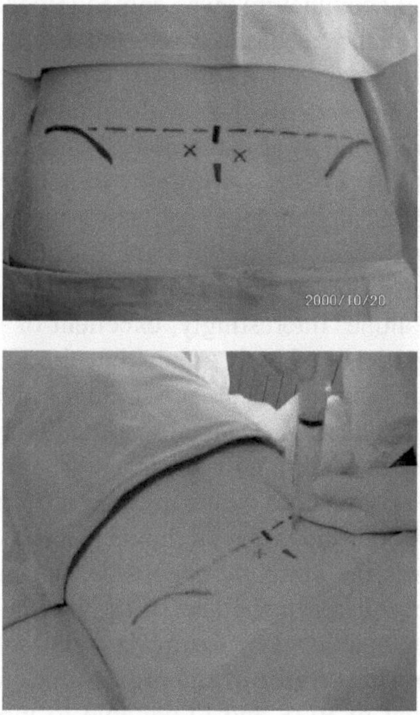

Figure 19.The iliac crests are palpated and the transiliac line is determined to identify the L4 spinous process, the interspinous spaces are identified by selecting the space corresponding to the herniated disc. Approximately 2 cm are calculated bilaterally to the spinous process (above). Once the needle is inserted through the fasciae, aspiration is carried while holding the needle still. We must be sure that we have not punctured a vein. Then ***a 10-20 mcg/ml concentration of an oxygen-ozone mixture is injected very slowly up to a maximum of 10 ml per infiltration.*** *Aspiration is repeated during infiltration (below) (Tabaracci, 2001).*

I repeat that injection of O_2-O_3 elicits a sharp pain lasting a few minutes and the injection must be done very slowly to avoid any risk of embolism. If we act carefully, we can avoid serious adverse effects, such as sudden hypotension, bradycardia, mydriasis, intense perspiration and cardiac arrest (vasovagal reflex). **Any serious ozonetherapist must be prepared for this emergency** (Cummins, 1994), **which is very rare but can happen.** ***A good***

experience with basic life support can save the patient ….and the ozonetherapists!

The results of a number of studies vary somewhat (Cinnella and Brayda-Bruno, 2001), but they can be summarized as: about 40% optimal, 35-40% marked improvement, 15-25% minimal or no result. Gionovich et al. (2001) compared three approaches:

A) Paravertebral injection of O_2-O_3: 75% good response
B) Peridural injections with desamethasone: 55% good response
C) Paravertebral injection of buvipicaine 0.25%: 70% good response

The term "chemical acupuncture" was coined (Bocci, 1998a) because we must clarify the role of the needle, oxygen and ozone. It was proposed to compare this procedure against a waiting-list control, two placebo controls (one with oxygen alone and another without any gas) and a standard-treatment control. Gionovich et al. have now shown that, as expected, even an anaesthetic has some effect. Owing to an unexpected, unintentional incorrect use of the medical generator (delivering medical oxygen only), we can now give a reasonable answer to the above-mentioned uncertainty. Torri et al. (1999) treated a group of 66 patients with ozone and a group of 30 patients with oxygen alone. Interestingly, excellent or good responses were observed in 86% of patients of both groups but the ozone group showed a statistically significant improvement of some clinical parameters. This suggests that the needle and oxygen together already have a therapeutic role, which is potentiated by the addition of ozone.

Then the question is: how does ozone injected intramuscularly work? The gas infiltrates the muscle and after 24 hours some gas bubbles (residual oxygen!) move towards the vertebral canal (as seen radiologically). It was postulated that ozone will reach the site of the herniated material and will lyse it. This is an untenable idea: **ozone is very soluble and dissolves rapidly into the interstitial water of the muscle** and, within 20-40 seconds, will generate a gradient of ROS and LOPs able to inhibit amyelinic fibres (C-nociceptors), which are able to elicit the elevation of pain threshold and an antalgic response via the descending antinociceptive system (Figure 20). As occurs during acupuncture (Ceccherelli et al., 1995), the introduction of the needle, reinforced by the pressure of the gas, induces strong inhibition of nociceptors, perhaps a prolonged stunning due to ROS and LOPs. It is known that an algic stimulation of the skin and muscles can reduce pain through the mechanism of diffuse noxious inhibitory control (DNIC). That is why the needle +ROS-LOPs + oxygen pressure can be translated into a chemical acupuncture.

This mechanism is likely correct because too low ozone concentrations (3-10 mcg/ml) or small gas volumes (1-2 ml) are ineffective, whereas too high concentrations or excessive gas volumes can cause lipothymia. It is unclear whether pre-infiltration with an anaesthetic reduces the effect of ozone but it likely does.

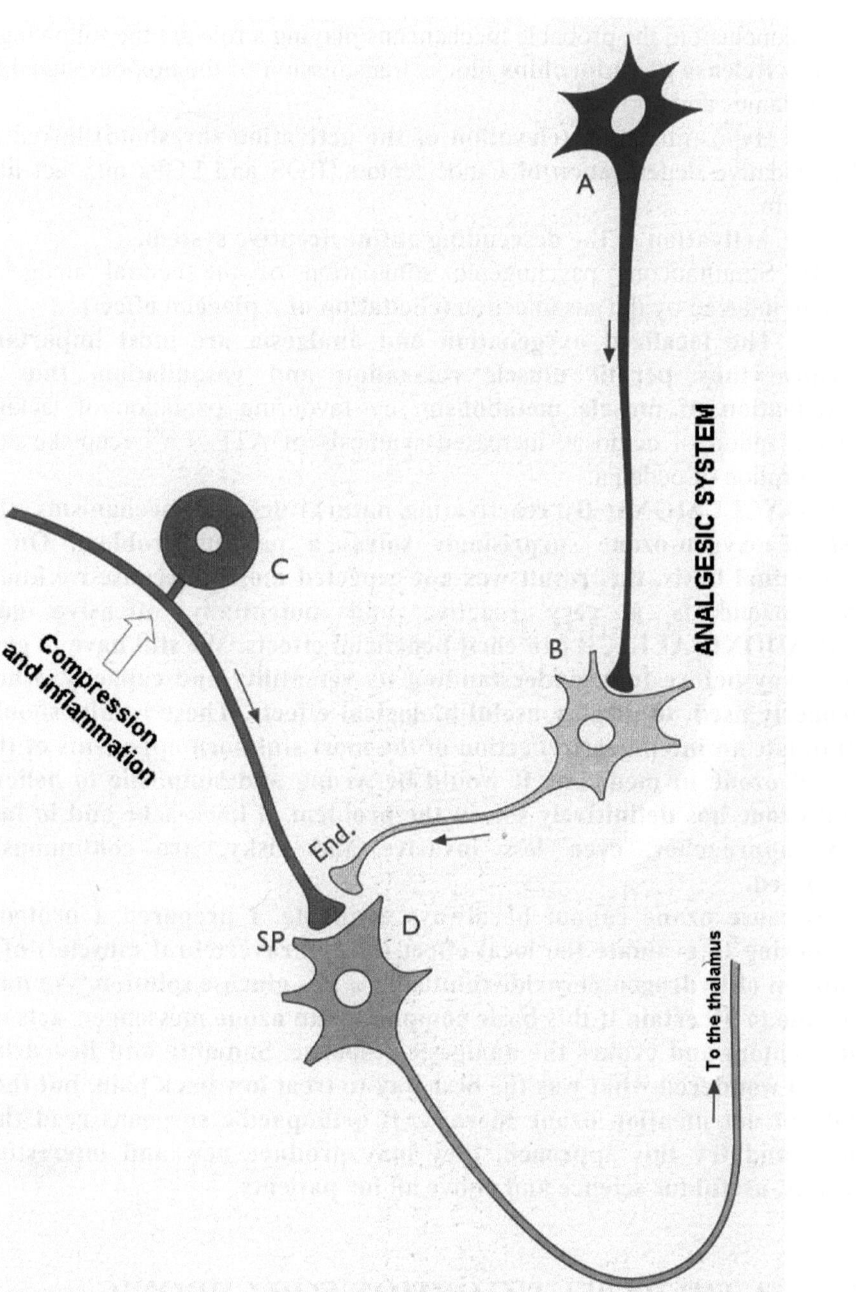

Figure 20.The scheme indicates the mechanisms for the control of algesic signals. By releasing endorphins (End.), the enkephalinergic interneuron may inhibit the presynaptic connection of a neurocyte (C) of a spinal ganglion which, under compression of a herniated disc, stimulates the release of substance P (SP). Endorphins can inhibit the transmission of the algesic signal to neuron D, hence to the ascending spinal-thalamic fibres. The monoaminergic or serotoninergic neuron A, as a component of antinociceptive descending fibres, can reinforce the analgesic effect of neuron B

In conclusion, the probable mechanisms playing a role are the following:

a) **Release of endorphins** blocks transmission of the noxious signal to the thalamus and cortex.

b) Hypostimulation **(elevation of the activation threshold)** linked to the oxidative degeneration of C-nociceptors. ROS and LOPs may act like capsaicin.

c) **Activation of the descending antinociceptive system.**

d) Simultaneous psychogenic stimulation of the central analgesic system induced by the gas injection **(elicitation of a placebo effect).**

e) **The localized oxygenation and analgesia are most important because they permit muscle relaxation and vasodilation, thus a reactivation of muscle metabolism,** by favouring oxidation of lactate, neutralization of acidosis, increased synthesis of ATP, Ca^{2+} reuptake and reabsorption of oedema.

CONCLUSIONS: By reactivating natural defence mechanisms, the use of oxygen-ozone surprisingly solves a painful problem. On a conceptual basis, this result was not expected mostly because we know that ozone is a very reactive and potentially offensive gas. PARADOXICALLY, it can elicit beneficial effects. We still have to go a long way before fully understanding its versatility and capacity, when properly used, to display useful biological effects. These results should stimulate an intelligent reflection of the most stubborn opponents of the use of ozone in medicine. It would be wrong and simplistic to believe that ozone has definitively solved the problem of back-ache and in fact new approaches, even less invasive and risky, are continuously proposed.

Because ozone cannot be always available, I prepared a protocol proposing to evaluate the local effect (into paravertebral muscles) of a solution of hydrogen peroxide diluted in a 5% glucose solution. We may be able to ascertain if this basic compound, an ozone messenger, acts on nociceptors and evokes the analgesic response. Samanta and Beardsley (1999) wondered what was the best way to treat low back pain, but they did not not mention ozone therapy. If orthopaedic surgeons read this book and try this approach, they may produce new and interesting results, useful for science and above all for patients

14. A THERAPEUTIC OPTION FOR CHRONIC FATIGUE SYNDROME (CFS) AND FIBROMYALGIA

Both diseases are frustrating illnesses characterized by a number of signs and symptoms among which severe fatigue and a flu-like syndrome predominate and profoundly disable patients (Natelson, 2001; Wessely, 2001).

CFS has also been named chronic mononucleosis syndrome, chronic Epstein-Barr virus (EBV) syndrome, myalgic encephalitis and postviral syndrome suggesting that the initial cause of the disease was believed to be a viral infection (Swartz, 1988; Cope et al., 1994). In spite of the fact that more than 4.000 papers have been published on CFS (Joyce et al., 1998), its aetiology and pathophysiology remain ambiguous, but it cannot be excluded that CFS is first triggered by an undefined viral or bacterial infection able to induce a chronic infection with a concomitant immunological dysregulation (Caligiuri et al., 1987; Landay et al., 1991; Konstatinov et al., 1996; Komaroff and Buchwald., 1998). Interestingly, De Meirleir et al., (2000) confirmed Suhadolnik's et al (1997) finding of an increased level of 2-5 A synthetase in lymphocytes of patients with CFS. This enzyme is an excellent biomarker of an underlying interferon (IFN) synthesis and IFN represents the prototypic cytokine causing a flu-like syndrome (Bocci, 1988, a).

However, we cannot say whether a primary infection is also responsible for the disturbance of the hypothalamic-pituitary-adrenal axis (HPA) characterized by low circulating cortisol, dysregulated secretion of central neurotransmitters (serotonin, opioids, arginine vasopressin) and growth hormone (Parker et al., 2001). Although the latter disturbance is controversial (Allain et al., 1997; Cleare et al., 2000), it must be kept in mind because growth hormone regulates the hepatic synthesis and release of somatomedin C, which is a mediator of muscle homeostasis possibly implicated in muscle pain. This aspect can be connected to the muscular alterations detected in patients with CSF (Fulle et al., 2000) characterized by mitochondrial dysfunction and oxidative damage documented by an increased level of 8-hydroxy-2-deoxyguanosine in nuclear DNA and malonyldialdehyde in supernatants of muscle homogenates. Relevant collateral findings have been an impaired oxygen delivery to muscle and a lower rate of creatinine phosphate resynthesis following high-intensity exercise in CFS patients compared to normal subjects (McCully and Natelson., 1999). Finally Georgiades et al., (2003) have emphasized the role of some CNS mechanisms gone astray in the pathogenesis of CFS, particularly the role of 5-hydroxytryptamine and dopaminergic systems.

Fibromyalgia has an obscure aetiology and it is also known as the Atlas's syndrome (nuchal or peripheral variety). In Italy, it is considered a disease causing considerable socio-economic problems, since it affects about six million people, predominantly women between the ages of 25 to 60 years. The disorder is characterized by musculoskeletal pain, stiffness, fatigue, exhaustion and frequent association with headache, unrefreshing sleep, irritable bowel syndrome and dysmenorrhea. Moldofsky et al., (1975) demonstrated that a disturbance of stage 4 non REM sleep characterized by alpha-wave intrusion into the delta rhythm may play a role in the development of fibromyalgia. The pulsatile secretion of growth hormone closely related to stage 4 sleep may therefore become impaired with consequent decreased release of somatomedin C and muscular damage (Fulle et al., 2000).

In relation to the various pathogenetic hypotheses, the following orthodox treatments have been tested:

Antivirals (acyclovir, IFN alpha, immunoglobulin G), **antidepressants** (fluoxetine, amitryptiline, hypericum extract), **anti-inflammatory drugs** (a variety of cyclo-oxygenase 1 inhibitors, corticosteroids) and **metabolic drugs** (vitamin B12, magnesium pidolate, Q10 coenzyme, carnitine, nicotinamide adenine dinucleotide). They have proved to be scarcely effective and some of them exert adverse effects (Reid et al., 2000). **Prolonged rest** similar to the deconditioning process occurring during ageing (Degens, 1998) **is ineffective or harmful.** On the contrary, **graded exercise therapy, GET**, (Powell et al., 2001) and **cognitive behavioural therapy, CBT,** (Prins et al., 2001) **administered by specialized therapists appears to be an effective intervention for CFS patients.** Prescribed graded aerobic exercise treatment has been found effective and widely available in fibromyalgia (Richards and Scott, 2002).

A working group set up in 1998 to review the management of CFS published a report in 2001 (Report of the Working Party, 2001) and has reached a fairly large consensus on the beneficial effects of GET and CBT. However, Clark et al., (2002) pointed out that *"none of the rehabilitation approaches is intended to be curative, no approach has been found to be beneficial for everyone, and all can be tainted by poor practice by therapists lacking proper understanding of the disorder".* Moreover, the report endorsed an additional approach known as **"pacing"** which consists in balancing activity and rest.

This state of uncertainty does not help patients and compels me to propose here **ozonetherapy** because, with little means, it has been quietly performed in the last few years **yielding a major, often definitive, improvement in most patients**. At least in theory, ozonetherapy may correct the most relevant deficiencies because it:

a) improves blood circulation and oxygen delivery to ischaemic tissues,

b) corrects the dysimmunity due to a possible primary infection,

c) corrects the endogenous chronic oxidative stress by upregulating the antioxidant system.

d) induces a release of hormones and neurotransmitters (probably a surge of serotonine).

e) induces a state of wellness and euphoria without adverse effects.

I shall describe a few open trials where patients were compared to those without treatment. The diagnosis of CFS was made on the basis of the definition of the disease made by the US Centers for Disease Control and

Prevention (Fukuda et al., 1994). This includes the manifestation of several physical symptoms such as severe fatigue during the last six months and at least four of the following symptoms: 1) sore throat, fever, muscle pain, multi-joint pain, frequent headaches, unrefreshing sleep, impaired memory and post-exertional malaise. The British criterion (Sharpe et al., 1991) which insists on the presence of mental fatigue, was also taken into consideration.

In 1988, at the hospital of Conegliano Veneto, a private ozonetherapist treated six patients diagnosed with CFS. AHT was carried out twice weekly for 8 weeks and this physician assured me that four patients showed a "remarkable improvement". He could not give information about the follow-up.

At Siena hospital (Rheumatology Department), Dr. Cosentino treated only one CSF-patient performing 14 AHTs but the patient reported only a slight improvement.

Our study (Borrelli and Bocci, 2002) included three patients with CFS (one man: age 47 and two women: age 51 and 55). Our patients reported fatigue, muscle weakness, sleep disturbance and two had frequent headaches. Pressure over tender sites very often elicited considerable but transitory pain. All of these patients had suspended medical treatments for at least three months. Two patients with depressive disorders taking antidepressants and other drugs were excluded. Before ozonetherapy, patients were informed that the treatment was experimental but it had a rational basis and did not yield toxic effects. All patients signed a specific informed consent form (Ernst and Cohen, 2001). The therapy consisted of two AHTs weekly but we decided that it would be wrong to predetermine a fixed schedule. Thus, they underwent 28, 32 and 40 initial treatments (during 3.5, 4 and 5 months, respectively) followed by three months rest.

Before starting the therapy, we tested the total antioxidant capacity (TAS) of the patient's plasma (Re et al., 1999; Ghiselli et al., 2000). Levels ranged within normal levels (1.3-1.8 mM). Occasionally we tested the TAS level after ozonation *ex vivo* and we found that it was decreased by no more than 15 %. Peroxidation levels (TBARS) barely increased with an ozone concentration of 20 mcg/ml but they were significantly increased with an ozone concentration of 40 mcg/ml indicating an effective blood ozonation. Haemolysis always remained at negligible (0.1-0.4%) levels.

The weakness of our work is due to the very limited number of patients. However, the compliance was excellent because, **as the patients slowly improved, they were enthusiastic to continue the therapy.** In these three CFS patients, most of the symptoms markedly decreased after 3.5, 3.8 and 4 months, respectively of continuous therapy. All of them were and felt practically normal six months after the initial treatment. No side effects were reported and all of them experienced a feeling of renewed energy and euphoria.

With regard to fibromyalgia, from 1988 to 2000, Dr. Salvatore Loconte (Andria, Bari) has treated 150 patients by infiltrating 5 ml gas directly on the trigger points (ozone concentration: 5-10 mcg/ml) and performing a cycle of AHT with about 150 ml blood and a total ozone dose of 4.5 mg (30 mcg/ml). He is a private ozonetherapist and cannot do a control but he has claimed to achieve total remission in about 60% of patients and partial improvement in 15%.

A RCT has been performed (1998-2000) in the Department of Rheumatology of our University on 40 women (age 30-50) diagnosed as having fibromyalgia on the basis of the ACR criteria. The scope of the study was to evaluate the effect of **A)** AHT with O_2-O_3 (20 patients, with ozone concentrations scaling up from 20 to 40 mcg/ml, twice per week for a total of 16 treatments), **B)** AHT with oxygen alone (10 patients as controls), and **C)** simple AHT without gas (10 patients as another control). Several standard end-points were tested before treatment, after 8 weeks and 1 month thereafter.

Patients of group C did not show any improvement and are now under ozone treatment.

Three **patients of group B (30%) showed good improvement.**

Seven patients of group A (35%) showed excellent improvement, while one (5%) had good improvement. Cosentino et al. (2000) concluded that ozonetherapy has therapeutic validity and no side effects

Dr. E. Borrelli and I (2002) evaluated five fibromyalgic patients. Four showed a definitive improvement after six months whereas one woman had very poor venous access and complained of blood extravasation. After four treatments, she was dissatisfied and dropped out. We suggested trying rectal insufflation of ozone but she did not accept. The problem of difficult venous puncture is rare but is real and now we can propose the option of quasi-total body exposure to ozone that is not invasive and quite practical. During the therapeutic session we take care to talk to the patient and explain the various biological effects resulting from the interaction of blood with ozone. Most of the patients appreciate the conversation and we believe that this is part of the treatment.

Our four fibromyalgic patients received between 24 and 36 treatments depending upon the response to the therapy. As Loconte (2000) previously reported, we performed careful infiltration of 5 ml of gas (O_3 concentrations: 5-15 mcg/ml) in some of the tender sites and trigger points, alternatively. The infiltration of ozone in both tender and trigger points of fibromyalgic muscles deserves a comment. Although they cause a transitory (3-5 minutes) pain, they usually elicit a diffuse analgesic effect after 5-8 infiltrations.

All patients throughout the therapy were advised to supplement their daily diet with vitamin C (0.5 mg), n-acetyl-cysteine (0.6 mg) and a multivitamin tablet (RD doses) including vitamin E, selenium and alpha-lipoic acid.

CONCLUSIONS: Orthodox medical care (antidepressants, corticosteroids, immunotherapy and metabolic drugs) is scarcely beneficial and with some side effects in CFS patients. Although GET and CBT appear to represent an effective intervention for CFS, they do not entirely solve the problem. We have been stimulated in evaluating ozonetherapy because, in other pathologies, most of the patients have reported a feeling of well-being and euphoria. This result is interesting and we can only speculate that the reasons for these positive effects are, at least in part, due to a functional restoration of hormonal and neurotransmitter functions. Moreover, ozonetherapy may interrupt the vicious circle due to a chronic oxidative stress and deranged muscle metabolism. The clinical results so far obtained appear to justify the use of ozone because it is able to activate simultaneously several metabolic pathways gone astray in these frustrating pathologies. This also explains why CBT, that certainly involves the psycho-neurohumoral system, is somehow more effective than using conventional drugs. Our data need to be expanded and compared with a group of patients treated with CBT. The use of a placebo (simple autotransfusion or only oxygenated blood) would be interesting, but these patients are severely distressed and randomisation appears unethical.

A few observations ought to be kept in mind for the future. Our schedule and the volume of blood exposed to O_2-O_3 may not have been optimal because the clinical improvement has progressed slowly. While we are insisting on the validity of the strategy "start low, go slow", we may have been too cautious. The schedule of two treatments per week appears valid and well accepted by patients but, while we should start with a 225 ml volume of blood and an ozone concentration of 20 mcg/ml, during a four week period, we should escalate the blood volume to the maximum of 270 ml and an ozone concentration of 40 mcg/ml. It also appeared clear that *a priori* we cannot fix a number of treatments (say 12 or 16 to be performed in 1.5 or 2 months) because, understandably, each patient responds differently to the therapy. In our case, among CFS patients, we noted one slow, one medium and one rapid responder. Consequently, *we must adjust the cycle and maintenance therapy to the single patient and not to a fixed, meaningless scheme.* This is an aspect that ought to be extended to other pathologies!

In the case of fibromyalgia, our statistics are very meagre compared to those reported by Loconte (2000) and Cosentino et al., (2000). The latter group determined a complete response in about 40% of patients while Loconte claimed to achieve total remission in 60% of patients. In our case, four patients (80%) had an excellent response and this is most likely due to our far longer treatment schedule. The direct infiltration of tender sites and trigger points can be compared with the "chemical acupuncture"performed in the paravertebral muscles for the problem

of backache and is interpreted to activate the anti-nociceptive system via the descending analgesic neuronal complex. It may be interesting to evaluate the local infiltration of a small volume of ozonated blood that may lead to a complete normalisation of nociceptors.

15. OZONE THERAPY IN EMERGENCY SITUATIONS, BEFORE TRANSPLANTATION AND ELECTIVE SURGERY.

There are several circumstances when **the use of ozone therapy combined with conventional therapies may improve prognosis.** I never managed to convince the chief doctor of intensive care medicine at Siena hospital of the potential usefulness of ozonated HAT performed at low concentrations (15-25 mcg/ml of blood) in seriously ill patients with permanently cannulated central or peripheral veins. He has been mostly concerned about the legal aspect of using a non-validated and somewhat controversial therapy in high-risk patients. When I visited Russian hospitals, I was told that they do not worry about it and use ozonetherapy to disinfect traumatic and war wounds, burns, radiation injuries and abdominal surgery after stomach or intestinal perforations. Disinfection with ozonated bidistilled water and application of ozonated oil has been found to be most useful in burns. It is unfortunate that they abundantly use ozonated saline instead of ozonated blood for systemic treatment. On this point, our opinions are greatly divergent. I cannot agree with their assertion that ozonated saline is as effective as blood, because firstly it contains sodium hypoclorite and secondly, because blood is far more efficacious

Serious trauma, burns and peritonitis lead more or less rapidly to systemic alterations and multiple organ failure particularly of the cardiopulmonary (ARDS), coagulative (DIC) and renal systems. Because of an adverse series of metabolic impairments, these alterations **frequently cause the patient's death.** Thus, using all the most appropriate conventional supporting therapies combined with a mild oxygen-ozone therapy (one AHT every 3-4 hours throughout the day), I "feel" that we could save some lives. Of the two million cases of nosocomial infections occurring each year in the USA about half are associated with indwelling devices and mortality is high among patients with cardiovascular implants, particularly prosthetic heart valves and aortic grafts (Darouiche, 2004). In section I, I mentioned already that these infections are often supported by antibiotic-resistant bacteria and/or by poor penetration of antibiotics into the infected area. Besides the fundamental role of surgery and medical treatment, both the parenteral ozonated AHT and the localized ozonation with either gas (if possible) or with strongly ozonated water could be useful for achieving a better outcome.

Unfortunately there is not yet the outlook that ozone therapy could represent a valid support!

The second topic is less tragic, but no less serious. I have often wondered if **a cardiac patient waiting for a heart transplant might gain increased resistance to infections and to immunesuppression (unavoidably linked to deep anaesthesia and surgery) if he could undergo three AHTs per week (at low ozone concentrations: 20 to 40 mcg/ml) for at least two weeks before transplantation.** This strategy is all too obvious and may induce a sort of ischaemic preconditioning or, to use language comprehensible to most people, the adaptation to chronic oxidative stress, that is present in these critical patients. During heart transplantation, all organs (particularly the CNS, retina and kidneys) undergo a bland ischaemia-reperfusion syndrome, which in unlucky cases may have dire consequences even if the operation is technically perfect. Thus **prophylactic ozonetherapy,** with little effort and expense, **might improve the outcome by reducing the risk of infections and shortening the hospitalization.**

The final point worth pursuing involves **the scheduled operation for application of prostheses, particularly joint implants.** In particular, as a precaution, coxo-femoral surgery requires the collection of 1 or 2 standard units of blood from the patient. Discussing this problem with several orthopaedic surgeons, I found that they are interested in evaluating whether performing at least four ozonated AHTs (ozone at low concentrations) during the two weeks before the operation and then every day immediately after it for 4-5 days (using the predeposits as well) would reduce the complications by enhancing healing and the patient's mood. I presented a protocol to our Ethical Committee, which was approved. However, unless we have appropriate fundings the trial cannot start because the orthopaedic surgeons do not have enough supporting personnel.

CONCLUSIONS: It is frustrating to have ideas that cannot be implemented owing to either incompetence, scepticism, lack of funds and possibly prosecution. In the supreme interest of the patient, Health Authorities should try to improve the situation but they remain entangled in economic and political problems.

16. OZONETHERAPHY IN DENTISTRY AND STOMATOLOGY

In spite of the dogma that "ozone is always toxic," a new development has stirred up great interest. The oral cavity normally hosts some 20 g of commensal bacteria, which are well kept in check by the MALT. However, they can become pathogenic and are mostly responsible for dental decay. As reported in the Introduction, Dr. E. Fisch (1899-1966) is considered the first dentist to use ozone in his practice and to have shown to Dr. E. Payr (1871-1946) the potent disinfectant activity of ozone. After several discussions with dentists, it has become clear to me that they have a vast armamentarium to fight oral and dental infections (Inaba et al., 1996; Dogan and Calt, 2001). Nonetheless, since 1995 in Germany, Filippi and Kirschner have used ozonated water under pressure, as a spray, during dental treatment and surgical operations. Obviously, one need an ozone generator and a reservoir of bidistilled water to freshly prepare ozonated water throughout the day but this is not a problem. Dr. Filippi is enthusiastic about this old-new possibility and has often asked me why ozonated water works so well. The jet of water removes all purulent material and disinfects the area. The ozone probably activates the local circulation and may stimulate the production of the usual cytokines, promoting the healing process. Indeed Filippi, at the 15th World Congress (IOA, 2001), reported that the application of ozonated water in the oral cavity significantly accelerated wound healing in comparison to placebo treatment. Although direct use of the gas is prohibited because one must never breathe ozone, a recent new invention has circumvented the problem. The application of ozone as a good disinfectant in Dentistry is not surprising, but this is a new approach. In a series of papers, Prof. E. Lynch's group (Baysan et al., 2000; Baysan A. and Lynch E., "Management of root caries using a novel ozone delivery system in vivo", submitted for publication) has shown that primary root carious lesions (PRCLs) can be successfully treated with a novel ozone delivery system able to avoid any toxic risk. The system includes a source of ozone and a dental handpiece with a removable silicon cup for exposing the tooth's lesion to the gas. Escape of ozone is prevented by a tightly fitting cup including a resilient edge for sealing the edge of the cup against the selected area of the tooth. The tooth's lesion is exposed to ozone for a period of 10-20 sec to a sort of ozone "hurricane" based on a low ozone concentration (about 4 mcg/ml) and a gas flow of about 600 ml/min. This treatment appears sufficient to kill all micro-organisms present in the PRCL and nobody has ever doubted the bactericidal potency of ozone. Particularly important is the protein denaturation and death of lactobacilli which, by normally acting on glucose, produce lactic acid favouring dental demineralization. **The ozone sterilized dental surface can be quickly (about an hour) remineralised by the calcium phosphate present in saliva, thus becoming hard and resistant to further bacterial attack for at least three months.** Previous suggestions using sodium hypochlorite do not appear as effective as ozone (Inaba et al., 1996; Dogan and Calt, 2001), According to Lynch's group, some 80% of PRCLs can be successfully treated by the quick, simple, inexpensive and painless use of ozone on root dentine carious lesions as an alternative to the conventional and painful "drilling and filling" management of PRCLs. To the best of my knowledge, this technique uses air and no medical oxygen to produce ozone. If I am correct, I am wondering how relevant the role of NOx generated with ozone is. This possible contamination imposes an extreme care to avoid breathing this gaseous mixture. I have heard that a new device, using only medical oxygen will deliver humidified ozone at higher concentration and therefore it will be more effective.

This technique has also stimulated the interest of the stomatologists. Indeed, around one in five people (frequent in children and women), or about ten million adults in the UK report each year the incidence of small, painful sores occurring on the tongue, lips and cheeks. Aphthous ulcers or, generically, cold sores have various aetiologies and tend to heal spontaneously in 8-10 days; there is not yet a good remedy to cure them and prevent recurrency. **I can foresee a great interest in developing ozone therapy with ozonated water and topical application of ozonated oil.** Since 1995, this treatment was proved to be very useful for treating herpetic lesions due to herpes virus type-I and II (Section I on herpetic infections). Moreover, for preventing recurrencies, the ozonetherapist CANNOT NEGLECT the use of very effective drugs (Acyclovir, Valacyclovir and Famciclovir) particularly for recurrent episodes or suppressive therapy (Corey et al., 2004; Crumpacker, 2004; Kimberlin and Rouse, 2004). **If the patient refuses these drugs, we can propose the use of ozone therapy: a cycle of minor ozonated AHT, presumably acting as an auto-vaccine is very effective and therefore I strongly recommend combining the parenteral and topical therapy.** Indeed the herpetic infection is not due to a simple local problem but is due to a recrudescence of the existing viral infection facilitated by immune depression caused by ageing, toxic drugs and various types of stress. The application of gas (oxygen and ozone only) appears to be more problematic for the risk of toxicity and should be used with utmost care. One possibility is to use a small sealing silicone cup on the lesion area and to insufflate the gas for a couple of minutes followed by the the suction of the residual gas. **However, the application of ozonated oil appears the most practical proposition and is rapidly risolutive as soon as the patient notes the typical prodromal symptoms.**

CONCLUSIONS: again, ozone has surprised us once more with its useful new applications in Dentistry and Stomatology. The obstinate opponents of ozone therapy should consider that this controversial gas can be intelligently and proficiently applied without procuring any side effect. However, in the case of a herpetic infection, the conscientious ozonetherapist cannot deceive the patient with the promise that a simple gas insufflation will be the "cure" but he must suggest the combination of the orthodox treatment with the parenteral and topical ozone therapy.

17. OZONETHERAPY IN COSMETOLOGY

Ironically, although ozonetherapy may eventually be accepted and used in important pathologies, in Italy, until recently it was mostly known for its application in cosmetology. This is due to the myopic vision of a few ozonetherapists, who have caused this approach to be discredited. This trend has been favoured during the last decade by the continuous opening of new beauty centres, making large profits. It is sad to think that, while every day in the world 600 million people are starving, in the so-called developed countries a huge amount of money is being spent to delay skin ageing or mask small imperfections. Since January 2002, the use of ozone in beauty centres has been correctly prohibited because it was performed by technicians without any medical qualification. Improper and excessive subcutaneous injection of gas is very dangerous!

There are two problems that mainly afflict women that require the attention of most ozonetherapists: one is the constantly increasing **obesity** and, particularly for aesthetic reasons, **localised lipomatosis**; the second is **chronic panniculitis**. The first problem can easily be prevented, in most cases, with an appropriate diet and healthy lifestyle. However, multiple symmetric lipomatosis is a real disease, found mainly in men. It is characterized by the formation of multiple lipomas, primarily present in the nape of the neck (Madelung collar) and in the supraclavicular, deltoid and abdominal regions. However, most women worry about localized layers of fat around the pelvis and on the thighs (steatopygic Venus). This excess of fat can now be removed in aesthetic medical centres by several techniques: surgery, but more frequently liposuction, carboxytherapy and ozonetherapy.

There is no doubt that ozone acts efficiently as a lipolytic agent because as soon as ozone dissolves in the interstitial water, lipids are the preferential substrate and they are broken down to a number of derivatives, such as lipoperoxides, hydroperoxides and small molecular weight LOPs.

The methodology is simple: injections of 2-4 ml O_2-O_3 (ozone concentrations range from 2-3 to a maximum of 5-6 mcg/ml) per site (abdomen, thighs, hips and gluteal areas) are carried out subcutaneously in the various areas as a mosaic, once a week. Five-eight sessions are generally sufficient to markedly and homogeneously dissolve the excessive fat. Using a disposable ozone-resistant (polypropylene, siliconated) 50 ml syringe, the gas can be applied in 10-25 sites at a time. Practical needles are the 26-27 G x 12 mm. During each session, **no more than 100 or 150 ml** (20-50 sites) **may be injected very slowly and with extreme care to avoid the risk of embolism**. Side effects may include a transitory slight burning sensation at the site of injection and occasional ecchymosis. After the treatment, the patient must rest for about 20 min and a gentle massage, possibly with slightly ozonated oil. may relieve possible pain. In Italy, apparently after receiving an excessive gas volume (up to 600 ml!) administerd via SC injections, three women died during the last five years.These episodes have been a backlash for ozone therapy and I appreciate that Prof. Cuccurullo, president of the National Health Committee, wrote that these deaths have been caused by malpractice or incompetence rather than to ozone. Indeed embolism is eventually caused by oxygen.

The total dose of ozone ranges from 200-2000 mcg/ml and does not elicit any toxicity and may give a sense of wellness. However, this aspect has not been evaluated. We have very successfully treated two male Madelung disease patients using the EBOO approach (Di Paolo et al., 2000). In discussing the therapy for HIV infection, I mentioned that a complication during HAART (due to protease inhibitors) is the appearance of lipodystrophy, so that there is a rational reason to use ozonetherapy in addition to HAART.

There are several types of pathological panniculitis. I would say that the least pathological is the chronic type, which today worries so many women who wish to remain sexually desirable. The etiopathogenesis remains unclear but hereditary factors, an excessively fat-rich diet, a sedentary life and smoking combine to produce an ugly cutaneous appearance (like an orange peel) on the thighs, hips and gluteal areas. It may start as a microvascular disturbance that slowly induces an uneven fibrosclerotic process, with intercellular oedema, frequent venous ectasis, occasional microhaemorrhages and abnormal lipocytes. It can be defined as an oedematous-fibro-sclerotic panniculitis (OFSP), according to Agostini and Agostini (1994). The skin is no longer smooth and the patient may report slight pain during palpation. It is a pathologic situation, which although not serious, embarrasses patients for its ugly appearance.

Ozonetherapy is performed with 20-40 SC injections of 2.5-4.0 ml gas each, respectively, for a total gas volume of 150 ml once a week for 5-8 weeks. **One must keep well in mind that gas volumes exceeding 20 ml represent a risk.** Depending on the stage of the panniculitis, the ozone concentration has been differentiated as: tough-type: 2 mcg/ml; soft-type: 1.5-2.6 mcg/ml; oedematous-type: 3-4 mcg/ml. However, the finesse of these details is superfluous, because I seriously doubt that cosmetologists have such precise ozone generators to select these concentrations. Most of them use portable generators of a firm that produces very poor quality apparatuses; they lack a photometric control and, even when new, produce very imprecise ozone concentrations. Every year at our course on ozonetherapy, several ozonetherapists come with their portable generators to check the real concentration on the basis of the iodometric method. Luckily, we always find far lower ozone concentrations than expected: 1-2 instead of 20 and 17-19 instead of 70 mcg/ml! I always tell them a true story: several years ago, after a lecture in which I had pointed out the serious problem of unreliability of ozone generators, one famous ozonetherapist working in Milan looked very worried. In a very reserved way, he asked me what might be the reason why, during the last year, he injected the gas as usual in many women but with no success at all. So I asked him: when did you last check your instrument? He said: I have never checked it! This means simply, I replied, that your generator does not produce ozone any longer and you inject only oxygen or air. He thanked me very profusely saying that I had saved his work just in time.

I have often said that ozonetherapy is vexed by several problems: the serious control and maintenance of generators is a crucial one and, only recently, after several warnings, some ozonetherapists have become aware of this. Health authorities do not understand and care about this problem either. Moreover, poor quality generators easily undergo corrosion and, if air mixes with oxygen, they may produce a very toxic mixture containing NOx.

Coming back to the treatment, **I insist that gas injections must be done very slowly with little pressure, taking care not to be inside a vein to avoid embolization.**

Always for cosmetic reasons, small superficial telangiectasis can be sclerotized by first blocking the blood flow and then slowly injecting 1-3 ml of gas (at high ozone concentration: 80 mcg/ml), remaining still for 30-60 sec. A compressive bandage must be left for one day. Almost needless to add, for the topical treatment of these unaesthetic features, there are many products prepared as gel or cream containing either ozonated oil or other substances, which are fairly effective and quite expensive.

CONCLUSION: therapy of panniculitis with ozone therapy has been popular in Italy but, owing to recent deaths, patients prefer now other approaches. It remains imperative that the ozonetherapist checks periodically his ozone generator and avoids injecting large volumes of gas.

18. MAY OZONE REPRESENT THE ELIXIR OF LIFE?

I thought better to end this chapter with a cheerful section discussing whether ozone may qualify as the elixir of life. In **"The Fountain of Youth",** Lucas Cranach painted a famous scenery (1546, State Museums Berlin), where crippled and old people, after reaching a pool, could bathe and swim in a magic water, which allowed them to reach the opposite side young, rejuvenated and ready to start a new life-cycle (Figure 21). **The ancient dream of overcoming the ageing process and extending life is today more actual than ever because well-off people, believing that the power of money is infinite, hope to buy extra time for our terrestrial life.** Everyone knows that the life expectancy in Europe has increased throughout the last century from an average of 47 to about 78. The advent of vaccines, antibiotics, anti-atherosclerosis drugs, vitamins, a low-fat and low-calorie diet rich in antioxidants, a regimen of moderate physical exercise and the avoidance of smoking and drinking have been the main factors in lengthening the life span and improving the quality of life.

Figure 21. In this beautiful painting, Lucas Cranach dreamt how mankind could overcome the ageing process by bathing in "The Fountain of Youth (1546, State Museums Berlin).

There has been an increasing knowledge of the ageing process (Harman,1956; Youngman et al., 1992; Ames et al., 1993; Beckman and Ames, 1998; Cadenas and Davies, 2000; Hamilton et al., 2001; Ames, 2004), and we have become aware that **chronic oxidative stress, the formation of advanced glycosylation end substances (AGES), shortening of telomeres, chronic exposure to pollutants, a stressful lifestyle, the physiologic decline of circulating hormones and immunological defences are all factors that, to different extents, play a role in ending life** (Sohal et al., 2002; Sastre et al., 2003).

During the last three decades, the theory that hormonal decline may be an important cause of ageing has gathered momentum, with the postulation that hormonal replacement may result in rejuvenating frail or ailing elderly people (Seeman and Robbins, 1994). Thus numerous hormones have been proposed and variably tested: **estrogen,** which produces numerous benefits in post-menopausal women (Grady et al., 1992; Peterson, 1998); **growth hormone** (Rudman et al., 1990); **dehydroepiandrosterone (DHEA)** and DHEA-sulphate, a sort of mother steroid (Bilger, 1995; Baulieu and Robel,

1998); **melatonin** (Reiter, 1991) and, last but not least, **testosterone** for androgen deficiency (Morley and Perry III, 2000) in ageing men.

Many experiments have been conducted in rodents, frequently using very high doses of hormones. However, it remains unclear whether the results obtained in these non-primate models can be extrapolated to human beings, also because rodents often have a different hormonal pattern from man. Several studies in humans have shown beneficial actions of some hormones: prevention of osteoporosis, improvement of memory and of the HDL/LDL ratio due to estrogen; increased energy and sex drive during testosterone replacement therapy; an apparent improvement of mental activities after DHEA, promoted to the role of a neurosteroid. Nevertheless, improvement of the quality of life is not a consistent finding and many questions remain to be explored, mostly because **long-term therapy may be associated with serious adverse effects.** Another problem is that, in order to achieve striking results, enthusiastic clinicians tend to administer pharmacological doses of a single hormone, thus possibly disrupting the physiological equilibrium with unforeseeable consequences. Moreover it remains unclear what is the optimal method of hormone replacement, although **slow-release patches and creams are probably better than oral administration or injection.** Without minimizing the importance of this approach, I must conclude that we have not yet reached the stage of an equilibrated and optimised exogenous therapy, which is conceptually difficult to individualize.

The justification of this prologue can be found in the following question: is there any possibility of inducing a harmonious and useful release of hormones and how might this can be achieved?

Throughout the book, **I have reported that most patients report a feeling of euphoria and a sense of wellness after ozonetherapy.** Is this simply due to faith in this medical treatment (the power of the mind!), in other words the power of the placebo effect (Benson and Friedman, 1996), or are the generated messengers actually able to modify the secretion and allow an orderly release of several hormones? If only we had enough money to pay ten volunteers and the testings, we could have answered this question a long time ago; indeed it would not be too difficult to evaluate, before and after ozonated autohaemotherapy, the complete hormonal pattern and cycling in the plasma throughout the day. This study would be very enlightening and might help to understand why the patient feels better after ozone therapy and to identify the best time of the day to perform it.

During the last two years, I have been able to examine a rather specific questionnaire distributed to ARMD men in the range of 67 to 78 years old. I could calculate that 47% of the younger patients (67-73 years old) reported an improvement of sexual desire and performance during and immediately after undergoing a cycle of 15 treatments of ozonated AHTs. This result is in line with a previous observation in a few pre-terminal vasculopathic patients, who informed us that, after a few EBOO treatments, they noticed a return of early morning penile erection. This may be due to improved oxygenation or/and enhanced DHEA secretion and is certainly preferable to fashionable pharmacological vasodilators.

Another thing that has always puzzled me is why and how ozonetherapy relieves pain. Is this due to a release of cortisol or is ozonetherapy able to enhance the effects of some endogenous neurotransmitters such as serotonin and dopamine, similar to the effects of endorphins observed after intense physical exercise (Viru and Tendzegolskis, 1995).

It has been postulated (Chapter 4) that ozonetherapy can paradoxically strengthen the antioxidant defences against a transitory and controlled oxidative stress. We have now good evidence that this hypothesis is correct. The exciting possibility is that, by performing two brief cycles (6-8 treatments per cycle) of ozonetherapy each year (around March and October), we may be able to delay ageing. In such a case low doses of ozone should be used for either ozonated – AHT (15-30 mcg/ml) or RI (5-15 mcg/ml) or BOEX (0.2-1 mcg/ml). One cycle of infusions of the "gluco-peroxide" solution at 0.03-0.09% concentrations every semester may also be a useful option. While I remain uncertain whether RI can perfectly substitute the ozonated AHT in chronic limb ischemia patients, I admit (on the basis of my own experience) that also patients undergoing frequent rectal insufflations report a feeling of euphoria and an increased stamina. **It is well known that the gut has an extensive neuronal system (our second brain!) releasing the bulk of serotonin and it is possible that the rectal insufflation of ozone enhances its release**.

Ageing is a multifactorial process and consequently administration of a single hormone, while temporarily beneficial, is unlikely to be useful in the long run. Longevity, and even better **"longevity free from disability and functional dependence"** as Hayflick (2000) has written, **may be more rationally achieved by the yearly repetition of a gentle, yet paradoxical, treatment like ozonetherapy, which is probably able to simultaneously reactivate several functions, such as antioxidant defences, T-cell mediated functions, the network of enzyme repair, a sustained and balanced hormonal and neuro-transmitter release, with the inherent benefits of more energy, improved mood and memory, prevention of cancer and atherosclerosis, and retention of sexual activity. However, for me to state that ozone will represent the eternal "fountain of youth"** (as it was hyped for melatonin) and that it will prolong the life-span by some 15-20 years so as to have an extra decade of a good and productive life will be necessary to have acquired clinical evidence in at least 10,000 people.

While I wish to everyone to have a long and happy life, I am also thinking that the earth already hosts 6 billion people and it is far better to give space and opportunity to young ENTERPRISING PEOPLE RATHER than to maintain too many almost mummified centenarians.

CONCLUSIONS: During the last three decades, the affluent society has frantically tried to remain beautiful and preserve a good health for a longer time. Interestingly, in a few villages, almost secluded in rural areas of the globe, clones of centenarians have been described in the medical literature (Mecocci et al., 2000). **Firstly, these people can thank their genes and then surely an unstressful life associated to a moderate, if not limited, dieting. After all, it has been well demonstrated that rats, kept for life to a low- caloric intake, live longer than controls fed *ad libitum.* The evaluation of the metabolic profile of 18 men and women who had been on self-imposed caloric restriction for 3-15 years is truly remarkable: it has shown significant beneficial effects on the major atherosclerosis risk factors and a decrease of inflammation** (Fontana et al., 2004). **Yu (1996) had also stressed the relevance of a dietary restriction for reducing oxidative stress and prolonging the life-time.**

Besides genes, which at the moment cannot be safely modified or substituted, today we can today try to prolong our life-time with a moderate, well-balanced diet, a daily physical exercise, a correct life-style, supplementary (but not excessive) antioxidants and, when necessary, good drugs for preserving the efficiency of the cardiovascular system. Prevention is the key of success. Exogenous administration of hormones can certainly yield an illusory period of youth but, in the long run, may have a boomerang effect. I would dare to say that for people closely observing the rules of prevention, ozone therapy may be helpful because ozone detains several fundamental requirements for maintaining active or revitalize critical physiological functions

GENERAL CONCLUSIONS FOR CHAPTER 9

Clinical results so far available have been objectively discussed showing that ozonetherapy is often more useful than orthodox treatments in a FIRST category of diseases such as:

1) Osteomyelitis, pleural empyema, abscesses with fistulae, infected wounds, bed sores, chronic ulcers, diabetic foot and burns.

2) Advanced ischaemic diseases (hind-limb ischemia and heart ischemia).

3) Age-related macular degeneration (atrophic form).

4) Orthopaedic diseases and localized osteoarthrosis.

5) Chronic fatigue syndrome and fibromyalgia.

6) Dentistry regarding primary root carious lesions, particularly in children.

7) Stomatology for chronic or recurrent infections in the oral cavity.

For these pathologies ozone is a real “wonder” drug.

In a SECOND category of diseases including:

1) Acute and chronic infectious diseases, particularly due to antibiotic or chemoresistant bacteria, virus and fungi (hepatitis,

herpetic infections and herpes zoster, papillomavirus infections, onychomycosis and candidiasis, giardiasis and cryptosporidiosis) and

2) Cancer-related fatigue, ozone therapy, associated with orthodox treatments, accelerates and improves the outcome.

There is a THIRD category of serious diseases such as:

1) Autoimmune diseases (multiple sclerosis, rheumatoid arthritis, Crohn's disease, psoriasis).

2) Senile dementias.

3) Pulmonary diseases (emphysema, asthma, chronic obstructive pulmonary disease, idiopathic pulmonary fibrosis and acute respiratory distress syndrome).

4) Skin diseases (psoriasis and atopic dermatitis).

5) Metastatic cancer.

6) Severe sepsis and multiple organ dysfunction,

where the combination of orthodox treatments and ozone therapy, at least on theoretical ground, may be helpful but clinical evidence is lacking. Whether ozone therapy with the advantages of low cost and no adverse effects, may equal the efficacy of current conventional treatments remains to be explored. I am doubtful, however, how and when we will be able to perform these investigations standing the actual situation of total disinterest of Health Authorities, lack of specific sponsors and the overwhelming power of pharmaceutical industries, which are only interested in pursuing their objectives. Ironically, it is possible that less developed countries with minimal budgets may have an interest in performing pilot trials that can give us precious infomations regarding the usefulness of ozone therapy.

I need to mention a FOURTH category of diseases such as retinitis pigmentosa, sudden hearing loss and tinnitus where ozone therapy has not yielded therapeutic results.

The next table reports tentative guidelines regarding ozone concentrations within the therapeutic window to be used in different pathologies with the classical ozonated AHT, twice weekly. Ozone concentrations are slowly upgraded no more than 5 mcg/ml at a time, to achieve the adaptation to COS in 2-3 weeks.

	PROPOSED O_3 CONCENTRATIONS (mcg/ml per ml of blood)	
	initial	final
Infectious diseases	20-25	70
Vascular diseases	20	40
Degenerative diseases	20	30-40
Respiratory diseases	10	30-40
Autoimmune diseases	50	80
Metastatic tumours	25	70-90

From the examination of the table, two facts emerge: firstly, the idea "more is better" is not always appropriate for ozone and its concentration must be calibrated in relation to the effector and target cells; secondly, the need for further experimentation with appropriate controls to generate definitive clinical data.

Clinical trials are demanding enterprises that require a concerted effort by official Medicine and government authorities. National Health Authorities, which are always complaining about the increasing costs of medical assistance, could have an economical advantage if ozonetherapy was widespread and organized in a systematic way in all public hospitals. Although I have no hard data to support my contention, I am convinced that the benefit of ozone therapy does outweigh its cost, particularly for the above mentioned first category of diseases. In a public hospital, as an example, ten nurses, under the supervision of an ozonetherapists could easily perform the therapy in about 15 patients per hour. As things are today, it is depressing to realize that ozone therapy will not be applied in public hospitals for years to come, thus depriving many patients of the possibility of restoring their health.

Chapter 10

THE DILEMMA BETWEEN HYPERBARIC OXYGEN THERAPY (HOT) AND OZONE THERAPY.

HOT is better known than ozone therapy because it is considered an orthodox approach and is widely used in the USA. This explains why many physicians and the layman often ask me if ozone therapy is a sort of HOT.

The latter is a medical procedure by which 100% medical oxygen (Kindwall, 1993; Tibbles and Edelsberg, 1996; Leach et al., 1998; Cianci, 2004) is delivered at 2-3 times (usually 2.6) the atmospheric pressure (1 atmosphere = 760 mmHg) at sea level. In physiological conditions, at this level with normal air, the pO_2 in the alveolar space (O_2:14%) is equivalent to 100 mmHg and the pO_2 of arterial blood is about 98 mmHg; Hb is fully saturated to Hb_4O_8 and there is about 0.3 ml per decilitre of O_2 solubilized in the plasma. Tissues at rest extract from blood an average of about 25% O_2 (i.e. 5-6 ml of O_2/dL), so that venous blood has a pO_2 of about 40 mmHg and Hb_4O_8, having released at least one molecule of O_2, becomes Hb_4O_6. Thus the amount of O_2 physically dissolved in the plasma is grossly insufficient for the requirements of the tissues and the necessary 5.5 ml of oxygen derive from deoxygenation of Hb_4O_8. **In the hyperbaric chamber, administering 100% O_2 at 3 atmospheres, the O_2 solubilized in plasma is as much as 6 ml/dL and the Hb is fully saturated with oxygen.** In this situation, the dissolved O_2 content is sufficient to satisfy the cellular requirements and Hb_4O_8 hardly release any oxygen.

Rapid decompression (say from 4-5 to 1-2 atmospheres) causes decompression sickness due to nitrogen dissolved in plasmatic water, which suddenly forms inert gas bubbles that cause disseminated embolism. The diver can be saved if rapidly placed in the hyperbaric chamber, because during slow decompression the nitrogen is replaced by oxygen and slowly expired while the oxygen is metabolized by the tissues.

Carbon monoxide (CO) poisoning is a cause of death all over the world (Ernst and Zibrak, 1998) due to the fact that CO binds to Hb with an affinity 240 times that of oxygen. In the presence of CO, the oxyhaemoglobin dissociation curve shifts to the left and changes to a more hyperbolic shape, with the result of impaired release of oxygen at the tissue level, where CO also binds to myoglobin.

The hyperbaric chamber can save the intoxicated subject by delivering oxygen dissolved in the plasma to anoxic tissues and by accelerating the dissociation of COHb: its half-life decreases from about 300 min while air is breathed, to about 20 min with hyperbaric 100% oxygen. Moreover, HOT allows the dissociation of CO from cytochrome C oxidase, thus improving the cellular energy state. **The immediate administration of normobaric oxygen to a CO-intoxicated patient** is certainly useful, because the half life of CO-Hb is only about 60 min and tissue oxygenation is improved, **but it is not as effective as HOT.**

On rare occasions, **haemorrhagic shock may cause intensive anaemia,** unable to satisfy the metabolic demands of tissues: if suitable blood is not available or blood transfusion is not allowed for religious reasons, **HOT may temporarily compensate for the lack of erythrocytes.** These three examples suffice to illustrate **the unique importance of HOT.**

Adverse effects are rare and partly due to typical oxygen toxicity (optic symptoms in about 20% of patients), which can be prevented by administration of antioxidants and by shortening the period of hyperoxia (DuBois, 1962). In addition to **the high cost of installing a HOT facility, the oxygen presents a fire hazard.** Indeed, in the last decade, owing to incompetence and negligence, there have been two tragic explosions in Italy: one in Naples in a single-place chamber and another in Milan in a multi-place chamber with several deaths. These accidents should never occur, as the chamber should be regularly filled with inert air. In comparison, **oxygen-ozone therapy does not present risks, unless a mad ozonetherapist directly injects the gas IV, a procedure that is prohibited.** Moreover, **the cost of the material for ozonetherapy is almost negligible.**

There are fundamental differences between HOT and ozonetherapy. Although the bulk of the gas mixture is represented by 95-99% oxygen, ozonetherapy does not aim to oxygenate blood directly. Indeed, with all the procedures (AHT, EBOO, BOEX and RI), the arterial pO_2 hardly increases in vivo. Yet **if ozone is used properly, it has many virtues:** disinfectant and immunomodulatory (cytokine release) activities, increased delivery of oxygen to hypoxic tissue through vasodilatation ($NO^•$, CO) and possibly a shift of the HbO_2 dissociation curve to the right (the venous pO_2 may fall to 20 mmHg), release of growth factors (PDGF, TGF-β1, etc.) thus enhancing tissue healing, possibly hormonal release due to a sudden homeostatic change and/or a placebo effect and, most importantly, a generalized metabolic improvement with enhancement of the antioxidant defence.

Another significant difference is that **ozonetherapy induces fairly long-lasting and interconnected metabolic changes, while the effects of HOT, being due mainly to a transitory oxygen hyperconcentration, are of shorter duration.** Interestingly, increased DNA damage was detected immediately at the end of the first HOT, while no effect was found one day later (Dennog et al., 1996). They also suggested that HOT, under the same conditions, may increase antioxidant defences. This suggestion is now supported by interesting experimental data (Kim et al., 2001). Cianci (2004) has provided evidence that HOT ie, oxygen favours cell replication *in vitro* and wound healing *in vivo*. The finding of significant oxidative base damage after the first HOT treatment reinforces my conviction that ozonetherapy should always start with a very low dose followed by a gradual increase to minimize any possible damage.

An objective comparison of the therapeutic efficacy of HOT versus ozonetherapy is not possible, mostly because valid RCTs of ozonetherapy are few and small, while there are many publications dealing with HOT. However, even though as many as 64 different disorders seemed to be improved with HOT, in most of them the evidence to warrant its clinical use was insufficient (Kindwall, 1993). There is only one paper comparing rheological parameters (but not clinical efficacy) between HOT and ozonetherapy: Verrazzo et al. (1995) claimed that only the latter approach caused a significant increase of erythrocyte filterability and a decrease of blood viscosity. On the basis of our data, these results need to be confirmed.

In Table 9, I attempt to summarize the diseases for which either HOT or ozonetherapy are used and to express an opinion, based on personal experience and not on hard data, about which of the two approaches seems more beneficial.

Table 8. Diseases for which HOT and ozonetherapy are used.

	HOT	OZONETHERAPY
1) Arterial gas embolism	+++	---
2) Decompression sickness	+++	---
3) Severe CO poisoning and smoke inhalation	+++	---
4) Severe blood-loss anaemia	+++	---
5) Clostridial myonecrosis (gas gangrene)	+++	++
6) Compromised skin grafts and flaps	+	+++
7) Prevention of osteo-radionecrosis	+	+++
8) Radiation damage	+	+++
9) Refractory osteomyelitis	+	+++
10) Necrotizing fascitis	+	+++
11) Traumatic ischaemic injury	+	+++
12) Thermal burns	+	+++
13) Chronic ulcers and failure of wound healing	+	+++
14) Multiple sclerosis	---	+?
15) Chronic fatigue syndrome	+	++
16) HIV-AIDS	+?	+
17) Senility	+	++

Legend : + little, ++ modest, +++ good activity, --- no activity

It may seem that I favour ozonetherapy and the reason is that, in some affections, ozonetherapy is very effective. In most cases, we can apply both parenteral administration, in the form of AHT, EBOO, BOEX and RI, and topical application, either as a gas mixture (bagging and dynamic insufflation) or ozonated water and oil. **The combination favours an incredible synergic effect, which acts on several targets.** Indeed this explains the efficacy of ozonetherapy where there are several components at work simultaneously (infection, inflammation, cell necrosis, ischaemia, dysmetabolism, impaired healing, etc.). Several of these affections have been discussed in the previous chapter (Sections I and VII).

Bevers et al. (1995) proposed HOT (20 sessions at 100% O2 at 3 bars for 90 min) for patients with severe radiation-induced haematuria. Dr. R. Dall'Aglio informed me to have solved this problem with only three intravesical applications of ozone gas (once weekly!).

HOT was proposed for patients with AIDS (Bocci, 1987a) and a subsequent study showed a transitory improvement of the quality of life ("Hyperbaric Oxygen Therapy for the Treatment of Debilitating Fatigue Associated with HIV/AIDS", Janac, vol. 4(3), July-September, 1993). **There is no doubt that HOT has a precise and unique rationale in affections no. 1 to 5. In all other diseases, the use of HOT is not well supported and the risk of transferring the patient, who often lives far away from the site of the chamber, discourages its use.**

The purpose of this chapter was to clarify that ozonetherapy is very versatile, practical, inexpensive, without side effects and quite beneficial in several affections. I would like to believe that orthodox physicians, rather than being biased against ozonetherapy, simply do neither know about it, nor how to perform the therapy.

CONCLUSIONS: the reader may find useful the objective comparison between OHT and ozone therapy. In my opinion, both approaches are important and basically use oxygen as the vital element for maintaining life and activating wound healing. However, while HOT uses oxygen under pressure, ozone therapy uses ozone as the compound able to generate messengers crucial for activating several biological functions. This fact DEEPLY differentiates their practical applications and, in order to maximize their usefulness, either HOT or ozone therapy must be used within their specific fields.

Chapter 11

THE PROMISING FUTURE OF OZONE THERAPY IN MEDICINE.

In this final chapter, I will try to ponder on the future of ozone therapy. The potent disinfectant activity of ozone against anaerobic bacteria was utilized during the World War I but, for the next six decades, there was no progress, which came only thanks to a few clinicians, who guessed its usefulness. A major advance came with the work of Dr. H. Wolff (1927-1980) and an Austrian surgeon, Dr. O. Rokitansky, who, in an empirical way, showed the efficacy of the ozonated autohaemotherapy in avoiding limb amputation in patients with chronic limb ischemia. However the lack of basic research and randomized clinical trials relegated ozone therapy in the field of complementary medicine with a few and nebulous ideas of how ozone could act.

Meantime, three negative aspects came about: **the first was the general awareness that ozone is a strong oxidant and a toxic gas for the respiratory tract never to be breathed.**

The second was the relevance of free radicals as determinants of ageing and of several human diseases and the knowledge that ozone is a master generator of free radicals. Even today this remains the easy objection raised by scientists and physicians, who do not know the progress that has been made on the biochemistry and pharmacology of ozone therapy.

The third problem arose with the spread of HIV and AIDS infection due to the lack of an appropriate therapeutic control until 1996 when, at long last, virologists understood the need to attack simultaneously the virus with a combination of different drugs (the HAART). In the early 90s, quacks around the world begun to inject the gas mixture oxygen-ozone directly into the blood stream, naively believing to disinfect blood, like dirty water flowing in an aqueduct. What is worse is that they exploited the desperate patients and claimed to "cure" the infection hiding the deleterious effect of the pulmonary embolism and possibly of a few deaths. It was easy and correct for orthodox medicine to condemn ozone therapy and these unforgivable mistakes did almost entomb ozone therapy. However in Cuba, owing to the embargo and lacks of medical drugs, by sheer necessity, a group of enterprising physicians started to use ozone in a meaningful way in several diseases confirming that ozone could be medically useful.

By pure serendipity, in 1988, we started our project and we tried to unravel the mechanisms of action when ozone dissolves in blood, hoping to explain the controversy between the too many opponents and the few proponents of ozone therapy. It has not been an easy job but we have started to see a faint light at the end of the tunnel. We were and are well aware of the intrinsic toxicity of ozone: any chemical compound can be a drug or a toxin and we realized the importance to differentiate the therapeutic dose from the toxic one. **Today we have clearly ascertained that OZONE RAPIDLY DISSOLVES in the water of plasma and biological fluids, IMMEDIATELY REACTS WITH BIOMOLECULES, GENERATES CRUCIAL MESSENGERS AND DISAPPEARS.** We know that the ozone-ROS-LOPs signalling cascade is not yet definitive and some aspects remains to be elucidated but it is clear that, among complementary approaches, **ozone therapy has emerged as the one that is well explainable with classical biochemical, physiological and pharmacological knowledge.** After some fifteen years, I feel that confused and wrong ideas have been dispelled and **this book presents the real first comprehensive framework for understanding and recommending ozone therapy**

Since 1992, we wanted to start clinical investigations and we realized how the scepticism and diffidence against ozone therapy was diffused in the academic world. The FDA, for several good reasons, had to prohibit the use of ozone in the USA. However, one reason was and still is based on the dogma that **"ozone is always toxic and should not be used in medicine".** *This is an absurd and antiscientific idea and today we have a million reasons for saying that it is totally wrong.* It is disappointing that some influent American scientists still BELIEVE that is correct. The FDA decision has negatively influenced the Health Authorities of other countries and this fact is not surprising because today only a few super-developed countries have a dominant (and not necessarily always positive) influence over the world's medical resources.

I still have to answer the question of the future of ozone therapy in medicine. As slowly we move on and explore this approach in new diseases, we are surprised to note the breadth of action of ozone and the lack of toxicity against the blackest prediction. Unfortunately lack of resources and of an efficient international organisation impede a rapid progress of basic and clinical researches. However the discovery that, paradoxically, ozone therapy can induce an adaptation to the chronic oxidative stress by upregulating the antioxidant system, and favour the release of oxidative stress proteins and probably of staminal cells suggest that ozone exerts multiform activities and has the capability of restoring health by reactivating wrecked biological functions.

As far as therapeutic activity is concerned (concisely summarized in the general conclusions of Chapter 9), it is a complex matter and there are relevant differences depending on the type of pathology. Indeed, against the sarcastic comment that ozone therapy is a *panacea*, we have clear evidence that for several diseases, ozone therapy represents only a useful approach, which must be combined with conventional therapy to achieve the best results. Moreover, as it was expected, **ozone therapy has failed to yield a result in retinitis pigmentosa, tinnitus and sudden hearing loss.**

This is a good opportunity of making a plea for exerting maximal objectivity and honesty: the competent ozonetherapist must present all possible options to the patient, who has the right to choose the treatment when she/he is fully informed about pros and cons of both conventional and orthodox treatments. Sheldon (2004) reported that the Netherlands, a very liberal and democratic nation, will crack down on six practitioners of complementary medicine after government health inspectors severely criticised the treatments offered to the brilliant actress Sylvia Millecam, who died of breast cancer. Apparently, although mainstream care was available, it seems that Sylvia was abducted to receive electroacupuncture, faith healing, salt therapy and psychic healing instead of a more appropriate therapy that may have procured a cure or a prolonged survival. Ozone therapy has been in the past already defamed with the label of dangerous quackery and today we do not want to deserve that label.

On the other hand, ozone therapy is extremely valid, often more than orthodox treatments, in vascular ischaemic diseases (caused by atherosclerosis, diabetes, uremia, smoking, etc,) and for HEALING chronic wounds, bed sores, chronic ulcers (the diabetic foot), burn injuries, intractable fistulae and an array of skin, mouth, vaginal and rectal infections. Ozone therapy is the only treatment that can restore some visual acuity in patients with the atrophic form of age-related macular degeneration. For all of these affections, ozone is a real "wonder" drug and it is even more wonderful because free of adverse effects and actually capable of generating a feeling of wellness and euphoria. Ironically, the highest percentage of patients with these diseases lives in countries obstructing ozone therapy.

I am absolutely convinced that the **the combination of parenteral ozone therapy carried out, when necessary, with the topical one (ozonated water and oil), in due time, will mark a medical revolution.** It remains difficult to foresee when it will happen because the pace of our research, in comparison to official medicine supported by colossal fundings, is too slow.

William James brilliantly described three famous phases characterizing new theories. It appears unavoidable that these blunders occur, from time to time, in the Sciences:

1) The new theory is attacked and declared absurd. We are at this phase!

2) Then it is admitted that it is true and OBVIOUS, but insignificant.

3) To the end, it is recognized the real importance and its detractors demand the honour to have discovered it.

We should not get discouraged and continue to work in spite of the antagonism and negligence of Health Authorities. I regret to say that prestigious scientific journals (FRBM and NEJM) have not given me the chance of opening a dialogue. Recently, the novelty that ozone may be produced in vivo and be responsible for athcrosclcrosis has been amply divulgated but **my letter stating that "ozone is NOT always toxic" was not published.** Similarly **the WHO Bulletin**, which should be responsible for health care of everyone, **has just rejected one of my recent reviews, where, provocatively, I discussed: "Why WHO does not promote the use of ozone therapy"?**

The antagonism of Health Authorithies is responsible for delaying the application of ozone therapy to billion of patients and we must do the maximal effort to break this situation. It may seem absurd but there is a hope that oxygen-ozone therapy will quickly extend in all the hospitals of poor or less developed countries before being recognized as a valid tool by the most advanced nations.

August 22nd 2004

REFERENCES

Abe, H., Ikebuchi K., Shimbo M., and Sekiguchi S., 1998, Hypotensive reactions with a white cell-reduction filter: activation of kallikrein-kinin cascade in a patient, *Transfusion* **38**:411-412.

Abraham, N. G., Drummond G. S., Lutton J. D., and Kappas A., 1996, The biological significance and physiological role of heme oxygenase, *Cell. Physiol. Biochem.* **6**:129-168.

Aejmelaeus, R. T., Holm P., Kaukinen U., Metsä-Ketelä T. J. A., Laippala P., Hervonen A. L. J., and Alho H. E. R., 1997, Age-related changes in the peroxyl radical scavenging capacity of human plasma, *Free rad. Biol. Med.* **23**:69-75.

Age-related Eye Disease Study Research Group, 2001, A randomized, placebo-controlled, clinical trial of high-dose supplementation with vitamins C and E, beta carotene, and zinc for age-related macular degeneration and vision loss: AREDS report no. 8, *Arch. Ophthalmol.* **119**:1417-1436.

Agostini, G., and Agostini S., 1994, Contributo alla conoscenza e al trattamento della parmiculopatia edemato-fibro-sclerotica, in *Proceedings: VII National Meeting of Ozonetherapy, Roma, 1994*.

Agus, D. B., Vera J. C., and Golde D. W., 1999, Stromal cell oxidation: a mechanism by which tumors obtain vitamin C, *Cancer Res.* **59**:4555-4558.

Aicher, A., Heeschen C., Mildner-Rihm C., Urbich C., Ihling C., Technau-Ihling K., Zeiher A.M., and Dimmeler S., 2003, Essential role of endothelial nitric oxide synthase for mobilization of stem and progenitor cells, *Nat. Med.* **9**: 1370-1376.

Aird, W. C., 2003, The role of the endothelium in severe sepsis and multiple organ dysfunction syndrome, *Blood* **101**: 3765-3777.

Aitken, C., and Jeffries D. J., 2001, Nosocomial spread of viral disease, *Clin. Microbiol. Rev.* **14**:528-546.

Akaike, T., Suga M., and Maeda H., 1998, Free radicals in viral pathogenesis: molecular mechanisms involving superoxide and NO, *Proc. Soc. Exp. Biol. Med.* **217**:64-73.

Akdis, C. A., and Blaser K., 2001, Mechanisms of interleukin-10-mediated immune suppression, *Immunology* **103**:136.

Akdis, C. A., Blesken T., Akdis M., Wüthrich B., and Blaser K., 1998, Role of interleukin 10 in specific immunotherapy, *J. Clin. Invest.* **102**:98-106.

Akey, D., and Walton T. E., 1985, Liquid-phase study of ozone inactivation of Venezuelan Equine Encephalomyelitis virus, *Appl. Environ. Microbiol.* **50**:882-886.

Al Dalain, S. M., Martinez G., Candelario-Jalil E., Menendez S., Re L., Giuliani A., and Leon O. S., 2001, Ozone treatment reduces markers of oxidative and endothelial damage in an experimental diabetes model in rats, *Pharmacol. Res.* **44**:391-396.

Al Sa'doni, H., and Ferro A., 2000, S-Nitrosothiols: a class of nitric oxide-donor drugs, *Clin. Sci. (Colch.)* **98**:507-520.

Alexander, H. R., Jr., 2003, Hyperthermia and its modern use in cancer treatment, *Cancer* **98**:219-221.

Alexandre, A., and Fumo G., 1998, Discolisi percutanea mediante O_2O_3 nell'ernia discale lombare, in *Lombalgie e lombosciatalgie. Criteri di diagnosi e cura* (F. Ceccherelli, and A. Ricciardi, Eds.), Edizioni Libreria Cortina, Torino, pp.367-377.

Alexandre, A., Buric J., Corò L., Rigobello L., and Scopetta S., 2000, Discolisi percutanea mediante O_2O_3 intradiscale, in *Proceedings: I Congresso IMOS, Italia, Siena, 2-4 novembre 2000*, pp.7-8.

Alexandre, A., Buric J., Paradiso R., Salgado H., Murga M., Corò L., Albarreal A., Scopetta S., Giocoli H., and Marin F., 2002, Intradiscal injection of O_2-O_3 to treat lumbar disc herniations: results at five years, *Rivista Italiana Di Ossigeno-Ozonoterapia* **1**:165-169.

Alexandre, A., Pentimalli L., Rigobello L., and Corò N., 1999, Amaurosi fugax in un caso di discolisi cervicale mediante O_2O_3, in *L'Ozonoterapia nel 2000* (F. Ceccherelli, and F. Giron, Eds.), Edizioni Libreria Cortina, Torino, pp.141-144.

Allain, T. J., Bearn J. A., Coskeran P., Jones J., Checkley A., Butler J., Wessely S., and Miell J. P., 1997, Changes in growth hormone, insulin, insulinlike growth factors (IGFs), and IGF-binding protein-1 in chronic fatigue syndrome, *Biol. Psychiatry* **41**:567-573.

Amato, G., 2000, Uso dell'ozonoterapia mediante grande autoemotrasfusione nella terapia dell'angina abdominis, in *Proceedings: I Congresso IMOS, Italia, Siena, 2-4 novembre 2000*, p.10.

Amato, G., Sacchetta A., Borrelli E., and Bocci V., 2000, Ruolo dell'ozonoterapia mediante grande autoemotrasfusione nel trattamento delle epatiti croniche post-epatite virale (II parte), in *Proceedings: I Congresso IMOS, Italia, Siena, 2-4 novembre 2000*, p.11.

Amersi, F., Buelow R., Kato H., Ke B., Coito A. J., Shen X. D., Zhao D., Zaky J., Melinek J., Lassman C. R., Kolls J. K., Alam J., Ritter T., Volk H. D., Farmer D. G., Ghobrial R. M., Busuttil R. W., and Kupiec-Weglinski J. W., 1999, Upregulation of heme oxygenase-1 protects genetically fat Zucker rat livers from ischemia/reperfusion injury, *J. Clin. Invest* **104**:1631-1639.

Ames, B. N., 2004, A role for supplements in optimizing health: the metabolic tune-up, *Arch. Biochem. Biophys.* **423**:227-234.

Ames, B. N., Shigenaga M. K., and Hagen T. M., 1993, Oxidants, antioxidants, and the degenerative diseases of aging, *Proc. Nat. Acad. Sci. USA* **90**:7915-7922.

Anderson, C., 1992, Gene therapy researcher under fire over controversial cancer trials, *Nature* **360**:399-400.

Anderson, M. M., Hazen S. L., Hsu F. F., and Heinecke J. W., 1997, Human neutrophils employ the myeloperoxidase-hydrogen peroxide-chloride system to convert hydroxy-amino acids into glycolaldehyde, 2-hydroxypropanal, and acrolein, *J. Clin. Invest.* **99**:424-432.

Andreula, C.F., Simonetti L., De Santis F., Agati R, Ricci R., and Leonardi M., 2003, Minimally Invasive Oxygen-Ozone Therapy for Lumbar Disk Herniation, *AJNR Am J Neuroradiol*, **24**: 996-1000.

Angelucci, E., Brittenham G. M., McLaren C. E., Ripalti M., Baronciani D., Giardini C., Galimberti M., Polchi P., and Lucarelli G., 2000, Hepatic iron concentration and total body iron stores in thalassemia major, *N. Engl. J. Med.* **343**:327-331.

Angelucci, E., Muretto P., Lucarelli G., Ripalti M., Baronciani D., Erer B., Galimberti M., Giardini C., Gaziev D., and Polchi P., 1997, Phlebotomy to reduce iron overload in patients cured of thalassemia by bone marrow transplantation. Italian Cooperative Group for Phlebotomy Treatment of Transplanted Thalassemia Patients, *Blood* **90**:994-998.

Antonelli, G., Bagnato F., Pozzilli C., Simeoni E., Bastianelli S., Currenti M., De Pisa F., Fieschi C., Gasperini C., Salvetti M., and Dianzani F., 1998, Development of neutralizing antibodies in patients with relapsing- remitting multiple sclerosis treated with IFN-beta1a, *J. Interferon Cytokine Res.* **18**:345-350.

Ardizzone, S. and Bianchi Porro G., 2002, Inflammatory bowel disease: new insights into pathogenesis and treatment, *J. Intern. Med*, **252**: 475-496.

Argiles, J. M., Moore-Carrasco R., Fuster G., Busquets S., and Lopez-Soriano F. J., 2003, Cancer cachexia: the molecular mechanisms, *Int. J. Biochem. Cell Biol.* **35**:405-409.

Aris, R. M., Christian D., Hearne P. Q., Kerr K., Finkbeiner W. E., and Balmes J. R., 1993, Ozone-induced airway inflammation in human subjects as determined by airway lavage and biopsy, *Amer. Rev. Respir. Dis.* **148**:1363-1372.

Arnason, B. G. W., 1993, Interferon beta in multiple sclerosis, *Neurology* **43**:641-643.

Arvin, A. M., and Prober C. G., 1997, Herpes simplex virus type 2 - a persistent problem, *N. Engl. J. Med.* **337**:1158-1159.

Aslan, M., Ryan T. M., Adler B., Townes T. M., Parks D. A., Thompson J. A., Tousson A., Gladwin M. T., Patel R. P., Tarpey M. M., Batinic-Haberle I., White C. R., and Freeman B. A., 2001, Oxygen radical inhibition of nitric oxide-dependent vascular function in sickle cell disease, *Proc. Natl. Acad. Sci. U. S. A* **98**:15215-15220.

Asplund, K., 2002, Antioxidant vitamins in the prevention of cardiovascular disease: a systematic review, *J. Intern. Med.* **251**:372-392.

Atherton, D. J., 2003, Topical corticosteroids in atopic dermatitis, *BMJ* **327**:942-943.

Aubourg, P., 1936, Colibacillose aigue, colibacillose cronique: ameliorations cliniques notables par un traitement d'ozone, *Bull. Med. Paris* **140**:644-654.

Aubourg, P., 1940, Ozon in der Chirurgie, *Mem. Acad. Chir* **65**:1183-1192.

Ault, J. G., and Lawrence D. A., 2003, Glutathione distribution in normal and oxidatively stressed cells, *Exp. Cell Res.* **285**:9-14.

Auphan, N., DiDonato J. A., Rosette C., Helmberg A., and Karin M., 1995, Immunosuppression by glucocorticoids: inhibition of NF-k B activity through induction of IkB synthesis, *Science* **270**:286-290.

Ayres, R. M., Stott R., Mara D. D., and Lee D. L., 1992, Wastewater reuse in agriculture and the risk of intestinal nematode infection, *Parasitol. Today* **8**:32-35.

Babior, B. M., 2000, Phagocytes and oxidative stress, *Am. J. Med.* **109**:33-44.

Babior, B. M., Takeuchi C., Ruedi J., Gutierrez A., and Wentworth P., Jr., 2003, Investigating antibody-catalyzed ozone generation by human neutrophils, *Proc. Natl. Acad. Sci. U. S. A* **100**:3031-3034.

Back, T., 1998, Pathophysiology of the ischemic penumbra--revision of a concept, *Cell Mol. Neurobiol.* **18**:621-638.

Badwey, J. A., and Karnovsky M. L., 1980, Active oxygen species and the functions of phagocytic leukocytes, *Annu. Rev. Biochem.* **49**:695-726.

Baert, F., Noman M., Vermeire S., Van Assche G., D' Haens G., Carbonez A., and Rutgeerts P., 2003, Influence of immunogenicity on the long-term efficacy of infliximab in Crohn's disease, *N. Engl. J. Med.* **348**:601-608.

Baeuerle, P. A., and Henkel T., 1994, Function and activation of NF-kB in the immune system, *Annu. Rev. Immunol.* **12**:141-179.

Bailar, J. C., III, and Gornik H. L., 1997, Cancer undefeated, *N. Engl. J. Med.* **336**:1569-1574.

Bak, I., Papp G., Turoczi T., Varga E., Szendrei L., Vecsernyes M., Joo F., and Tosaki A., 2002, The role of heme oxygenase-related carbon monoxide and ventricular fibrillation in ischemic/reperfused hearts, *Free Radic. Biol. Med.* **33**:639-648.

Baker, K. H., Hegarty J. P., Redmond B., Reed N. A., and Herson D. S., 2002, Effect of oxidizing disinfectants (chlorine, monochloramine, and ozone) on Helicobacter pylori, *Appl. Environ. Microbiol.* **68**: 981-984.

Barakat, S., Seif-El Nasr A., Ardel-Maksoud N., El-Ebiary F., Amer H., Zaghloul A., and Thabet S., 2004, Induktion der angiogenese durch medizinisches ozon, in *Ozon-handbuch.*

Grundlagen pravention, therapie (R. Viebahn-Hansler, and H. G. Knoch, Eds.), Landsberg in press.

Barber, E., Menéndez S., León O. S., Barber M. O., Merino N., Calunga J. L., Cruz E., and Bocci V., 1999, Prevention of renal injury after induction of ozone tolerance in rats submitted to warm ischaemia., *Mediat. Inflamm.* **8**:37-41.

Barnes, P. J., 2000, Chronic obstructive pulmonary disease, *N. Engl. J. Med.* **343**:269-280.

Barnes, P. J., and Karin M., 1997, Nuclear factor-kappaB: a pivotal transcription factor in chronic inflammatory diseases, *N. Engl. J. Med.* **336**:1066-1071.

Barnes, P. J., and Liew F. Y., 1995, Nitric oxide and asthmatic inflammation, *Immunol. Today* **16**:128-130.

Basu, S., 2004, Isoprostanes: novel bioactive products of lipid peroxidation, *Free Radic. Res.* **38**:105-122.

Baulieu, E.-E., and Robel P., 1998, Dehydroepiandrosterone (DHEA) and dehydroepiandrosterone sulfate (DHEAS) as neuroactive neurosteroids, *Proc. Nat. Acad. Sci. USA* **95**:4089-4091.

Baykal, Y., Yilmaz M. I., Celik T., Gok F., Rehber H., Akay C., and Kocar I. H., 2003, Effects of antihypertensive agents, alpha receptor blockers, beta blockers, angiotensin-converting enzyme inhibitors, angiotensin receptor blockers and calcium channel blockers, on oxidative stress, *J. Hypertens.* **21**:1207-1211.

Baynes, J. W., 1991, Role of oxidative stress in development of complications in diabetes, *Diabetes* **40**:405-412.

Baysan, A., Whiley R. A., and Lynch E., 2000, Antimicrobial effect of a novel ozone- generating device on micro- organisms associated with primary root carious lesions in vitro, *Caries Res.* **34**:498-501.

Beal, M. F., 2002, Oxidatively modified proteins in aging and disease, *Free Radic. Biol. Med.* **32**:797-803.

Beck, L. S., DeGuzman L., Lee W. P., Xu Y., Siegel M. W., and Amento E. P., 1993, One systemic administration of transforming growth factor- b1 reverses age- or glucocorticoid-impaired wound healing, *J. Clin. Invest.* **92**:2841-2849.

Beckman, K. B., and Ames B. N., 1998, The free radical theory of aging matures, *Physiol Rev.* **78**:547-581.

Bell, D. S., 2004a, Type 2 diabetes mellitus: what is the optimal treatment regimen?, *Am. J. Med.* **116 Suppl 5A**:23S-29S.

Bell, D. S., 2004b, Advantages of a third-generation beta-blocker in patients with diabetes mellitus, *Am. J. Cardiol.* **93**:49B-52B.

Bell, S., and Kamm M. A., 2000, Antibodies to tumour necrosis factor alpha as treatment for Crohn's disease, *Lancet* **355**:858-860.

Belluzzi, A., Brignola C., Campieri M., Pera A., Boschi S., and Miglioli M., 1996, Effect of an enteric -coated fish-oil preparation on relapses in Crohn's disease, *N. Engl. J. Med.* **334**:1557-1560.

Beltrani, V. S., 1999, The clinical spectrum of atopic dermatitis, *J. Allergy Clin. Immunol.* **104**:S87-S98.

Bender, D.A., 2002, Daily doses of multivitamin tablets, *B. M. J.* **325**: 173-174.

Bennett, S. P., Griffiths G. D., Schor A. M., Leese G. P., and Schor S. L., 2003, Growth factors in the treatment of diabetic foot ulcers, *Br. J. Surg.* **90**:133-146.

Benson, H., and Friedman R., 1996, Harnessing the power of the placebo effect and renaming it "remembered wellness", *Annu. Rev. Med.* **47**:193-199.

Bergamini, A., Capozzi M., Ghibelli L., Dini L., Salanitro A., Milanese G., Wagner T., Beninati S., Delfina Pesce C., Amici C., and Rocchi G., 1994, Cystamine potently suppresses in vitro HIV replication in acutely and chronically infected human cells, *J. Clin. Invest.* **93**:2251-2257.

Bergo, G. W., and Tyssebotn I., 1999, Cardiovascular effects of hyperbaric oxygen with and without addition of carbon dioxide, *Eur. J. Appl. Physiol Occup. Physiol* **80**:264-275.

Bergofsky, E. H., and Bertun P., 1966, Response of regional circulations to hyperoxia, *J. Appl. Physiol* **21**:567-572.

Bergqvist, D., 1999, Salvage of critically ischaemic limbs, *Lancet* **354**:1920-1921.

Bernier, J., Denekamp J., Rojas A., Minatel E., Horiot J., Hamers H., Antognoni P., Dahl O., Richaud P., van Glabbeke M., and Pi inverted question m. M., 2000, ARCON: accelerated radiotherapy with carbogen and nicotinamide in head and neck squamous cell carcinomas. The experience of the Co-operative group of radiotherapy of the european organization for research and treatment of cancer (EORTC), *Radiother. Oncol.* **55** :111-119.

Berson, E. L., Remulla J. F. C., Rosner B., Sandberg M. A., and Weigel-DiFranco C., 1996, Evaluation of patients with retinitis pigmentosa receiving electric stimulation, ozonated blood, and ocular surgery in Cuba, *Arch. Ophthalmol.* **114**:560-563.

Bevers, R. F. M., Bakker D. J., and Kurth K. H., 1995, Hyperbaric oxygen treatment for haemorrhagic radiation cystitis, *Lancet* **346**:803-805.

Beyerle, 1996, cited by Null, 1996 (Ozone: a wide-spectrum realer, *Penthouse Magazine* **January**).

Biedunkiewicz, B., Tylicki L., Nieweglowski T., Burakowski S., and Rutkowski B., 2004, Clinical efficacy of ozonated autohemotherapy in hemodialyzed patients with intermittent claudication: an oxygen-controlled study, *Int. J. Artif. Organs* **27**:29-34.

Bilger, B., 1995, Forever young, *The Sciences* **September/October**:26-30.

Bishop, G. A., Ramirez L. M., Baccam M., Busch L. K., Pederson L. K., and Tomai M. A., 2001, The immune response modifier resiquimod mimics CD40-induced B cell activation, *Cell Immunol.* **208**:9-17.

Block, J. A., and Sequeira W., 2001, Raynaud's phenomenon, *Lancet* **357**:2042-2048.

Bocchi, L., Cervelli C., and Ferrata P., 1998, La nucleoaspirazione, in *Lombalgie e lombosciatalgie. Criteri di diagnosi e cura* (F. Ceccherelli, and A. Ricciardi, Eds.), Edizioni Libreria Cortina, Torino, pp.285-293.

Bocchi, L., Cervelli C., and Ferrata P., 2000, L'ossigeno-ozono terapia nel trattamento delle patologie vertebrali lombari, in *Proceedings: I Congresso IMOS, Italia, Siena, 2-4 novembre 2000*, p.15.

Bocci, V., 1981a, Determinants of erythrocyte ageing: a reappraisal., *Brit. J. Haematol.* **48**:515-522.

Bocci, V., 1981b, Pharmacokinetic studies of interferons, *Pharmacol. Ther.* **13 (3)**:421-440.

Bocci, V., 1981c, Production and role of interferon in physiological conditions, *Biol. Rev.* **56**:49-85.

Bocci, V., 1985a, Immunomodulators as local hormones: new insights regarding their clinical utilization., *J. Biol. Resp. Modif.* **4**:340-352.

Bocci, V., 1985b, Administration of interferon at night may increase its therapeutic index., *Cancer Drug Del.* **2**:313-318.

Bocci, V., 1987a, May hyperbaric oxigenation be useful to patients with AIDS?, *J. Biol. Regulat. Homeost. Agent.* **1**:201.

Bocci, V., 1987b, Metabolism of protein anticancer agents. (Updated and reprinted in 1994 in Int. Encyclopedia of Pharmacology and Therapeutics, Sec.140 Anticancer Drugs, Oxford: Pergamon Press, pp. 387-436), *Pharmacol. Ther.* **34**:1-49.

Bocci, V., 1988, Roles of interferon produced in physiological conditions. A speculative review, *Immunology* **64**:1-9.

Bocci, V., 1988a, Central nervous system toxicity of interferons and other cytokines, *J. Biol. Regulat. Homeost. Agent.* **2**:107-118.

Bocci, V., 1988b, Roles of interferon produced in physiological conditions. A speculative review, *Immunology* **64**:1-9.

Bocci, V., 1990a, Catabolism of therapeutic proteins and peptides with implications for drug delivery., *Advan. Drug Delivery Rev.* **4**:149-169.

Bocci, V., 1990b, Tumor therapy with biological response modifiers. Why is progress slow?, *EOS-J. Immunol. Immunopharmacol.* **10**:79-82.

Bocci, V., 1991a, Absorption of cytokines via oropharyngeal-associated lymphoid tissues. Does an unorthodox route improve the therapeutic index of interferon?, *Clin. Pharmacokinet.* **21**:411-417.

Bocci, V., 1991b, Interleukins. Clinical pharmacokinetics and practical implications, *Clin. Pharmacokinet.* **21**:274-284.

Bocci, V., 1992a, Ozonization of blood for the therapy of viral diseases and immunodeficiencies. A hypothesis, *Med. Hypotheses* **39**:30-34.

Bocci, V., 1992b, Physicochemical and biologic properties of interferons and their potential uses in drug delivery systems., *Crit. Rev. Ther. Drug Carr. Syst.* **9**:91-133.

Bocci, V., 1992c, The neglected organ: bacterial flora has a crucial immunostimulatory role, *Perspect. Biol. Med.* **35**:251-260.

Bocci, V., 1993a, *Interferon. Una storia recente ed antichissima. Fisiopatologia e clinica del sistema interferon*, Antea Edizioni, pp.1-205.

Bocci, V., 1993b, Mistletoe (viscum album) lectins as cytokine inducers and immunoadjuvant in tumor therapy. A review, *J. Biol. Regulat. Homeost. Agent.* **7**:1-6.

Bocci, V., 1994a, A reasonable approach for the treatment of HIV infection in the early phase with ozonetherapy (autohemotherapy). How inflammatory cytokines may have a therapeutic role, *Mediat. Inflamm.* **3**:315-321.

Bocci, V., 1994b, Autohaemotherapy after treatment of blood with ozone. A reappraisal., *J. Int. Med. Res.* **22**:131-144.

Bocci, V., 1996a, Does ozone therapy normalize the cellular redox balance?, *Med. Hypotheses* **46**:150-154.

Bocci, V., 1996b, Ozone as a bioregulator. Pharmacology and toxicology of ozonetherapy today, *J. Biol. Regulat. Homeost. Agent.* **10**:31-53.

Bocci, V., 1996c, Ozone: a mixed blessing. New mechanisms of the action of ozone on blood cells make ozonated major autohaemotherapy (MAH) a rational approach, *Forsch. Komplementärmed.* **3**:25-33.

Bocci, V., 1998a, Ipotetici meccanismi di azione dell'ozono nel trattamento del conflitto discoradicolare, in *Lombalgie e lombosciatalgie. Criteri di diagnosi e cura* (F. Ceccherelli, and A. Ricciardi, Eds.), Edizioni Libreria Cortina, Torino, pp.331-340.

Bocci, V., 1998b, Is ozonetherapy therapeutic ?, *Perspect. Biol. Med.* **42**:131-143.

Bocci, V., 1998c, Ozonetherapy as a possible biological response modifier in cancer, *Forsch. Komplementärmed.* **5**:54-60.

Bocci, V., 1999a, Biological and clinical effects of ozone. Has ozonetherapy a future in medicine ?, *Brit. J. Biomed. Sci.* **56**:270-279.

Bocci, V., 1999b, Ozonetherapy as a complementary medical approach. Where are we and where do we need to go?, in *Proceedings of the Int. Ozone Symposium, 21 and 22 October 1999, Basel, Switzerland* (IOA – EA_3G Ed.), Bauer Druck AG, Basel, pp.353-374.

Bocci, V., 2000, *Ossigeno-ozono terapia*, Casa Editrice Ambrosiana, Milano, pp.1-324.

Bocci, V., 2002, *Oxygen-ozone therapy. A critical evaluation*, Kluwer Academic Publischer.

Bocci, V., 2004, Ozone as Janus: this controversial gas can be either toxic or medically useful, *Mediators. Inflamm.* **13**:3-11.

Bocci, V., Aldinucci C., Borrelli E., Corradeschi F., Diadori A., Fanetti G., and Valacchi G., 2001a, Ozone in medicine, *Ozone-Sci. Eng.* **23**:207-217.

Bocci, V., and Di Paolo N., 2004, Oxygenation-ozonization of blood during extracorporeal circulation (EBOO). PartIII: A new medical approach, *Ozone:Sci. Enginnering* **26**:195-205.

Bocci, V., and Paulesu L., 1990, Studies on the biological effects of ozone 1. Induction of interferon gamma on human leucocytes, *Haematologica* **75**:510-515.

Bocci, V., Borrelli E., Corradeschi F., and Valacchi G., 2000, Systemic effects after colorectal insufflation of oxygen-ozone in rabbits, *Int. J. Med. Biol. Environ.* **28**:109-113.

Bocci, V., Borrelli E., Valacchi G., and Luzzi E., 1999, Quasi-total-body exposure to an oxygen-ozone mixture in a sauna cabin, *Eur. J. Appl. Physiol. Occup. Physiol.* **80**: 549-554.

Bocci, V., Carraro F., Naldini A., Paulesu L., and Pessina G. P., 1990, Roles of interferons in physiological conditions and for the control of viral diseases, in *Microbiological, chemotherapeutical and immunological problems in high risk patients* (E. Garaci, G. Renzini, F. Filadoro, A. L. Goldstein, and J. Verhoef, Eds.), Serono Symposia Publication from Raven Press, New York, pp.243-250.

Bocci, V., Di Paolo N., Borrelli E., Larini A., and Cappelletti F., 2001c, Ozonation of blood during extracorporeal circulation II. Comparative analysis of several oxygenators-ozonators and selection of one type., *Int. J. Artif. Organs* **24**:890-897.

Bocci, V., Di Paolo N., Garosi G., Aldinucci C., Borrelli E., Valacchi G., Cappelli F., Guerri L., Gavioli G., Corradeschi F., Rossi R., Giannerini F., and Di Simplicio P., 1999b, Ozonation of blood during extracorporeal circulation. I. Rationale, methodology and preliminary studies, *Int. J. Artif. Organs* **22**:645-651.

Bocci, V., Luzzi E., Corradeschi F., and Paulesu L., 1994a, Studies on the biological effects of ozone: 5. Evaluation of immunological parameters and tolerability in normal volunteers receiving ambulatory autohaemotherapy., *Biotherapy* **7**:83-90.

Bocci, V., Luzzi E., Corradeschi F., and Silvestri S., 1994b, Studies on the biological effects of ozone: 6. Production of transforming growth factor b1 by human blood after ozone treatment, *J. Biol. Regulat. Homeost. Agent.* **8**:108-112.

Bocci, V., Luzzi E., Corradeschi F., Paulesu L., and Di Stefano A., 1993a, Studies on the biological effects of ozone: 3. An attempt to define conditions for optimal induction of cytokines, *Lymphokine Cytokine Res.* **12**:121-126.

Bocci, V., Luzzi E., Corradeschi F., Paulesu L., Rossi R., Cardaioli E., and Di Simplicio P., 1993b, Studies on the biological effects of ozone: 4. Cytokine production and glutathione levels in human erythrocytes., *J. Biol. Regulat. Homeost. Agent.* **7**:133-138.

Bocci, V., Pessina G. P., Paulesu L., Muscettola M., and Valeri A., 1988, The lymphatic route. V. Distribution of human natural interferon-□ in rabbit plasma and lymph., *J. Interferon Res.* **8**:633-640.

Bocci, V., Pogni R., Corradeschi F., Busi E., Cervelli C., Bocchi L., and Basosi R., 2001b, Oxygen-ozone in orthopaedics: EPR detection of hydroxyl free radicals in ozone-treated "nucleus pulposus" material, *Riv. Neuroradiol.* **14**:55-59.

Bocci, V., Russi M., and Rita G., 1967, Recovery and identification of interferon in the rabbit urine, *Experientia* **23**:1-5.

Bocci, V., Valacchi G., Corradeschi F., Aldinucci C., Silvestri S., Paccagnini E., and Gerli R., 1998a, Studies on the biological effects of ozone: 7. Generation of reactive oxygen species (ROS) after exposure of human blood to ozone, *J. Biol. Regulat. Homeost. Agent.* **12**:67-75.

Bocci, V., Valacchi G., Corradeschi F., and Fanetti G., 1998b, Studies on the biological effects of ozone: 8. Effects on the total antioxidant status and on interleukin-8 production, *Mediat. Inflamm.* **7**:313-317.

Bocci, V., Valacchi G., Rossi R., Giustarini D., Paccagnini E., Pucci A. M., and Di Simplicio P., 1999a, Studies on the biological effects of ozone: 9. Effects of ozone on human platelets, *Platelets* **10**:110-116.

Bocci, V., Venturi G., Catucci M., Valensin P. E., and Zazzi M., 1998c, Lack of efficacy of ozone therapy in HIV infection, *Clin. Microbiol. Infec.* **4**:667-669.

Boehm, T., Folkman J., Browder T., and O'Reilly M. S., 1997, Antiangiogenic therapy of experimental cancer does not induce acquired drug resistance, *Nature* **390**:404-407.

Bondy, S. C., 1995, The relation of oxidative stress and hyperexcitation to neurological disease, *Proc. Soc. Exp. Biol. Med.* **208**:337-345.

Boneschi, F. M., Rovaris M., Johnson K. P., Miller A., Wolinsky J. S., Ladkani D., Shifroni G., Comi G., and Filippi M., 2003, Effects of glatiramer acetate on relapse rate and accumulated disability in multiple sclerosis: meta-analysis of three double-blind, randomized, placebo-controlled clinical trials, *Mult. Scler.* **9**:349-355.

Bonetti, M., Cotticelli B., Valdenassi L., and Richelmi P., 2001, Analisi dei risultati dopo trattamento con O_2-O_3 nelle ernie intra ed extra foraminali lombari, *Riv. Neuroradiol.* **14**:89-92.

Boni, C., Bertoletti A., Penna A., Cavalli A., Pilli M., Urbani S., Scognamiglio P., Boehme R., Panebianco R., Fiaccadori F., and Ferrari C., 1998, Lamivudine treatment can restore T cell responsiveness in chronic hepatitis B, *J. Clin. Invest* **102**:968-975.

Bonifati, C., and Ameglio F., 1999, Cytokines in psoriasis, *Int. J. Dermatol.* **38**:241-251.

Booth, S., and Wade R., 2003, Oxygen or air for palliation of breathlessness in advanced cancer, *J. R. Soc. Med.* **96**:215-218.

Borrego, A., Zamora Z. B., Gonzalez R., Romay C., Menendez S., Hernandez F., Montero T., and Rojas E., 2004, Protection by ozone preconditioning is mediated by the antioxidant system in cisplatin-induced nephrotoxicity in rats, *Mediators. Inflamm.* **13**:13-19.

Borrelli, E., and Bocci V., 2002, A novel therapeutic option for chronic fatigue syndrome and fibromyalgia, *Rivista Italiana Di Ossigeno-Ozonoterapia* **1**:149-153.

Bosch-Morell, F., Flohé L., Marìn N., and Romero F. J., 1999, 4-hydroxynonenal inhibits glutathione peroxidase: protection by glutathione, *Free rad. Biol. Med.* **26**:1383-1387.

Boxer, L. A., and Smolen J. E., 1988, Neutrophil granule constituents and their release in health and disease, *Hematol. Oncol. Clin. N. Amer.* **2**:101-134.

Brahimi-Horn, C., Berra E., and Pouyssegur J., 2001, Hypoxia: the tumor's gateway to progression along the angiogenic pathway, *Trends Cell Biol.* **11**:S32-S36.

Brandes, M. E., Allen J. B., Ogawa Y., and Wahl S. M., 1991, Transforming growth factor b1 suppresses acute and chronic arthritis in experimental animals., *J. Clin. Invest.* **87**:1108-1113.

Brayda-Bruno, M., and Cinnella P., 1998, Il trattamento dell'ernia discale con infiltrazioni di ossigeno-ozono in paravertebrale, in *Lombalgie e lombosciatalgie. Criteri di diagnosi e cura* (F. Ceccherelli, and A. Ricciardi, Eds.), Edizioni Libreria Cortina, Torino, pp.361-365.

Bressler, N. M., Bressler S. B., and Fine S. L., 1988, Age-related macular degeneration, *Surv. Ophthalmol.* **32**:375-413.

Bridgeman, M. M., Marsden M., MacNee W., Flenley D. C., and Ryle A. P., 1991, Cysteine and glutathione concentrations in plasma and bronchoalveolar lavage fluid after treatment with N-acetylcysteine, *Thorax* **46**:39-42.

Brizel, D. M., Scully S. P., Harrelson J. M., Layfield L. J., Bean J. M., Prosnitz L. R., and Dewhirst M. W., 1996, Tumor oxygenation predicts for the likelihood of distant metastases in human soft tissue sarcoma, *Cancer Res.* **56**:941-943.

Broeckaert, F., Arsalane K., Hermans C., Bergamaschi E., Brustolin A., Mutti A., and Bernard A., 1999, Lung epithelial damage at low concentrations of ambient ozone, *Lancet* **353**:900-901.

Brouard, S., Otterbein L. E., Anrather J., Tobiasch E., Bach F. H., Choi A. M., and Soares M. P., 2000, Carbon monoxide generated by heme oxygenase 1 suppresses endothelial cell apoptosis, *J. Exp. Med.* **192**:1015-1026.

Brownlee, M., 2001, Biochemistry and molecular cell biology of diabetic complications, *Nature* **414**:813-820.

Brugnara, C., Gee B., Armsby C. C., Kurth S., Sakamoto M., Rifai N., Alper S. L., and Platt O. S., 1996, Therapy with oral clotrimazole induces inhibition of the Gardos channel and reduction of erythrocyte dehydration in patients with sickle cell disease, *J. Clin. Invest* **97**:1227-1234.

Bubenìk, J., 1996, Cytokine gene-modified vaccines in the therapy of cancer, *Pharmacol. Ther.* **69**:1-14.

Buege, J. A., and Aust S. D., 1994, Microsomal lipid peroxidation, *Meth. Enzymol.* **233**:302-310.

Bulinin, V. I., Solod N. V., and Moshurov I. P., 1995, The first experience of chronic abscesses and pleura emphyemas treatment by the method of ozonization, in *The ozone in biology and medicine. 2nd all Russian scientific-practical conference, September 6-8, 1995. Russian association of ozonetherapy, Reshetnikovskaya street 2, Nizhni Novgorod, 603006 Russia*, p.20.

Burkey, K. O., and Eason G., 2002, Ozone tolerance in snap bean is associated with elevated ascorbic acid in the leaf apoplast, *Physiol Plant* **114**:387-394.

Burstein, H. J., Gelber S., Guadagnoli E., and Weeks J. C., 1999, Use of alternative medicine by women with early-stage breast cancer, *N. Engl. J. Med.* **340**:1733-1739.

Bush, R. S., Jenkin R. D., Allt W. E., Beale F. A., Bean H., Dembo A. J., and Pringle J. F., 1978, Definitive evidence for hypoxic cells influencing cure in cancer therapy, *Br. J. Cancer Suppl* **37**:302-306.

Bustamante, J., Lodge J. K., Marcocci L., Tritschler H., Packer L., and Rihn B. H., 1998, a-lipoic acid in liver metabolism and disease, *Free rad. Biol. Med.* **24**:1023-1039.

Butterfield, D. A., and Lauderback C. M., 2002, Lipid peroxidation and protein oxidation in Alzheimer's disease brain: potential causes and consequences involving amyloid beta-peptide-associated free radical oxidative stress, *Free Radic. Biol. Med.* **32**:1050-1060.

Cadenas, E., and Davies K. J., 2000, Mitochondrial free radical generation, oxidative stress, and aging, *Free rad. Biol. Med.* **29**:222-230.

Calabrese, E. J., 2002, Hormesis: changing view of the dose-response, a personal account of the history and current status, *Mutat. Res.* **511**:181-189.

Calabrese, E. J., and Baldwin L. A., 2001, Hormesis: U-shaped dose responses and their centrality in toxicology, *Trends Pharmacol. Sci.* **22**:285-291.

Calder, P. C., 1998, Fat chance of immunomodulation, *Immunol. Today* **19**:244-247.

Caligiuri, M., Murray C., Buchwald D., Levine H., Cheney P., Peterson D., Komaroff A. L., and Ritz J., 1987, Phenotypic and functional deficiency of natural killer cells in patients with chronic fatigue syndrome, *J. Immunol.* **139**:3306-3313.

Callahan, J. T., Collecutt M. F., Lightbody J. R., and Faragher B. S., 1982, Alteration of human red blood cells stored in plastic packs, *Transfusion* **22**:154-157.

Campbell, D. E., Fryga A. S., Bol S., and Kemp A. S., 1999, Intracellular interferon-gamma (IFN-g) production in normal children and children with atopic dermatitis, *Clin. Exp. Immunol.* **115**:377-382.

Cannistra, S. A., and Niloff J. M., 1996, Cancer of the uterine cervix, *N. Engl. J. Med.* **334**:1030-1038.

Cardile, V., Jiang X., Russo A., Casella F., Renis M., and Bindoni M., 1995, Effects of ozone on some biological activities of cells *in vitro*, *Cell Biol. Toxicol.* **11**:11-21.

Carette, S., Leclaire R., Marcoux S., Morin F., Blaise G. A., St.-Pierre A., Truchon R., Parent F., Lévesque J., Bergeron V., Montminy P., and Blanchette C., 1997, Epidural corticosteroid injections for sciatica due to herniated nucleus pulposus, *N. Engl. J. Med.* **336**:1634-1640.

Carlsson, L. M., Jonsson J., Edlund T., and Marklund S. L., 1995, Mice lacking extracellular superoxide dismutase are more sensitive to hyperoxia, *Proc. Nat. Acad. Sci. USA* **92**:6264-6268.

Carmeliet, P., and Jain R. K., 2000, Angiogenesis in cancer and other diseases, *Nature* **407**:249-257.

Carmeliet, P., Dor Y., Herbert J. M., Fukumura D., Brusselmans K., Dewerchin M., Neeman M., Bono F., Abramovitch R., Maxwell P., Koch C. J., Ratcliffe P., Moons L., Jain R. K., Collen D., Keshert E., and Keshet E., 1998, Role of HIF-1alpha in hypoxia-mediated apoptosis, cell proliferation and tumour angiogenesis, *Nature* **394**:485-490.

Carpendale, M. T. F., and Freeberg J. K., 1991, Ozone inactivates HIV at noncytotoxic concentrations, *Antivir. Res.* **16**:281-292.

Carpendale, M. T., Freeberg J., and McLeod Griffiss J., 1993, Does Ozone alleviate AIDS diarrhea?, *J. Clin. Gastroenterol.* **17**:142-145.

Cassileth, B. R., and Chapman C. C., 1996, Alternative and complementary cancer therapies, *Cancer* **77**:1026-1034.

Castagnola, E., Molinari A. C., Fratino G., Viscoli C., 2003, Conditions associated with infections of indwelling central venous catheters in cancer patients: a summary, *Br. J. Haematol.* **121**: 233-239.

Ceballos-Picot, I., Merad-Boudia M., Nicole A., Thevenin M., Hellier G., Legrain S., and Berr C., 1996a, Peripheral antioxidant enzyme activities and selenium in elderly subjects and in dementia of Alzheimer's type-place of the extracellular glutathione peroxidase, *Free rad. Biol. Med.* **20**:579-587.

Ceballos-Picot, I., Witko-Sarsat V., Merad-Boudia M., Nguyen A. T., Thévenin M., Jaudon M. C., Zingraff J., Verger C., Jungers P., and Descamps-Latscha B., 1996b, Glutathione antioxidant system as a marker of oxidative stress in chronic renal failure, *Free rad. Biol. Med.* **21**:845-853.

Ceccherelli, F., Gagliardi G., Matterazzo G., Rossato M., and Giron G., 1995, La riflessoterapia per agopuntura, in *La riflessoterapia per agopuntura* (P. Procacci, A. Di Massa, F. Ceccherelli, and R. Casale, Eds.), Edizioni A.I.R.A.S., Padova, pp.49-77.

Chader, G. J., 2001, PEDF: Raising both hopes and questions in controlling angiogenesis, *Proc. Natl. Acad. Sci. U. S. A* **98**:2122-2124.

Chae, H. Z., Kim K., and Kim I.-H., 1999, The novel antioxidant enzyme, thioredoxin peroxidase, and mammalian peroxiredoxins, in *Redox regulation of cell signaling and its clinical application* (L. Packer, and J. Yodoi, Eds.), Marcel Dekker, Inc., New York, pp.85-92.

Chan WM, Lam DS, Wong TH, Lai TY, Kwok AK, Tam BS, Li KK., 2003, Photodynamic therapy with verteporfin for subfoveal idiopathic choroidal neovascularization: one-year results from a prospective case series., *Ophthalmology.* **110**(12):2395-402.

Chanock, S. J., El Benna J., Smith R. M., and Babior B. M., 1994, The respiratory burst oxidase, *J. Biol. Chem.* **269**:24519-24522.

Chen, Z., Oberley T. D., Ho Y., Chua C. C., Siu B., Hamdy R. C., Epstein C. J., and Chua B. H., 2000, Overexpression of CuZnSOD in coronary vascular cells attenuates myocardial ischemia/reperfusion injury, *Free rad. Biol. Med.* **29**:589-596.

Cherkin, D. C., Deyo R. A., Battie M., Street J., and Barlow W., 1998, A comparison of physical therapy, chiropractic manipulation, and provision of an educational booklet for the treatment of patients with low back pain, *N. Engl. J. Med.* **339**:1021-1029.

Cho, H. Y., Zhang L. Y., and Kleeberger S. R., 2001, Ozone-induced lung inflammation and hyperreactivity are mediated via tumor necrosis factor-alpha receptors, *Am. J. Physiol Lung Cell Mol. Physiol* **280**:L537-L546.

Choi, B.M., Pae H.O., Kim Y.M., Chung H.T., 2003, Nitric oxide-mediated cytoprotection of hepatocytes from glucose deprivation-induced cytotoxicity: involvement of heme oxygenase-1, *Hepatology* **37**: 810-823.

Chopdar, A., Chakravarthy U., and Verma D., 2003, Age related macular degeneration, *BMJ* **326**:485-488.

Chow, C. K., and Kaneko J. J., 1979, Influence of dietary vitamin E on the red cells of ozone-exposed rats, *Environ. Res.* **19**:49-55.

Christian, D. L., Chen L. L., Scannell C. H., Ferrando R. E., Welch B. S., and Balmes J. R., 1998, Ozone-induced inflammation is attenuated with multiday exposure, *Amer. J. Respir. Crit. Care Med.* **158**:532-537.

Chun, T. W., and Fauci A. S., 1999, Latent reservoirs of HIV: obstacles to the eradication of virus, *Proc. Natl. Acad. Sci. U. S. A* **96**:10958-10961.

Cianci, P., 2004, Advances in the treatment of the diabetic foot: Is there a role for adjunctive hyperbaric oxygen therapy?, *Wound. Repair Regen.* **12**:2-10.

Cighetti, G., Duca L., Bortone L., Sala S., Nava I., Fiorelli G., and Cappellini M. D., 2002, Oxidative status and malondialdehyde in beta-thalassaemia patients, *Eur. J. Clin. Invest* **32 Suppl 1**:55-60.

Cinatl, J., Morgenstern B., Bauer G., Chandra P., Rabenau H., and Doerr H. W., 2003, Treatment of SARS with human interferons, *Lancet* **362**:293-294.

Cinnella, P., and Brayda-Bruno M., 2001, La nostra esperienza nel trattamento dei conflitti disco-radicolari e delle radicolopatie post-chirurgiche con ossigeno-ozono terapia infiltrativa paravertebrale, *Riv. Neuroradiol.* **14**:75-79.

Clark, C., Buchwald D., MacIntyre A., Sharpe M., and Wessely S., 2002, Chronic fatigue syndrome: a step towards agreement, *Lancet* **359**:97-98.

Clavo, B., Català L., Pérez J. L., Rodrìguez V., and Robaina F., 2004, Effect of ozone therapy on cerebral blood flow: A preliminary report, *Evid. Based. Complement Alternat. Med.,* in press.

Clavo, B., Perez J. L., Lopez L., Suarez G., Lloret M., Rodriguez V., Macias D., Santana M., Morera J., Fiuza D., Robaina F., and Gunderoth M., 2003, Effect of ozone therapy on muscle oxygenation, *J. Altern. Complement Med.* **9**:251-256.

Clavo, B., Perez J. L., Lopez L., Suarez G., Lloret M., Rodriguez V., Macias D., Santana M., Hernandez M. A., Martin-Oliva R., and Robaina F., 2004a, Ozone Therapy for Tumor Oxygenation: a Pilot Study, *Evid. Based. Complement Alternat. Med.* **1**:93-98.

Clavo, B., Ruiz A., Lloret M., Lopez L., Suarez G., Macias D., Rodriguez V., Hernandez M. A., Martin-Oliva R., Quintero S., Cuyas J. M., and Robaina F., 2004b, Adjuvant ozonetherapy in advanced head and neck tumors: A comparative study, *J. Complement. Altern. Med.,* in press.

Cleare, A. J., Sookdeo S. S., Jones J., O'Keane V., and Miell J. P., 2000, Integrity of the growth hormone/insulin-like growth factor system is maintained in patients with chronic fatigue syndrome, *J. Clin. Endocrinol. Metab* **85**:1433-1439.

Clinton, S. K., 1998, Lycopene: chemistry, biology, and implications for human health and disease, *Nutr. Rev.* **56**:35-51.

Cohen, J., 2002, The immunopathogenesis of sepsis, *Nature* **420**: 885-891

Cohen, S. M., Olin K. L., Feuer W. J., Hjelmeland L., Keen C. L., and Morse L. S., 1994, Low glutathione reductase and peroxidase activity in age-related macular degeneration, *Brit. J. Ophthalmol.* **78**:791-794.

Coleman, C. N., 1988, Hypoxia in tumors: a paradigm for the approach to biochemical and physiologic heterogeneity, *J. Natl. Cancer Inst.* **80**:310-317.

Cooke, E. D., Pockley A. G., Tucker A. T., Kirby J. D. T., and Bolton A. E., 1997, Treatment of severe Raynaud's syndrome by injection of autologous blood pretreated by heating, ozonation and exposure to ultraviolet light (H-O-U) therapy, *Int. Angiol.* **16**:250-254.

Cope, H., David A., Pelosi A., and Mann A., 1994, Predictors of chronic "postviral" fatigue, *Lancet* **344**:864-868.

Corey, L., Wald A., Patel R., Sacks S. L., Tyring S. K., Warren T., Douglas J. M., Jr., Paavonen J., Morrow R. A., Beutner K. R., Stratchounsky L. S., Mertz G., Keene O. N., Watson H. A., Tait D., and Vargas-Cortes M., 2004, Once-daily valacyclovir to reduce the risk of transmission of genital herpes, *N. Engl. J. Med.* **350**:11-20.

Cosentino, R., Manca S., De Stefano R., Frati E., Hammoud M., Manganelli S., and Marcolongo R., 2000, Efficacia dell'ozonoterapia nella sindrome fibromialgica, in *Proceedings: I Congresso IMOS, Italia, Siena, 2-4 novembre 2000*, p.30.

Courbat, R., Urfer D., Walther J. L., and Mironova T. A., 2001, Optimisation of disinfection with ozone at full-scale in Nizhny Novgorod, Russia, in *Proceedings of the 15th Ozone World Congress, London, UK, 11th-15th September 2001, Volume I* (IOA 2001, Ed.), Speedprint MacMedia Ltd, Ealing, London, UK, pp.235-249.

Crabb, J. W., Miyagi M., Gu X., Shadrach K., West K. A., Sakaguchi H., Kamei M., Hasan A., Yan L., Rayborn M. E., Salomon R. G., and Hollyfield J. G., 2002, Drusen proteome analysis: an approach to the etiology of age-related macular degeneration, *Proc. Natl. Acad. Sci. U. S. A* **99**:14682-14687.

Cracowski, J. L., Devillier P., Durand T., Stanke-Labesque F., and Bessard G., 2001, Vascular biology of the isoprostanes, *J. Vasc. Res.* **38**:93-103.

Cruickshanks, K. J., Klein R., and Klein B. E., 1993, Sunlight and age-related macular degeneration. The Beaver Dam Eye Study, *Arch. Ophthalmol.* **111**:514-518.

Crumpacker, C. S., 2004, Use of antiviral drugs to prevent herpesvirus transmission, *N. Engl. J. Med.* **350**:67-68.

Csonka, C., Pataki T., Kovacs P., Muller S. L., Schroeter M. L., Tosaki A., and Blasig I. E., 2000, Effects of oxidative stress on the expression of antioxidative defense enzymes in spontaneously hypertensive rat hearts, *Free rad. Biol. Med.* **29**:612-619.

Csonka, C., Varga E., Kovacs P., Ferdinandy P., Blasig I. E., Szilvassy Z., and Tosaki A., 1999, Heme oxygenase and cardiac function in ischemic/reperfused rat hearts, *Free rad. Biol. Med.* **27**:119-126.

Cummins, R. O., 1994, *Textbook of advanced cardiac life support*, Scientific Publishing American Heart Association, Dallas,Tx.

Curran, S. F., Amoruso M. A., Goldstein B. D., and Berg R. A., 1984, Degradation of soluble collagen by ozone or hydroxyl radicals, *FEBS Lett.* **176**:155-160.

Curtis-Prior, P., Vere D., and Fray P., 1999, Therapeutic value of Ginkgo biloba in reducing symptoms of decline in mental function, *J. Pharm. Pharmacol.* **51**:535-541.

Dale, J. J., Ruckley C. V., Harper D. R., Gibson B., Nelson E. A., and Prescott R. J., 1999, Randomised, double blind placebo controlled trial of pentoxifylline in the treatment of venous leg ulcers, *Brit. Med. J.* **319**:875-878.

Daly, M. E., Makris A., Reed M., and Lewis C. E., 2003, Hemostatic regulators of tumor angiogenesis: a source of antiangiogenic agents for cancer treatment?, *J. Natl. Cancer Inst.* **95**:1660-1673.

D'Ambrosio, C. M., 2002a, Trattamento delle malattie infiammatorie croniche dell'intestino mediante ossigeno-ozonoterapia, *Rivista Italiana Di Ossigeno-Ozonoterapia* **1**:155-158.

D'Ambrosio, C. M., 2002b, Terapia delle IBD mediante ozonoterapia per via rettale, *Rivista Italiana Di Ossigeno-Ozonoterapia* **1**:159-163.

D'Amico, D. J., 1994, Diseases of the retina, *N. Engl. J. Med.* **331**:95-106.

Darouiche, R. O., 2004, Treatment of infections associated with surgical implants, *N. Engl. J. Med.* **350**:1422-1429.

Darzins, P., Mitchell P., and Heller R. F., 1997, Sun exposure and age-related macular degeneration. An Australian case- control study, *Ophthalmology* **104**:770-776.

Das, D., Bandyopadhyay D., Bhattacharjee M., and Banerjee R. K., 1997, Hydroxyl radical is the major causative factor in stress-induced gastric ulceration, *Free rad. Biol. Med.* **23**:8-18.

Das, U. N., 2003, Folic acid says NO to vascular diseases, *Nutrition* **19**:686-692.

Day, R., 2002, Adverse reactions to TNF-alpha inhibitors in rheumatoid arthritis, *Lancet* **359**:540-541.

De Capua, B., De Felice C., D'Onza M., De Lauretis A., Monaco G., Cosentino G., Tassi R., Gistri M., and Passali D., 2001, [Idiopathic sudden hearing loss: role of the posterior communicating cerebral arteries of the Willis' circle], *Acta Otorhinolaryngol. Ital.* **21**:144-150.

De Maio, A., 1999, Heat shock proteins: facts, thoughts, and dreams, *Shock* **11**:1-12.

De Maria, N., Colantoni A., Fagiuoli S., Liu G.-J., Rogers B. K., Farinati F., van Thiel D. H., and Floyd R. A., 1996, Association between reactive oxygen species and disease activity in chronic hepatitis C, *Free rad. Biol. Med.* **21**:291-295.

De Meirleir, K., Bisbal C., Campine I., De Becker P., Salehzada T., Demettre E., and Lebleu B., 2000, A 37 kDa 2-5A binding protein as a potential biochemical marker for chronic fatigue syndrome, *Am. J. Med.* **108**:99-105.

De Monte, A., van der Zee H., and Bocci V., 2004, Major ozonated auto-haemotherapy in chronic limb ischemia with ulcerations, *J. Complement. Altern. Med.,* in press.

Dedon, P. C., and Tannenbaum S. R., 2004, Reactive nitrogen species in the chemical biology of inflammation, *Arch. Biochem. Biophys.* **423**:12-22.

Degens, H., 1998, Age-related changes in the microcirculation of skeletal muscle, *Adv. Exp. Med. Biol.* **454**:343-348.

Delgado, J., 1991, Tratamiento con ozono del herpes zoster, *CENIC Ciencias Biologicas* **20**:160-162.

Denko, N. C., and Giaccia A. J., 2001, Tumor hypoxia, the physiological link between Trousseau's syndrome (carcinoma-induced coagulopathy) and metastasis, *Cancer Res.* **61**:795-798.

Dennog, C., Hartmann A., Frey G., and Speit G., 1996, Detection of DNA damage after hyperbaric oxygen (HBO) therapy, *Mutagenesis* **11**:605-609.

Dernek, S., Tunerir B., Sevin B., Aslan R., Uyguc O., and Kural T., 1999, The effects of methylprednisolone on complement, immunoglobulins and pulmonary neutrophil sequestration during cardiopulmonary bypass, *Cardiovasc. Surg.* **7**: 414-418.

Devlin, R. B., McDonnell W. F., Mann R., Becker S., House D. E., Schreinemachers D., and Koren H. S., 1991, Exposure of humans to ambient levels of ozone for 6.6. hours causes cellular and biochemical changes in the lung, *Amer. J. Respir. Cell Molec. Biol.* **4**:72-81.

Dewey, W. C., Hopwood L. E., Sapareto S. A., and Gerweck L. E., 1977, Cellular responses to combinations of hyperthermia and radiation, *Radiology* **123**:463-474.

Di Mascio, P., Kaiser S., and Sies H., 1989, Lycopene as the most efficient biological carotenoid singlet oxygen quencher, *Arch. Biochem. Biophys.* **274**:532-538.

Di Paolo, N., Bocci V., Cappelletti F., Petrini G., and Gaggiotti E., 2002, Necrotizing fasciitis successfully treated with extracorporeal blood oxygenation and ozonization (EBOO), *Int. J. Artif. Organs* **25**:1194-1198.

Di Paolo, N., Bocci V., Garosi G., Borrelli E., Bravi A., Bruci A., Aldinucci C., and Capotondo L., 2000, Extracorporeal blood oxigenation and ozonation (EBOO) in man. Preliminary report., *Int. J. Artif. Organs* **23**:131-141.

Dianzani, F., 1999, Chronic hepatitis B, biological basis for new therapeutic strategies, *J. Biol. Regulat. Homeost. Agent.* **13**:71-79.

Dianzani, M. U., 1998, 4-Hydroxynonenal and cell signalling, *Free rad. Res.* **28**:553-560.

Diaz, S., Menendez S., Eng L., and Fernandez I., 1995, No increase in sister chromatid exchanges and micronuclei frequencies in human lymphocytes exposed to ozone in vitro, in *Proceedings Ozone in Medicine 12th World Congress of the International Ozone Association, 15th to 18th May 1995, Lille France* (International Ozone Association, Ed.), Instaprint S.A., Tours, pp.43-52.

Diaz-Llera, S., Gonzalez-Hernandez Y., Prieto-Gonzalez E. A., and Azoy A., 2002, Genotoxic effect of ozone in human peripheral blood leukocytes, *Mutat. Res.* **517**:13-20.

Didier, C., Pouget J. P., Cadet J., Favier A., Beani J. C., and Richard M. J., 2001, Modulation of exogenous and endogenous levels of thioredoxin in human skin fibroblasts prevents DNA damaging effect of ultraviolet A radiation, *Free rad. Biol. Med.* **30**:537-546.

Dinarello, C. A., 1999, IL-18: a T_{H1}-inducing, proinflammatory cytokine and new member of the IL-1 family, *J. Allergy Clin. Immunol.* **103**:11-24.

Dische, S., Anderson P. J., Sealy R., and Watson E. R., 1983, Carcinoma of the cervix--anaemia, radiotherapy and hyperbaric oxygen, *Br. J. Radiol.* **56**:251-255.

Dobie, R. A., Sakai C. S., Sullivan M. D., Katon W. J., and Russo J., 1993, Antidepressant treatment of tinnitus patients: report of a randomized clinical trial and clinical prediction of benefit, *Am. J. Otol.* **14**:18-23.

Dockrell, H. M., and Playfair J. H., 1983, Killing of blood-stage murine malaria parasites by hydrogen peroxide, *Infect. Immun.* **39**:456-459.

Dogan, H., and Qalt S., 2001, Effects of chelating agents and sodium hypochlorite on mineral content of root dentin, *J. Endod.* **27**:578-580.

Dong, Z., Lavrovsky Y., Venkatachalam M. A., and Roy A. K., 2000, Heme oxygenase-1 in tissue pathology: the Yin and Yang, *Am. J. Pathol.* **156**:1485-1488.

Dore, S., 2002, Decreased activity of the antioxidant heme oxygenase enzyme: implications in ischemia and in Alzheimer's disease, *Free Radic. Biol. Med.* **32**:1276-1282.

Doroshow, J. H., 1995, Glutathione peroxidase and oxidative stress, *Toxicol. Lett.* **82/83**:395-398.

Dreher, D., and Junod A. F., 1996, Role of oxygen free radicals in cancer development, *Eur. J. Cancer* **32A**:30-38.

DuBois, A. B., 1962, Oxygen toxicity, *Anestesiology* **23**:473-477.

Duckers, H. J., Boehm M., True A. L., Yet S. F., San H., Park J. L., Clinton W. R., Lee M. E., Nabel G. J., and Nabel E. G., 2001, Heme oxygenase-1 protects against vascular constriction and proliferation, *Nat. Med.* **7**:693-698.

Duda, P. W., Schmied M. C., Cook S. L., Krieger J. I., and Hafler D. A., 2000, Glatiramer acetate (Copaxone) induces degenerate, Th2-polarized immune responses in patients with multiple sclerosis, *J. Clin. Invest* **105**:967-976.

Dumaswala, U. J., Wilson M. J., Wu Y. L., Wykle J., Zhuo L., Douglass L. M., and Daleke D. L., 2000, Glutathione loading prevents free radical injury in red blood cells after storage, *Free rad. Res.* **33**:517-529.

Durante, W., 2003, Heme oxygenase-1 in growth control and its clinical application to vascular disease, *J Cell Physiol.* **195**: 373-382.

Durelli, L., Verdun E., Barbero P., Bergui M., Versino E., Ghezzi A., Montanari E., and Zaffaroni M., 2002, Every-other-day interferon beta-1b versus once-weekly interferon beta-1a for multiple sclerosis: results of a 2-year prospective randomised multicentre study (INCOMIN), *Lancet* **359**:1453-1460.

Dworkin, R. H., 1999, Prevention of postherpetic neuralgia, *Lancet* **353**:1636-1637.

Eaton, L., 2003, World cancer rates set to double by 2020, *BMJ* **326**:728.

Edmunds, L.H., Jr., 1998, Inflammatory response to cardiopulmonary bypass, *Ann Thorac Surg.* **66**: S12-S16.

Eliakim, R., Karmeli F., Rachmilewitz D., Cohen P., and Zimran A., 2001, Ozone enema: a model of microscopic colitis in rats, *Dig Dis Sci.* **46**: 2515-2520.

Emanuel, E. J., Schnipper L. E., Kamin D. Y., Levinson J., and Lichter A. S., 2003, The costs of conducting clinical research, *J. Clin. Oncol.* **21**:4145-4150.

Emery, P., and Buch M., 2002, Treating rheumatoid arthritis with tumour necrosis factor alpha blockade, *BMJ* **324**:312-313.

Engelhart, M. J., Geerlings M. I., Ruitenberg A., Van Swieten J. C., Hofman A., Witteman J. C., and Breteler M. M., 2002, Dietary intake of antioxidants and risk of Alzheimer disease, *JAMA* **287**:3223-3229.

Enwonwu, J. W., 1989, Increased metabolic demand for arginine in sickle cell anemia, *Med. Sci. Res.* **17**:997-998.

Ernst, A., and Zibrak J. D., 1998, Carbon monoxide poisoning, *N. Engl. J. Med.* **339**:1603-1608.

Ernst, E., 1997, Thymus therapy for cancer? A criteria-based, systematic review, *Eur. J. Cancer* **33**:531-535.

Ernst, E., 2001, Mistletoe for cancer?, *Eur. J. Cancer* **37**:9-11.

Ernst, E., 2003, The current position of complementary/alternative medicine in cancer, *Eur. J. Cancer* **39**:2273-2277.

Ernst, E., and Cohen M. H., 2001, Informed consent in complementary and alternative medicine, *Arch. Intern. Med.* **161**:2288-2292.

Ernst, E., and Resch K. L., 1996, Evaluating specific effectiveness of complementary therapies - a position paper, *Forsch. Komplementärmed.* **3**:35-38.

Esterbauer, H., Schaur R. J., and Zollner H., 1991, Chemistry and biochemistry of 4-hydroxynonenal, malonaldehyde and related aldehydes, *Free rad. Biol. Med.* **11**:81-128.

Evans, H., Bauer M., Luckman I., and Page M., 2001, An assessment of the benefits afforded by the continuous versus intermittent operation of ozone for drinking water treatment, in *Proceedings of the 15th Ozone World Congress, London, UK, 11th-15th September 2001, Volume I* (IOA 2001, Ed.), Speedprint MacMedia Ltd, Ealing, London, UK, pp.219-234.

Evans, J. L., Goldfine I. D., Maddux B. A., and Grodsky G. M., 2002, Oxidative stress and stress-activated signaling pathways: a unifying hypothesis of type 2 diabetes, *Endocr. Rev.* **23**:599-622.

Evans, J. R., 2001, Risk factors for age-related macular degeneration, *Prog. Retin. Eye Res.* **20**:227-253.

Fabris, G., Tommasini G., Petralia B., Lavaroni A., De Nardi F., De Luca G., Biasizzo E., and Iaiza F., 2001, L'ossigeno-ozono terapia intra-foraminale, *Riv. Neuroradiol.* **14**:61-66.

Falk, S. J., Ward R., and Bleehen N. M., 1992, The influence of carbogen breathing on tumour tissue oxygenation in man evaluated by computerised p02 histography, *Br. J. Cancer* **66** :919-924.

Falm, E., 2004, Angiogenesis inhibitor in clinical development; where are we now and where are we going?, *Br. J. Cancer* **90**:1-7.

Farber, J. L., Kyle M. E., and Coleman J. B., 1990, Biology of disease. Mechanisms of cell injury by activated oxygen species, *Lab. Invest.* **62**:670-679.

Farquharson, C. A., Butler R., Hill A., Belch J. J., and Struthers A. D., 2002, Allopurinol improves endothelial dysfunction in chronic heart failure, *Circulation* **106**:221-226.

Farr, C. H., 1993, *Protocol for the intravenous administration of hydrogen peroxide*, International Bio-Oxidative Medicine Foundation, Oklahoma City, pp.29-31.

Feldmann, M., and Maini R. N., 2001, Anti-TNF alpha therapy of rheumatoid arthritis: what have we learned?, *Annu. Rev. Immunol.* **19**:163-196.

Fields, H. L., 1986, La generazione ectopica di impulsi negli afferenti primari, in *Il dolore: meccanismi di insorgenza e trattamento terapeutico* McGraw-Hill, pp.126-129.

Figueroa, M. S., Regueras A., and Bertrand J., 1996, Laser photocoagulation to treat macular soft drusen in age-related macular degeneration, *Retina* **14**:391-396.

Filippi, A., 2001, The influence of ozonised water on the epithelial wound healing process in the oral cavity, in *Proceedings of the 15th Ozone World Congress, London, UK, 11th-15th September 2001, Medical Therapy Conference* (IOA 2001, Ed.), Speedprint MacMedia Ltd, Ealing, London, UK, pp.109-116.

Filippi, A., and Kirschner H., 1995, Ozoniertes Wasser zur Desinfektion und Prophylaxe in der Zahn-Mund-Kieferheilkunde, in *Ozon-Handbuch. Grundlagen. Prävention. Therapie* (E. G. Beck, and R. Viebahn-Hänsler, Eds.), Ecomed, Landsberg, p.V-12.2 1-V-12.2 12.

Filippini, G., Munari L., Incorvaia B., Ebers G. C., Polman C., D'Amico R., and Rice G. P., 2003, Interferons in relapsing remitting multiple sclerosis: a systematic review, *Lancet* **361**:545-552.

Finger PT, Gelman YP, Berson AM, Szechter A., 2003, Palladium-103 plaque radiation therapy for macular degeneration: results of a 7 year study., *Br J Ophthalmol.* **87**:1497-503.

Fiocchi, C., 1998, Inflammatory bowel disease: etiology and pathogenesis, *Gastroenterology* **115**:182-205.

Fiocchi, C., 1999, From immune activation to gut tissue injury: the pieces of the puzzle are coming together, *Gastroenterology* **117**:1238-1241.

Fiocchi, C., 2004, Closing fistulas in Crohn's disease--should the accent be on maintenance or safety?, *N. Engl. J. Med.* **350**:934-936.

Fishman, A., Martinez F., Naunheim K., Piantadosi S., Wise R., Ries A., Weinmann G., and Wood D. E., 2003, A randomized trial comparing lung-volume-reduction surgery with medical therapy for severe emphysema, *N. Engl. J. Med.* **348**:2059-2073.

FitzGerald, G. A., and Patrono C., 2001, The coxibs, selective inhibitors of cyclooxygenase-2, *N. Engl. J. Med.* **345**:433-442.

Flach, J., and Seachrist L., 1994, Mind-body meld may boost immunity, *J. Nat. Cancer Inst.* **86**:256-258.

Fleischer, A. B., Jr., 1999, Treatment of atopic dermatitis: role of tacrolimus ointment as a topical noncorticosteroidal therapy, *J. Allergy Clin. Immunol.* **104**:S126-S130.

Floyd, R. A., 1999, Neuroinflammatory processes are important in neurodegenerative diseases: an hypothesis to explain the increased formation of reactive oxygen and nitrogen species as major factors involved in neurodegenerative disease development, *Free rad. Biol. Med.* **26**:1346-1355.

Foksinski, M., Bialkowski K., Skiba M., Ponikowska I., Szmurlo W., and Olinski R., 1999, Evaluation of 8-oxodeoxyguanosine, typical oxidative DNA damage, in lymphocytes of ozone-treated arteriosclerotic patients, *Mutat. Res.* **438**:23-27.

Fontana, L., McNeill K. L., Ritter J. M., and Chowienczyk P. J., 1999, Effects of vitamin C and of a cell permeable superoxide dismutase mimetic on acute lipoprotein induced endothelial dysfunction in rabbit aortic rings, *Br. J. Pharmacol.* **126**:730-734.

Fontana, L., Meyer T. E., Klein S., and Holloszy J. O., 2004, Long-term calorie restriction is highly effective in reducing the risk for atherosclerosis in humans, *Proc. Natl. Acad. Sci. U. S. A* **101**:6659-6663.

Frank RN., 2004, Diabetic retinopathy, *N Engl J Med.* **350**(1):48-58

Freeman, B. A., Miller B. E., and Mudd J. B., 1979, Reaction of ozone with human erythrocytes, in *Assessing toxic effects of environmental pollutants* (S. D. Lee, and J. B. Mudd, Eds.), Ann Arbor Science Publishers, Ann Arbor, Mich, pp.151-171.

Frei, B., 1999, On the role of vitamin C and other antioxidants in atherogenesis and vascular dysfunction, *Proc. Soc. Exp. Biol. Med.* **222**:196-204.

Fridovich, I., 1995, Superoxide radical and superoxide dismutases, *Annu. Rev. Biochem.* **64**:97-112.

Friedman-Kien, A. E., Eron L. J., Conant M., Growdon W., Badiak H., Bradstreet P. W., Fedorczyk D., Trout J. R., and Plasse T. F., 1988, Natural interferon alfa for treatment of condylomata acuminata., *JAMA-J. Am. Med. Assn.* **259**:533-538.

Fuchs, J., and Kern H., 1998, Modulation of UV-light-induced skin inflammation by D-alpha-tocopherol and L-ascorbic acid: a clinical study using solar simulated radiation, *Free rad. Biol. Med.* **25**:1006-1012.

Fukuda, K., Straus S. E., Hickie I., Sharpe M. C., Dobbins J. G., and Komaroff A., 1994, The chronic fatigue syndrome: a comprehensive approach to its definition and study. International Chronic Fatigue Syndrome Study Group, *Ann. Intern. Med.* **121**:953-959.

Fukunaga, K., Nakazono N., Suzuki T., and Takama K., 1999, Mechanism of oxidative damage to fish red blood cells by ozone, *IUBMB. Life* **48**:631-634.

Fulle, S., Mecocci P., Fano G., Vecchiet I., Vecchini A., Racciotti D., Cherubini A., Pizzigallo E., Vecchiet L., Senin U., and Beal M. F., 2000, Specific oxidative alterations in vastus lateralis muscle of patients with the diagnosis of chronic fatigue syndrome, *Free rad. Biol. Med.* **29**:1252-1259.

Fung, W. E., 1991, Interferon alpha 2a for treatment of age-related macular degeneration, *Amer. J. Ophthalmol.* **112**:349-350.

Fyles, A., Milosevic M., Hedley D., Pintilie M., Levin W., Manchul L., and Hill R. P., 2002, Tumor hypoxia has independent predictor impact only in patients with node-negative cervix cancer, *J. Clin. Oncol.* **20**:680-687.

Gabriel, C., Blauhut B., Greul R., Schneewels B., and Roggendorf M., 1996, Transmission of hepatitis C by ozone enrichment of autologous blood, *Lancet* **347**:541.

Galbraith, R., 1999, Heme oxygenase: who needs it?, *Proc. Soc. Exp. Biol. Med.* **222**:299-305.

Galleano, M., and Puntarulo S., 1995, Role of antioxidants on the erythrocytes resistance to lipid peroxidation after acute iron overload in rats, *Biochim. Biophys. Acta* **1271**:321-326.

Garber, G. E., Cameron D. W., Hawley-Foss N., Greenway D., and Shannon M. E., 1991, The use of ozone-treated blood in the therapy of HIV infection and immune disease: a pilot study of safety and efficacy., *AIDS* **5**:981-984.

Garg, A., 2004, Acquired and inherited lipodystrophies, *N. Engl. J. Med.* **350**:1220-1234.

Gatenby, R. A., Kessler H. B., Rosenblum J. S., Coia L. R., Moldofsky P. J., Hartz W. H., and Broder G. J., 1988, Oxygen distribution in squamous cell carcinoma metastases and its relationship to outcome of radiation therapy, *Int. J. Radiat. Oncol. Biol. Phys.* **14**:831-838.

Georgiades, E., Behan W. M., Kilduff L. P., Hadjicharalambous M., Mackie E. E., Wilson J., Ward S. A., and Pitsiladis Y. P., 2003, Chronic fatigue syndrome: new evidence for a central fatigue disorder, *Clin. Sci. (Lond)* **105**:213-218.

Ghiselli, A., Serafini M., Natella F., and Scaccini C., 2000, Total antioxidant capacity as a tool to assess redox status: critical view and experimental data, *Free rad. Biol. Med.* **29**:1106-1114.

Ghosh, S., Goldin E., Gordon F. H., Malchow H. A., Rask-Madsen J., Rutgeerts P., Vyhnalek P., Zadorova Z., Palmer T., and Donoghue S., 2003, Natalizumab for active Crohn's disease, *N. Engl. J. Med.* **348**:24-32.

Gionchetti, P., Rizzello F., Ferrieri A., Venturi A., Brignola C., Ferretti M., Peruzzo S., Miglioli M., and Campieri M., 1999, Rifaximin in patients with moderate or severe ulcerative colitis refractory to steroid-treatment: a double-blind, placebo-controlled trial, *Dig. Dis. Sci.* **44**:1220-1221.

Giunta, R., Coppola A., Luongo C., Sammartino A., Guastafierro S., Grassia A., Giunta L., Mascolo L., Tirelli A., and Coppola L., 2001, Ozonized autohemotransfusion improves hemorheological parameters and oxygen delivery to tissues in patients with peripheral occlusive arterial disease, *Ann. Hematol.* **80**:745-748.

Gjonovich, A., Sattin G. F., Girotto L., Bordin M., Gallo L., and Preciso G., 2001, Lombalgie ribelli: l'ossigeno-ozono terapia a confronto con altre metodiche, *Riv. Neuroradiol.* **14**:35-38.

Gladwin, M. T., Crawford J. H., and Patel R. P., 2004, The biochemistry of nitric oxide, nitrite, and hemoglobin: role in blood flow regulation, *Free Radic. Biol. Med.* **36**:707-717.

Gladwin, M. T., Schechter A. N., Shelhamer J. H., Pannell L. K., Conway D. A., Hrinczenko B. W., Nichols J. S., Pease-Fye M. E., Noguchi C. T., Rodgers G. P., and Ognibene F. P., 1999, Inhaled nitric oxide augments nitric oxide transport on sickle cell hemoglobin without affecting oxygen affinity, *J. Clin. Invest* **104**:937-945.

Glover, R. E., Ivy E. D., Orringer E. P., Maeda H., and Mason R. P., 1999, Detection of nitrosyl hemoglobin in venous blood in the treatment of sickle cell anemia with hydroxyurea, *Mol. Pharmacol.* **55**:1006-1010.

Goldman, M., 1996, Cancer risk of low-level exposure, *Science* **271**:1821-1822.

Goldstein, B. D., and Balchum O. J., 1967, Effect of ozone on lipid peroxidation in the red blood cell., *Proc. Soc. Exp. Biol. Med.* **126**:356-359.

Gomez, M., Espinosa E., and Caplan J. A., 1995, Application of medicinal ozone/oxygen in patients with sickle cell anemia, *Townsend Letter for Doctors* **January**:48-52.

Gonzalez, R., Borrego A., Zamora Z., Romay C., Hernandez F., Menendez S., Montero T., and Rojas A., 2004, Reversion by ozone treatment of acute nephrotoxicity induced by cisplatin in rats, *Mediators. Inflamm.,* in press.

Gooch, P. C., Creasia D. A., and Brewen J. G., 1976, The cytogenetic effects of ozone: inhalation and in vitro exposures, *Environ. Res.* **12**:188-195.

Gorre, M. E., Mohammed M., Ellwood K., Hsu N., Paquette R., Rao P. N., and Sawyers C. L., 2001, Clinical resistance to STI-571 cancer therapy caused by BCR-ABL gene mutation or amplification, *Science* **293**:876-880.

Grady, D., Rubin S. M., Petitti D. B., Fox C. S., Black D., Ettinger B., Ernster V. L., and Cummings S. R., 1992, Hormone therapy to prevent disease and prolong life in postmenopausal women, *Ann. Intern. Med.* **117**:1016-1037.

Graham, N., and Paraskeva P., 2001, Recent studies of ozone disinfection of secondary municipal effluents, in *Proceedings of the 15th Ozone World Congress, London, UK, 11th-15th September 2001, Volume I* (IOA 2001, Ed.), Speedprint MacMedia Ltd, Ealing, London, UK, pp.276-291.

Griffin, R. J., Okajima K., Barrios B., and Song C. W., 1996, Mild temperature hyperthermia combined with carbogen breathing increases tumor partial pressure of oxygen (pO2) and radiosensitivity, *Cancer Res.* **56**:5590-5593.

Grimaud, E., Heymann D., and Redini F., 2002, Recent advances in TGF-beta effects on chondrocyte metabolism. Potential therapeutic roles of TGF-beta in cartilage disorders, *Cytokine Growth Factor Rev.* **13**:241-257.

Grisham, M. B., Granger D. N., and Lefer D. J., 1998, Modulation of leukocyte-endothelial interactions by reactive metabolites of oxygen and nitrogen: relevance to ischemic heart disease, *Free rad. Biol. Med.* **25**:404-433.

Gross, T. J., and Hunninghake G. W., 2001, Idiopathic pulmonary fibrosis, *N. Engl. J. Med.* **345**:517-525.

Gubitz, G., and Sandercock P., 2000, Prevention of ischaemic stroke, *Brit. Med. J.* **321**:1455-1459.

Guerrero, A., Torres P., Duran M. T., Ruiz-Diez B., Rosales M., and Rodriguez-Tudela J. L., 2001, Airborne outbreak of nosocomial Scedosporium prolificans infection, *Lancet* **357**:1267-1268.

Gutstein, H. B., 2001, The biologic basis of fatigue, *Cancer* **92**:1678-1683.

Haas, A. F., Wong J. W., Iwahashi C. K., Halliwell B., Cross C. E., and Davis P. A., 1998, Redox regulation of wound healing? NF-kappaB activation in cultured human keratinocytes upon wounding and the effect of low energy HeNe irradiation, *Free rad. Biol. Med.* **25**:998-1005.

Hack, V., Breitkreutz R., Kinscherf R., Rohrer H., Bartsch P., Taut F., Benner A., and Droge W., 1998, The redox state as a correlate of senescence and wasting and as a target for therapeutic intervention, *Blood* **92**:59-67.

Hahn, M., Fennerty M. B., Corless C. L., Magaret N., Lieberman D. A., and Faigel D. O., 2000, Noninvasive tests as a substitute for histology in the diagnosis of Helicobacter pylori infection, *Gastrointest. Endosc.* **52**:20-26.

Halliwell, B., 1994, Free radicals and antioxidants: a personal view, *Nutr. Rev.* **52**:253-265.

Halliwell, B., 1996, Antioxidants in human health and disease, *Annu. Rev. Nutr.* **16**:33-50.

Halliwell, B., 1999a, Antioxidant defence mechanisms: from the beginning to the end (of the beginning), *Free rad. Res.* **31**:261-272.

Halliwell, B., 1999b, Vitamin C: poison, prophylactic or panacea?, *Trends Biochem. Sci.* **24**:255-259.

Halliwell, B., 2001, Role of free radicals in the neurodegenerative diseases: therapeutic implications for antioxidant treatment, *Drugs Aging* **18**:685-716.

Halliwell, B., 2003, Oxidative stress in cell culture: an under-appreciated problem?, *FEBS Lett.* **540**:3-6.

Halliwell, B., Zhao K., and Whiteman M., 2000, The gastrointestinal tract: a major site of antioxidant action?, *Free rad. Res.* **33**:819-830.

Hamilton, M. L., Van Remmen H., Drake J. A., Yang H., Guo Z. M., Kewitt K., Walter C. A., and Richardson A., 2001, Does oxidative damage to DNA increase with age?, *Proc. Natl. Acad. Sci. U. S. A* **98**:10469-10474.

Hanauer, S. B., and Dassopoulos T., 2001, Evolving treatment strategies for inflammatory bowel disease, *Annu. Rev. Med.* **52**:299-318.

Hanauer, S. B., Feagan B. G., Lichtenstein G. R., Mayer L. F., Schreiber S., Colombel J. F., Rachmilewitz D., Wolf D. C., Olson A., Bao W., and Rutgeerts P., 2002, Maintenance infliximab for Crohn's disease: the ACCENT I randomised trial, *Lancet* **359**:1541-1549.

Haniflin, J. M., and Tofte S. J., 1999, Update on therapy of atopic dermatitis, *J. Allergy Clin. Immunol.* **104**:S123-S125.

Hannuksela, M. L., and Ellahham S., 2001, Benefits and risks of sauna bathing, *Am. J. Med.* **110**:118-126.

Hardwick, C., 1940, The indications for and technique of whole-blood inkijections, *Practitioner* **144**:79-82.

Harman, D., 1956, A theory based on free radical and radiation chemistry, *J. Gerontol.* **11**:298-300.

Harris, A. L., 2002, Hypoxia--a key regulatory factor in tumour growth, *Nat. Rev. Cancer* **2**:38-47.

Harris, J. P., Weisman M. H., Derebery J. M., Espeland M. A., Gantz B. J., Gulya A. J., Hammerschlag P. E., Hannley M., Hughes G. B., Moscicki R., Nelson R. A., Niparko J. K., Rauch S. D., Telian S. A., and Brookhouser P. E., 2003, Treatment of corticosteroid-responsive autoimmune inner ear disease with methotrexate: a randomized controlled trial, *JAMA* **290**:1875-1883.

Hasselwander, O., and Young I. S., 1998, Oxidative stress in chronic renal failure, *Free Radic. Res.* **29**:1-11.

Hatoum, O. A., Miura H., and Binion D. G., 2003, The vascular contribution in the pathogenesis of inflammatory bowel disease, *Am. J. Physiol Heart Circ. Physiol* **285**:H1791-H1796.

Hawkins, C. L., and Davies M. J., 1996, Direct detection and identification of radicals generated during the hydroxyl radical-induced degradation of hyaluronic acid and related materials, *Free rad. Biol. Med.* **21**:275-290.

Hayflick, L., 2000, The future of ageing, *Nature* **408**:267-269.

Head, C. A., Brugnara C., Martinez-Ruiz R., Kacmarek R. M., Bridges K. R., Kuter D., Bloch K. D., and Zapol W. M., 1997, Low concentrations of nitric oxide increase oxygen affinity of sickle erythrocytes in vitro and in vivo, *J. Clin. Invest* **100**:1193-1198.

Helczynska, K., Kronblad A., Jogi A., Nilsson E., Beckman S., Landberg G., and Pahlman S., 2003, Hypoxia promotes a dedifferentiated phenotype in ductal breast carcinoma in situ, *Cancer Res.* **63**:1441-1444.

Heng, H., Rucker R. B., Crotty J., and Dubick M. A., 1987, The effects of ozone on lung, heart, and liver superoxide dismutase and glutathione peroxidase activities in the protein-deficient rat, *Toxicol. Lett.* **38**:225-237.

Henke, M., Laszig R., Rube C., Schafer U., Haase K. D., Schilcher B., Mose S., Beer K. T., Burger U., Dougherty C., and Frommhold H., 2003, Erythropoietin to treat head and neck cancer patients with anaemia undergoing radiotherapy: randomised, double-blind, placebo-controlled trial, *Lancet* **362**:1255-1260.

Hennekens, C. H., Buring J. E., and Peto R., 1994, Antioxidant vitamins - benefits not yet proved, *N. Engl. J. Med.* **330**:1080-1081.

Hernandez, F., Alvarez I., Corcho I., and Gonzalez M., 2004, Changes in glutathione antioxidant pathway components, HLA-DR and IgE in blood from asthma patients treated with ozone therapy, *Ozone in Science,* in press.

Hernandez, F., Menendez S., and Wong R., 1995, Decrease of blood cholesterol and stimulation of antioxidative response in cardiopathy patients treated with endovenous ozone therapy, *Free rad. Biol. Med.* **19**:115-119.

Herrmann, M., Voll R. E., and Kalden J. R., 2000, Etiopathogenesis of systemic lupus erythematosus, *Immunol. Today* **21**:424-426.

Herzenberg, L. A., De Rosa S. C., Dubs J. G., Roederer M., Anderson M. T., Ela S. W., Deresinski S. C., and Herzenberg L. A., 1997, Glutathione deficiency is associated with impaired survival in HIV disease, *Proc. Natl. Acad. Sci. U. S. A* **94**:1967-1972.

Hijnen, W. A. M., Bosklopper Th. G. J., Hofman J. A. M. H., Bosch A. D., and Medema G. J., 2001, Improvement of the disinfection efficiency of the full-scale ozonation of the River-Iake waterworks of Amsterdam water supply, in *Proceedings of the 15th Ozone World Congress, London, UK, 11th-15th September 2001, Volume I* (IOA 2001, Ed.), Speedprint MacMedia Ltd, Ealing, London, UK, pp.250-261.

Hillerdal, G., 1997, New principles for the treatment of diffuse pulmonary emphysema, *J. Intern. Med.* **242**:441-448.

Hirsch, K. R., and Wright T. L., 2000, The dilemma of disease progression in hepatitis C patients with normal serum aminotransferase levels, *Am. J. Med.* **109**:66-67.

Ho, D. D., 1997, Perspectives series: host/pathogen interactions. Dynamics of HIV-1 replication in vivo, *J. Clin. Invest* **99**:2565-2567.

Hockel, M., and Vaupel P., 2001, Tumor hypoxia: definitions and current clinical, biologic, and molecular aspects, *J. Natl. Cancer Inst.* **93**:266-276.

Hockel, M., Schlenger K., Aral B., Mitze M., Schaffer U., and Vaupel P., 1996, Association between tumor hypoxia and malignant progression in advanced cancer of the uterine cervix, *Cancer Res.* **56**:4509-4515.

Hodgson, H. J., 1996, Keeping Crohn's disease quiet, *N. Engl. J. Med.* **334**:1599-1600.

Holmboe, E. S., 2002, Oral antihyperglycemic therapy for type 2 diabetes: clinical applications, *JAMA* **287**:373-376.

Holmgren, A., 1989, Thioredoxin and glutaredoxin systems, *J. Biol. Chem.* **264**:13963-13966.

Honess, D. J., Andrews M. S., Ward R., and Bleehen N. M., 1995, Pentoxifylline increases RIF-1 tumour pO2 in a manner compatible with its ability to increase relative tumour perfusion, *Acta Oncol.* **34**:385-389.

Hooper, D. C., Scott G. S., Zborek A., Mikheeva T., Kean R. B., Koprowski H., and Spitsin S. V., 2000, Uric acid, a peroxynitrite scavenger, inhibits CNS inflammation, blood- CNS barrier permeability changes, and tissue damage in a mouse model of multiple sclerosis, *FASEB J.* **14**:691-698.

Hooper, L. V., and Gordon J. I., 2001, Commensal host-bacterial relationships in the gut, *Science* **292**:1115-1118.

Hoppe, C. C., and Walters M. C., 2001, Bone marrow transplantation in sickle cell anemia, *Curr. Opin. Oncol.* **13**:85-90.

Horsman, M. R., Chaplin D. J., and Brown J. M., 1989, Tumor radiosensitization by nicotinamide: a result of improved perfusion and oxygenation, *Radiat. Res.* **118**:139-150.

Houston, M., Estevez A., Chumley P., Aslan M., Marklund S., Parks D. A., and Freeman B. A., 1999, Binding of xanthine oxidase to vascular endothelium. Kinetic characterization and oxidative impairment of nitric oxide-dependent signaling, *J. Biol. Chem.* **274**:4985-4994.

Hrinczenko, B. W., Alayash A. I., Wink D. A., Gladwin M. T., Rodgers G. P., and Schechter A. N., 2000, Effect of nitric oxide and nitric oxide donors on red blood cell oxygen transport, *Br. J. Haematol.* **110**:412-419.

Hruz, P. W., Murata H., and Mueckler M., 2001, Adverse metabolic consequences of HIV protease inhibitor therapy: the search for a central mechanism, *Am. J. Physiol Endocrinol. Metab* **280**:E549-E553.

Hu, M.-L., 1994, Measurement of protein thiol groups and glutathione in plasma, *Meth. Enzymol.* **233**:380-385.

Huang, L. E., and Bunn H. F., 2003, Hypoxia-inducible factor and its biomedical relevance, *J. Biol. Chem.* **278**:19575-19578.

Hui, C. K., Yuen M. F., Sablon E., Chan A. O., Wong B. C., and Lai C. L., 2003, Interferon and ribavirin therapy for chronic hepatitis C virus genotype 6: a comparison with genotype 1, *J. Infect. Dis.* **187**:1071-1074.

Humayun, M. S., 2001, Intraocular retinal prosthesis, *Trans. Am. Ophthalmol. Soc.* **99**:271-300.

Ikeda, F., Shimomura H., Miyake M., Fujioka S. I., Itoh M., Takahashi A., Iwasaki Y., Sakaguchi K., Yamamoto K., Higashi T., and Tsuji T., 2000, Early clearance of circulating hepatitis C virus enhanced by induction therapy with twice-a-day intravenous injection of IFN-beta, *J. Interferon Cytokine Res.* **20**:831-836.

Imray, C. H., Walsh S., Clarke T., Tiivas C., Hoar H., Harvey T. C., Chan C. W., Forster P. J., Bradwell A. R., and Wright A. D., 2003, Effects of breathing air containing 3% carbon dioxide, 35% oxygen or a mixture of 3% carbon dioxide/35% oxygen on cerebral and peripheral oxygenation at 150 m and 3459 m, *Clin. Sci. (Lond)* **104**:203-210.

Inaba, D., Ruben J., Takagi O., and Arends J., 1996, Effect of sodium hypochlorite treatment on remineralization of human root dentine in vitro, *Caries Res.* **30**:218-224.

Inch, W. R., McCredie J. A., and Sutherland R. M., 1970, Effect of duration of breathing 95 percent oxygen plus 5 percent carbon dioxide before x-irradiation on cure of C3H mammary tumor, *Cancer* **25**:926-931.

Inui, Y., and Ichiyanagi I., 2001, *"Ozone Cleaner for Bedding, Bedclothes, etc.", Japanese Patent 2001 161797 A2 (Assignee: Yasunaga K.K.).*

Inzucchi, S. E., 2002, Oral antihyperglycemic therapy for type 2 diabetes: scientific review, *JAMA* **287**:360-372.

Iuliano, L., Colavita A. R., Leo R., Praticò D., and Violi F., 1997, Oxygen free radicals and platelet activation, *Free rad. Biol. Med.* **22**:999-1006.

Jackson, K. A., Majka S. M., Wang H., Pocius J., Hartley C. J., Majesky M. W., Entman M. L., Michael L. H., Hirschi K. K., and Goodell M. A., 2001, Regeneration of ischemic cardiac muscle and vascular endothelium by adult stem cells, *J. Clin. Invest* **107**:1395-1402.

Jacobs, M.-T., 1982, Untersuchung uber Zwischenfalle und typische Komplikationen in der Ozon-Sauerstoff-Therapie., *OzoNachrichten* **1**:5.

Jacobson, M. D., 1996, Reactive oxygen species and programmed cell death, *Trends Biochem Sci* **21**:83-86.

Jaeschke, H., 1995, Mechanisms of oxidant stress-induced acute tissue injury, *Proc. Soc. Exp. Biol. Med.* **209**:104-111.

Jang, M., Cai L., Udeani G. O., Slowing K. V., Thomas C. F., Beecher C. W., Fong H. H., Farnsworth N. R., Kinghorn A. D., Mehta R. G., Moon R. C., and Pezzuto J. M., 1997, Cancer chemopreventive activity of resveratrol, a natural product derived from grapes, *Science* **275**:218-220.

Jeffcoate, W. J., and Harding K. G., 2003, Diabetic foot ulcers, *Lancet* **361**:1545-1551.

Jenner, P., 1994, Oxidative damage in neurodegenerative disease, *Lancet* **344**:796-798.

Jia, L., Bonaventura C., Bonaventura J., and Stamler J. S., 1996, S-nitrosohaemoglobin: a dynamic activity of blood involved in vascular control, *Nature* **380**:221-226.

Jiang, J., Jordan S. J., Barr D. P., Gunther M. R., Maeda H., and Mason R. P., 1997, In vivo production of nitric oxide in rats after administration of hydroxyurea, *Mol. Pharmacol.* **52**:1081-1086.

Jindal, N., and Dellinger R. P., 2000, Inhalation of nitric oxide in acute respiratory distress syndrome, *J. Lab Clin. Med.* **136**:21-28.

Johnson, P. W., Dixon R., and Ross A. D., 1998, An in-vitro test for assessing the viability of Ascaris suum eggs exposed to various sewage treatment processes, *Int. J. Parasitol.* **28**:627-633.

Johnson, R. J., Willson R., Yamabe H., Couser W., Alpers C. E., Wener M. H., Davis C., and Gretch D. R., 1994, Renal manifestations of hepatitis C virus infection, *Kidney Int.* **46**:1255-1263.

Jolly, C., and Morimoto R. I., 2000, Role of the heat shock response and molecular chaperones in oncogenesis and cell death, *J. Natl. Cancer Inst.* **92**:1564-1572.

Jordan, L., Beaver K., and Foy S., 2002, Ozone treatment for radiotherapy skin reactions: is there an evidence base for practice?, *Eur. J. Oncol. Nurs.* **6**:220-227.

Jornot, L., Mirault M. E., and Junod A. F., 1991, Differential expression of hsp70 stress proteins in human endothelial cells exposed to heat shock and hydrogen peroxide, *Amer. J. Respir. Cell Molec. Biol.* **5**:265-275.

Joyce, J., Rabe-Hesketh S., and Wessely S., 1998, Reviewing the reviews: the example of chronic fatigue syndrome, *JAMA* **280**:264-266.

Joyner, M. J., and Dietz N. M., 1997, Nitric oxide and vasodilation in human limbs, *J. Appl. Physiol.* **83**:1785-1796.

Jucopilla, N., Ferrarese C., Tirapelle G., Battista R., Mazzo G., and Robert A., 2000, Infiltrazioni disco-foraminali con O2-O3 nelle SDR da conflitto disco-radicolari lombari, in *Proceedings: I Congresso IMOS, Italia, Siena, 2-4 novembre 2000*, p. 38.

Kadokawa, N., Morioka T., Motoyama N., Hashino M., Mori Y., Nishijima W., Okada M., and Moniwa T., 2001, Advanced water treatment using ozone resistant microfiltration membrane, in *Proceedings of the 15th Ozone World Congress, London, UK, 11th-15th September 2001, Volume I* (IOA 2001, Ed.), Speedprint MacMedia Ltd, Ealing, London, UK, pp.125-133.

Kalebic, T., Kinter A., Poli G., Anderson M. E., Meister A., and Fauci A. S., 1991, Suppression of human immunodeficiency virus expression in chronically infected monocytic cells by glutathione, glutathione ester, and N-acetylcysteine, *Proc. Nat. Acad. Sci. USA* **88**:986-990.

Kamp, D. W., 2003, Idiopathic pulmonary fibrosis: the inflammation hypothesis revisited, *Chest* **124** :1187-1190.

Kang, H.J., Kim, H.S., Zhang, S.Y., Park, K.W., Cho, H.J., Koo, B.K., Kim, Y.J., Soo Lee, D., Sohn, D.W., Han, K.S., Oh, B.H., Lee, M.M., Park, Y.B., 2004, Effects of intracoronary infusion of peripheral blood stem-cells mobilised with granulocyte-colony stimulating factor on left ventricular systolic function and restenosis after coronary stenting in myocardial infarction: the MAGIC cell randomised clinical trial., *Lancet,* **363**:751-756.

Karandikar, N. J., Crawford M. P., Yan X., Ratts R. B., Brenchley J. M., Ambrozak D. R., Lovett-Racke A. E., Frohman E. M., Stastny P., Douek D. C., Koup R. A., and Racke M. K., 2002, Glatiramer acetate (Copaxone) therapy induces CD8(+) T cell responses in patients with multiple sclerosis, *J. Clin. Invest* **109**:641-649.

Karp, C. L., Biron C. A., and Irani D. N., 2000, Interferon beta in multiple sclerosis: is IL-12 suppression the key?, *Immunol. Today* **21**:24-28.

Karppinen, J., Korhonen T., Malmivaara A., Paimela L., Kyllonen E., Lindgren K. A., Rantanen P., Tervonen O., Niinimaki J., Seitsalo S., and Hurri H., 2003, Tumor necrosis factor-alpha monoclonal antibody, infliximab, used to manage severe sciatica, *Spine* **28** :750-753.

Kashiba, M., Kasahara E., Chien K. C., and Inoue M., 1999, Fates and vascular action of S-nitrosoglutathione and related compounds in the circulation, *Arch. Biochem. Biophys.* **363**:213-218.

Kasumjan, S. A., Lelyanov A. D., Guseva E. D., and Alexeev B. P., 1995, The ozonetherapy of the acute suppurative infection, in *The ozone in biology and medicine. 2nd all Russian scientific-practical conference, September 6-8, 1995. Russian association of ozonetherapy, Reshetnikovskaya street 2, Nizhni Novgorod, 603006 Russia*, p.16.

Katano, H., Pesnicak L., and Cohen J. I., 2004, Simvastatin induces apoptosis of Epstein-Barr virus (EBV)-transformed lymphoblastoid cell lines and delays development of EBV lymphomas, *Proc. Natl. Acad. Sci. U. S. A* **101**:4960-4965.

Kaul, D. K., Tsai H. M., Liu X. D., Nakada M. T., Nagel R. L., and Coller B. S., 2000, Monoclonal antibodies to alphaVbeta3 (7E3 and LM609) inhibit sickle red blood cell-endothelium interactions induced by platelet-activating factor, *Blood* **95**:368-374.

Keane, J., Gershon S., Wise R. P., Mirabile-Levens E., Kasznica J., Schwieterman W. D., Siegel J. N., and Braun M. M., 2001, Tuberculosis associated with infliximab, a tumor necrosis factor alpha- neutralizing agent, *N. Engl. J. Med.* **345**:1098-1104.

Keane, M. P., and Strieter R. M., 2002, The importance of balanced pro-inflammatory and anti-inflammatory mechanisms in diffuse lung disease, *Respir. Res.* **3**:5.

Keegan, B. M., and Noseworthy J. H., 2002, Multiple sclerosis, *Annu. Rev. Med.* **53**:285-302.

Kelly, F. J., Mudway I., Krishna M. T., and Holgate S. T., 1995, The free radical basis of air pollution: focus on ozone, *Resp. Med.* **89**:647-656.

Kiang, J. G., and Tsokos G. C., 1998, Heat shock protein 70 kDa: molecular biology, biochemistry, and physiology, *Pharmacol. Ther.* **80**:183-201.

Kief, H., 1993, The treatment of malignant diseases with AHIT, in *Proceedings: IOA Congress, San Francisco, Ca, USA* (International Ozone Association, Ed.), pp.26-31.

Kim, C. H., Choi H., Chun Y. S., Kim G. T., Park J. W., and Kim M. S., 2001, Hyperbaric oxygenation pretreatment induces catalase and reduces infarct size in ischemic rat myocardium, *Pflugers Arch.* **442**:519-525.

Kimberlin, D. W., and Rouse D. J., 2004, Clinical practice. Genital herpes, *N. Engl. J. Med.* **350**:1970-1977.

Kimura, I., Shinoda K., Tanino T., Ohtake Y., Mashima Y., and Oguchi Y., 2003, Scanning laser Doppler flowmeter study of retinal blood flow in macular area of healthy volunteers, *Br. J. Ophthalmol.* **87**:1469-1473.

Kindwall, E. P., 1993, Hyperbaric Oxygen, *Brit. Med. J.* **307**:515-516.

King, G. L., and Suzuma K., 2000, Pigment-epithelium-derived factor--a key coordinator of retinal neuronal and vascular functions, *N. Engl. J. Med.* **342**:349-351.

Kinnula, V. L., and Crapo J. D., 2004, Superoxide dismutases in malignant cells and human tumors, *Free Radic. Biol. Med.* **36**:718-744.

Kipnis, J., Yoles E., Porat Z., Cohen A., Mor F., Sela M., Cohen I. R., and Schwartz M., 2000, T cell immunity to copolymer 1 confers neuroprotection on the damaged optic nerve: possible therapy for optic neuropathies, *Proc. Natl. Acad. Sci. U. S. A* **97**:7446-7451.

Kirby, P. K., Kiviat N., Beckman A., Wells D., Sherwin S., and Corey L., 1988, Tolerance and efficacy of recombinant human interferon gamma in the treatment of refractory genital warts., *Amer. J. Med.* **85**:183-188.

Kleeberger, S. R., Levitt R. C., Zhang L. Y., Longphre M., Harkema J., Jedlicka A., Eleff S. M., DiSilvestre D., and Holroyd K. J., 1997, Linkage analysis of susceptibility to ozone-induced lung inflammation in inbred mice, *Nat. Genet.* **17**:475-478.

Klein, R., Klein B. E., and Franke T., 1993, The relationship of cardiovascular disease and its risk factors to age- related maculopathy. The Beaver Dam Eye Study, *Ophthalmology* **100**:406-414.

Klein, R., Klein B. E., Jensen S. C., and Meuer S. M., 1997, The five-year incidence and progression of age-related maculopathy: the Beaver Dam Eye Study, *Ophthalmology* **104**:7-21.

Klomp, H. M., Spincemaille G. H. J. J., Steyerberg E. W., Habbema J. D. F., van Urk H., and for the ESES study group, 1999, Spinal-cord stimulation in critical limb ischaemia: a randomised trial., *Lancet* **353**:1040-1044.

Knoch, H.-G., and Klug W., 1990, Ozon-Sauerstoff-Therapie der Proktitis., *Med. Welt* **41**:371-374.

Knoch, H.-G., Roschke W., and Klug W., 1987, Ozone/oxygen therapy in proctology, *OzoNachrichten* **6**:51-70.

Knudsen, P. J., Leon J., Ng A. K., Shaldon S., Floege J., and Koch K. M., 1989, Hemodialysis-related induction of beta-2-
microglobulin and interleukin-1 synthesis and release by mononuclear phagocytes, *Nephron* **53**:188-193.

Kohner, E., 2003a, Extracts from "concise clinical evidence". Commentary: treatment of diabetic retinopathy, *BMJ* **326**:1023-1025. Kohner, E. M., 2003b, Aspirin for diabetic retinopathy, *BMJ* **327**:1060-1061.

Kokura, S., Yoshida N., and Yoshikawa T., 2002, Anoxia/reoxygenation-induced leukocyte-endothelial cell interactions, *Free Radic. Biol. Med.* **33**:427-432.

Kollef, M. H., and Fraser V. J., 2001, Antibiotic resistance in the intensive care unit, *Ann. Intern. Med.* **134**:298-314.

Komaroff, A. L., 2000, The biology of chronic fatigue syndrome, *Am. J. Med.* **108**:169-171.

Komaroff, A. L., and Buchwald D. S., 1998, Chronic fatigue syndrome: an update, *Annu. Rev. Med.* **49** :1-13.

Kondo, S., Toyokuni S., Iwasa Y., Tanaka Y., Onodera H., Hiai H., and Imamura M., 1999, Persistent oxidative stress in human colorectal carcinoma, but not in adenoma, *Free rad. Biol. Med.* **27**:401-410.

Konrad, H., 1995, Ozone therapy for herpes simplex and herpes zoster, in *Proceedings Ozone in Medicine, 12th World Congress of the International Ozone Association, 15th to 18th May 1995, Lille, France* (International Ozone Association, Ed.), Instaprint S.A., Tours, pp.187-194.

Konrad, H., 2001, Ozone therapy for post-herpetic neuralgia. A retrospective study of 55 cases, in *Proceedings of the 15th Ozone World Congress, London, UK, 11th-15th September 2001, Medical Therapy Conference* (IOA 2001, Ed.), Speedprint MacMedia Ltd, Ealing, London, UK, pp.85-88.

Konstantinov, K., von Mikecz A., Buchwald D., Jones J., Gerace L., and Tan E. M., 1996, Autoantibodies to nuclear envelope antigens in chronic fatigue syndrome, *J. Clin. Invest* **98**:1888-1896.

Kotler, D. P., 2003, HIV infection and lipodystrophy, *Prog. Cardiovasc. Dis.* **45**:269-284.

Kraft, K., Stenkamp E., Sachinidis A., Seewald S., and Vetter H., 1998, Effect of autohemotherapy with ozone on cardiovascular risk factors in patients with mild hypertension, *Perfusion* **11**:216-219.

Kramer, B. S., and Klausner R. D., 1997, Grappling with cancer, Defeatism versus the reality of progress, *N. Engl. J. Med.* **337**:931-934.

Kremer, J. M., Westhovens R., Leon M., Di Giorgio E., Alten R., Steinfeld S., Russell A., Dougados M., Emery P., Nuamah I. F., Williams G. R., Becker J. C., Hagerty D. T., and Moreland L. W., 2003, Treatment of rheumatoid arthritis by selective inhibition of T-cell activation with fusion protein CTLA4Ig, *N. Engl. J. Med.* **349**:1907-1915.

Krinsky, N. I., Landrum J. T., and Bone R. A., 2003, Biologic mechanisms of the protective role of lutein and zeaxanthin in the eye, *Annu. Rev. Nutr.* **23**:171-201.

Kudravcev, B. P., Miroshin S. J., and Semyonov S. V., 1995, The use of ozonized solutions in complex treatment of peritonitis, in *The ozone in biology and medicine. 2nd all Russian Scientific-Practical Conference, September 6-8, 1995, Russian Association of Ozonetherapy, Reshetnikovskaya stret, 2 - Nizhni Novgorod, 603006 Russia*, Nizhni Novgorod, p.20.

Kumaraguruparan, R., Subapriya R., Viswanathan P., and Nagini S., 2002, Tissue lipid peroxidation and antioxidant status in patients with adenocarcinoma of the breast, *Clin. Chim. Acta* **325**:165-170.

Kume, M., Yamamoto Y., Saad S., Gomi T., Kimoto S., Shimabukuro T., Yagi T., Nakagami M., Takada Y., Morimoto T., and Yamaoka Y., 1996, Ischemic preconditioning of the liver in rats: implications of heat shock protein induction to increase tolerance of ischemia-reperfusion injury, *J. Lab Clin. Med.* **128**:251-258.

Kupper, T. S., 2003, Immunologic targets in psoriasis, *N. Engl. J. Med.* **349**:1987-1990.

Kuruvilla, A. P., Shah R., Hochwald G. M., Liggitt H. D., Palladino M. A., and Thorbecke G. J., 1991, Protective effect of transforming growth factor b_1 on experimental autoimmune diseases in mice., *Proc. Nat. Acad. Sci. USA* **88**:2918-2921.

Labow, R. S., Tocchi M., and Rock G., 1986, Contamination of platelet storage bags by phthalate esters, *Toxicol. Environ. Health* **19**:591-598.

Laitinen, M., Mäkinen K., Manninen H., Matsi P., Kossila M., Agrawal R. S., Pakkanen T., Luoma J. S., Viita H., Hartikainen J., Alhava E., Laakso M., and Ylä-Herttuala S., 1998, Adenovirus-mediated gene transfer to lower limb artery of patients with chronic critical leg ischemia, *Hum. Gene Ther.* **9**:1481-1486.

Lalezari, J. P., Henry K., O'Hearn M., Montaner J. S., Piliero P. J., Trottier B., Walmsley S., Cohen C., Kuritzkes D. R., Eron J. J., Jr., Chung J., DeMasi R., Donatacci L., Drobnes C., Delehanty J., and Salgo M., 2003, Enfuvirtide, an HIV-1 fusion inhibitor, for drug-resistant HIV infection in North and South America, *N. Engl. J. Med.* **348**:2175-2185.

Lambeth, J. D., 2004, NOX enzymes and the biology of reactive oxygen, *Nat. Rev. Immunol.* **4**:181-189.

Landay, A. L., Jessop C., Lennette E. T., and Levy J. A., 1991, Chronic fatigue syndrome: clinical condition associated with immune activation, *Lancet* **338**:707-712.

Lane, R., and Phillips M., 2003, Rhabdomyolysis, *BMJ* **327**:115-116.

Lang, A. E., and Lozano A. M., 1998a, Parkinson's disease. First of two parts, *N. Engl. J. Med.* **339**:1044-1053.

Lang, A. E., and Lozano A. M., 1998b, Parkinson's disease. Second of two parts, *N. Engl. J. Med.* **339**:1130-1143.

Langen, R. C., Korn S. H., and Wouters E. F., 2003, ROS in the local and systemic pathogenesis of COPD, *Free Radic. Biol. Med.* **35**:226-235.

Larini, A., and Bocci V., 2004, Effects of ozone on isolated peripheral blood mononuclear cells, *Toxicol. Vitro,* in press.

Larini, A., Bianchi L., and Bocci V., 2003, The ozone tolerance: I) Enhancement of antioxidant enzymes is ozone dose-dependent in Jurkat cells, *Free Radic. Res.* **37**: 1163-1168.

Larini, A., Bianchi L., and Bocci V., 2004, Effect of 4-hydroxynonenal on antioxidant capacity and apoptosis induction in Jurkat T cells, *Free Radic. Res.* **38**:509-516.

Larrea, E., Beloqui O., Muñoz-Navas M. A., Civeira M. P., and Prieto J., 1998, Superoxide dismutase in patients with chronic hepatitis C virus infection, *Free rad. Biol. Med.* **24**:1235-1241.

Last, J. A., Warren D. L., Pecquet-Goad E., and Witschi H., 1987, Modification by ozone of lung tumor development in mice, *J. Nat. Cancer Inst.* **78**:149-154.

Lawrence, W. H., 1978, Phthalate esters: the question of safety, *Clin. Toxicol.* **13**:89.

Leach, R. M., Rees P. J., and Wilmshurst P., 1998, Hyperbaric oxygen therapy, *Brit. Med. J.* **317**:1140-1143.

Lebwohl, M., Tyring S. K., Hamilton T. K., Toth D., Glazer S., Tawfik N. H., Walicke P., Dummer W., Wang X., Garovoy M. R., and Pariser D., 2003, A novel targeted T-cell modulator, efalizumab, for plaque psoriasis, *N. Engl. J. Med.* **349**:2004-2013.

Lederman, R. J., Mendelsohn F. O., Anderson R. D., Saucedo J. F., Tenaglia A. N., Hermiller J. B., Hillegass W. B., Rocha-Singh K., Moon T. E., Whitehouse M. J., and Annex B. H., 2002, Therapeutic angiogenesis with recombinant fibroblast growth factor-2 for intermittent claudication (the TRAFFIC study): a randomised trial, *Lancet* **359**:2053-2058.

Lee, T. S., and Chau L. Y., 2002, Heme oxygenase-1 mediates the anti-inflammatory effect of interleukin-10 in mice, *Nat. Med.* **8**:240-246.

Leist, M., Raab B., Maurer S., and Brigelius-Flohé R., 1996, Conventional cell culture media do not adequately supply cells with antioxidants and thus facilitate peroxide-induced genotoxicity, *Free rad. Biol. Med.* **21**:297-306.

León, O. S., Menéndez S., Merino N., Castillo R., Sam S., Pérez L., Cruz E., and Bocci V., 1998, Ozone oxidative preconditioning: a protection against cellular damage by free radicals, *Mediat. Inflamm.* **7**:289-294.

Leonardi, C. L., Powers J. L., Matheson R. T., Goffe B. S., Zitnik R., Wang A., and Gottlieb A. B., 2003, Etanercept as monotherapy in patients with psoriasis, *N. Engl. J. Med.* **349**:2014-2022.

Leonardi, M., Simonetti L., and Barbara C., 2001b, Effetti dell'ozono sul nucleo polposo: reperti anatomo-patologici su un caso operato, *Riv. Neuroradiol.* **14**:57-59.

Letterio, J. J., and Roberts A. B., 1998, Regulation of immune responses by TGF-b, *Annu. Rev. Immunol.* **16**:137-161.

Leung, D. Y., 1999, Pathogenesis of atopic dermatitis, *J. Allergy Clin. Immunol.* **104**:S99-108.

Levi, F., Lucchini F., Negri E., Boyle P., and La Vecchia C., 1999, Cancer mortality in Europe, 1990-1994, and an overview of trends from 1955 to 1994, *Eur. J. Cancer* **35**:1477-1516.

Levine, M., Conry-Cantilena C., Wang Y., Welch R. W., Washko P. W., Dhariwal K. R., Park J. B., Lazarev A., Graumlich J. F., King J., and Cantilena L. R., 1996, Vitamin C pharmacokinetics in health volunteers: evidence for a recommended dietary allowance, *Proc. Nat. Acad. Sci. USA* **93**:3704-3709.

Levine, M., Daruwala R. C., Park J. B., Rumsey S. C., and Wang Y., 1998, Does vitamin C have a pro-oxidant effect?, *Nature* **395**:231.

Levine, R. L., 2002, Carbonyl modified proteins in cellular regulation, aging, and disease, *Free Radic. Biol. Med.* **32**:790-796.

Lewin, N., Craik S., Li H., Smith D. W., and Belosevic M., 2001, Sequential inactivation of *Cryptosporidium* using ozone followed by free chlorine in natural water, *Ozone-Sci. Eng.* **23**:411-420.

Lewis, L. M., Flechtner T. W., Kerkay J., Pearson K. H., Chen W. T., Popowniak K. L., and Nakamoto S., 1977, Determination of plasticizer levels in serum of hemodialysis patients, *Trans. Am. Soc. Artif. Intern. Organs* **23**:566-572.

Li, C. K., Chan P. K., Ling S. C., and Ha S. Y., 2002, Interferon and ribavirin as frontline treatment for chronic hepatitis C infection in thalassaemia major, *Br. J. Haematol.* **117**:755-758.

Li, P. A., Liu G. J., He Q. P., Floyd R. A., and Siesjo B. K., 1999, Production of hydroxyl free radical by brain tissues in hyperglycemic rats subjected to transient forebrain ischemia, *Free rad. Biol. Med.* **27**:1033-1040.

Liao, J. K., 2002, Isoprenoids as mediators of the biological effects of statins, *J. Clin. Invest* **110**:285-288.

Liaw, K.-L., Glass A. G., Manos M. M., Greer C. E., Scott D. R., Sherman M., Burk R. D., Kurman R. J., Wacholder S., Rush B. B., Cadell D. M., Lawler P., Tabor D., and Schiffman M., 1999, Detection of human papillomavirus DNA in cytologically normal women and subsequent cervical squamous intraepithelial lesions, *J. Nat. Cancer Inst.* **91**:954-960.

Lilienfeld, D. E., and Perl D. P., 1993, Projected neurodegenerative disease mortality in the United States, 1990-2040, *Neuroepidemiology* **12**:219-228.

Lindner, A., Charra B., Sherrard D. J., and Scribner B. H., 1974, Accelerated atherosclerosis in prolonged maintenance hemodialysis, *N. Engl. J. Med.* **290**:697-701.

Liou, C. T., Wang J. S., Ooi H. K., 2002, Effect of ozone treatment on Eimeria colchici oocysts, *J. Parasitol.* **88**: 159-162.

Lippman, M., 1989, Health effects of ozone, a critical review, *J. Am. Air Pollut. Control Assoc.* **39**:672-695.

Littlewood, T. J., Bajetta E., Nortier J. W., Vercammen E., and Rapoport B., 2001, Effects of epoetin alfa on hematologic parameters and quality of life in cancer patients receiving nonplatinum chemotherapy: results of a randomized, double-blind, placebo-controlled trial, *J. Clin. Oncol.* **19**:2865-2874.

Liu, H., Ren J. G., Cooper W. L., Hawkins C. E., Cowan M. R., and Tong P. Y., 2004, Identification of the antivasopermeability effect of pigment epithelium-derived factor and its active site, *Proc. Natl. Acad. Sci. U. S. A* **101**:6605-6610.

Livrea, M. A., Tesoriere L., Pintaudi A. M., Calabrese A., Maggio A., Freisleben H. J., D'Arpa D., D'Anna R., and Bongiorno A., 1996, Oxidative stress and antioxidant status in beta-thalassemia major: iron overload and depletion of lipid-soluble antioxidants, *Blood* **88**:3608-3614.

Llevadot, J., Murasawa S., Kureishi Y., Uchida S., Masuda H., Kawamoto A., Walsh K., Isner J. M., and Asahara T., 2001, HMG-CoA reductase inhibitor mobilizes bone marrow--derived endothelial progenitor cells, *J. Clin. Invest* **108**:399-405.

Lockwood, A. H., Salvi R. J., and Burkard R. F., 2002, Tinnitus, *N. Engl. J. Med.* **347**:904-910.

Loconte, S., 2000, La sindrome fibromialgica primaria, in *Proceedings: I Congresso IMOS, Italia, Siena, 2-4 novembre 2000*, p.40.

Loebstein, R., Lehotay D. C., Luo X., Bartfay W., Tyler B., and Sher G. D., 1998, Diabetic nephropathy in hypertransfused patients with beta-thalassemia. The role of oxidative stress, *Diabetes Care* **21**:1306-1309.

Long, N. C., Suh J., Morrow J. D., Schiestl R. H., Murthy G. G., Brain J. D., and Frei B., 2001, Ozone causes lipid peroxidation but little antioxidant depletion in exercising and nonexercising hamsters, *J. Appl. Physiol* **91**:1694-1700.

Loprete, F., 1999, Utilizzo dell'ossigeno-ozonoterapia nel trattamento della malattia varicosa e sue complicanze, in *L'Ozonoterapia nel 2000* (F. Ceccherelli, and F. Giron, Eds.), Edizioni Libreria Cortina, Torino, pp.129-135.

Los, M., Dröge W., Stricker K., Baeuerle P. A., and Schulze-Osthoff K., 1995, Hydrogen peroxide as a potent activator of T lymphocyte functions, *Eur. J. Immunol.* **25**:159-165.

Love, I. N., 1888, Peroxide of hydrogen as a remedial agent, *JAMA* :262-265.

Lovell, D. J., Giannini E. H., Reiff A., Cawkwell G. D., Silverman E. D., Nocton J. J., Stein L. D., Gedalia A., Ilowite N. T., Wallace C. A., Whitmore J., and Finck B. K., 2000, Etanercept in children with polyarticular juvenile rheumatoid arthritis. Pediatric Rheumatology Collaborative Study Group, *N. Engl. J. Med.* **342**:763-769.

Lusis, A. J., 2000, Atherosclerosis, *Nature* **407**:233-241.

Mach, F., 2003, Statins as novel immunomodulators: from cell to potential clinical benefit, *Thromb. Haemost.* **90**: 607-10.

Macular photocoagulation group, 1991, Argon laser photocoagulation for neovascular maculopathy after five years: results for randomized clinical trials, *Arch. Ophthalmol.* **109**:1109-1114.

Maddox, K., and Back R. F., 1935, An enquiry into the value of autohaemotherapy in juvenile asthma, *Arch. Dis. Child.* **10**:381-388.

Maestrelli, P., Paska C., Saetta M., Turato G., Nowicki Y., Monti S., Formichi B., Miniati M., and Fabbri L. M., 2003, Decreased haem oxygenase-1 and increased inducible nitric oxide synthase in the lung of severe COPD patients, *Eur. Respir. J.* **21**:971-976.

Maini, R., St Clair E. W., Breedveld F., Furst D., Kalden J., Weisman M., Smolen J., Emery P., Harriman G., Feldmann M., and Lipsky P., 1999, Infliximab (chimeric anti-tumour necrosis factor alpha monoclonal antibody) versus placebo in rheumatoid arthritis patients receiving concomitant methotrexate: a randomised phase III trial. ATTRACT Study Group, *Lancet* **354**:1932-1939.

Makino, Y., Okamoto K., Yoshikawa N., Aoshima M., Hirota K., Yodoi J., Umesono K., Makino I., and Tanaka H., 1996, Thioredoxin: a redox-regulating cellular cofactor for glucocorticoid hormone action. Cross talk between endocrine control of stress response and cellular antioxidant defense system, *J. Clin. Invest* **98**:2469-2477.

Mallozzi, C., Di Stasi A. M., and Minetti M., 1997, Peroxynitrite modulates tyrosine-dependent signal transduction pathway of human erythrocyte band 3, *FASEB J.* **11**:1281-1290.

Manu, P., 2000, Chronic fatigue syndrome: the fundamentals still apply, *Am. J. Med.* **108**:172-173.

Markesbery, W. R., 1997, Oxidative stress hypothesis in Alzheimer's disease, *Free rad. Biol. Med.* **23**:134-147.

Marrades, R. M., Roca J., Campistol J. M., Diaz O., Barbera J. A., Torregrosa J. V., Masclans J. R., Cobos A., Rodriguez-Roisin R., and Wagner P. D., 1996, Effects of erythropoietin on muscle O2 transport during exercise in patients with chronic renal failure, *J. Clin. Invest* **97**:2092-2100.

Martin, P., 1997, Wound healing-aiming for perfect skin regeneration, *Science* **276**:75-81.

Martindale, W., and Capper K. T., 1952, *The extra pharmacopoeia*, The Pharmaceutical Press, London, pp.1-816.

Masschelein, W. J., 1996, Iodometric method for the determination of ozone in a process gas, in *Ozon-Handbuch. Grundlagen. Prävention. Therapie* (E. G. Beck, and R. Viebahn-Hänsler, Eds.), Ecomed, Landsberg, pp.1-3.

Matos, H. R., Di Mascio P., and Medeiros M. H., 2000, Protective effect of lycopene on lipid peroxidation and oxidative DNA damage in cell culture, *Arch. Biochem. Biophys.* **383**:56-59.

Matsumoto, A., Sakurai S., Shinriki N., Suzuki S., and Miura T., 2001, Therapeutic effects of ozonized olive oil in the treatment of intractable fistula and wound after surgical operation, in *Proceedings of the 15th Ozone World Congress, London, UK, 11th-15th September 2001, Medical Therapy Conference* (IOA 2001, Ed.), Speedprint MacMedia Ltd, Ealing, London, UK, pp.77-84.

Mattassi, R., Bassi P., D'Angelo F., Franchina A., and Sbrascini S., 1985, Ozone as therapy in herpes simplex and herpes zoster diseases, in *Medical applications of ozone* (J. LaRaus, Ed.), International Ozone Association, Norwalk, pp.134-137.

Mattassi, R., D'Angelo F., Bisetti P., Colombo R., and Vaghi M., 1987, Terapia con ozono per via parenterale nelle arteriopatie obliteranti periferiche: meccanismo biochimico e risultati clinici, *Il Giornale Di Chirurgia* **VIII**:109-111.

Mattox, D. E., and Simmons F. B., 1977, Natural history of sudden sensorineural hearing loss, *Ann. Otol. Rhinol. Laryngol.* **86**:463-480.

Mawsouf, N., Tanbouli T. T., and El-Tayar W. I., 2004, Ozonetherapy in HCV infection, in *Ozon-Handbuch. Grundlagen Prävention, therapie* (R. von Viebahn-Hänsler, and H. G. Knoch, Eds.), Landsberg in press.

Mayer, R. J., 2004, Two steps forward in the treatment of colorectal cancer, *N. Engl. J. Med.* **350**:2406-2408.

McCall, M. R., and Frei B., 1999, Can antioxidant vitamins materially reduce oxidative damage in humans?, *Free rad. Biol. Med.* **26**:1034-1053.

McCarey, D.W., McInnes, I.B., Madhok, R., Hampson, R., Scherbakov, O., Ford, I., Capell, H.A., Sattar, N., 2004, Trial of Atorvastatin in Rheumatoid Arthritis (TARA): double-blind, randomised placebo-controlled trial., *Lancet* **363**:2015-2021

McConnell, R., Berhane K., Gilliland F., London S. J., Islam T., Gauderman W. J., Avol E., Margolis H. G., and Peters J. M., 2002, Asthma in exercising children exposed to ozone: a cohort study, *Lancet* **359**:386-391.

McConnell, R., Berhane K., Gilliland F., London S. J., Islam T., Gauderman W. J., Avol E., Margolis H. G., and Peters J. M., 2002, Asthma in exercising children exposed to ozone: a cohort study, *Lancet* **359**:386-391.

McCord, J. M., 1974, Free radicals and inflammation: protection of synovial fluid by superoxide dismutase, *Science* **185**:529-531.

McCully, K. K., and Natelson B. H., 1999, Impaired oxygen delivery to muscle in chronic fatigue syndrome, *Clin. Sci. (Lond)* **97**:603-608.

McDonnell, W. F., 1991, Intersubject variability in human acute ozone responsiveness, *Pharmacogenetics* **1**:110-113.

McInnes, I. B., and Liew F. Y., 1998, Interleukin 15: a proinflammatory role in rheumatoid arthritis synovitis, *Immunol. Today* **19**:75-79.

Mecocci, P., Polidori M. C., Troiano L., Cherubini A., Cecchetti R., Pini G., Straatman M., Monti D., Stahl W., Sies H., Franceschi C., and Senin U., 2000, Plasma antioxidants and longevity: a study on healthy centenarians, *Free rad. Biol. Med.* **28**:1243-1248.

Meewes, C., Brenneisen P., Wenk J., Kuhr L., Ma W., Alikoski J., Poswig A., Krieg T., and Scharffetter-Kochanek K., 2001, Adaptive antioxidant response protects dermal fibroblasts from UVA- induced phototoxicity, *Free rad. Biol. Med.* **30**:238-247.

Mellor, A. L., and Munn D. H., 1999, Tryptophan catabolism and T-cell tolerance: immunosuppression by starvation?, *Immunol. Today* **20**:469-473.

Mendiratta, S., Qu Z.-C., and May J. M., 1998a, Erythrocyte ascorbate recycling: antioxidant effects in blood, *Free rad. Biol. Med.* **24**:789-797.

Mendiratta, S., Qu Z.-C., and May J. M., 1998b, Enzyme-dependent ascorbate recycling in human erythrocytes: role of thioredoxin reductase, *Free rad. Biol. Med.* **25**:221-228.

Menendez, S., et al, 1995, Application of ozonized oil in the treatment of infantile giardiasis, in *Proceedings Ozone in Medicine. 12th World Congress of the International Ozone Association, 15th to 18th May 1995, Lille, France* (International Ozone Association, Ed.), Instaprint S.A., Tours, pp.297-300.

Menendez, S.., Falcon L, Simon D.R., Landa N., 2002, Efficacy of ozonized sunflower oil in the treatment of tinea pedis, *Mycoses* **45**: 329-332.

Merz, T., Bender M. A., Kerr H. D., and Kulle T. J., 1975, Observations of aberrations in chromosomes of lymphocytes from human subjects exposed at a concentration of 0.5 ppm for 6 and 10 hours, *Mutat. Res.* **3**:299-302.

Mezey, E., Key S., Vogelsang G., Szalayova I., Lange G. D., and Crain B., 2003, Transplanted bone marrow generates new neurons in human brains, *Proc. Natl. Acad. Sci. U. S. A* **100**:1364-1369.

Micheli, V., Ricci C., Taddeo A., and Gili R., 1985, Centrifugal fractionation of human erythrocytes according to age: comparison between Ficoll and Percoll density gradients, *Quad. Sclavo. Diagn.* **21**:236-248.

Miller, D. H., 2003, Commentary: Evaluating disease modifying tretments in multiple sclerosis, *BMJ* **326**:525.

Miller, D. H., Khan O. A., Sheremata W. A., Blumhardt L. D., Rice G. P., Libonati M. A., Willmer-Hulme A. J., Dalton C. M., Miszkiel K. A., and O'Connor P. W., 2003, A controlled trial of natalizumab for relapsing multiple sclerosis, *N. Engl. J. Med.* **348**:15-23.

Miller, N. J., Rice-Evans C., Davies M. J., Gopinathan V., and Milner A., 1993, A novel method for measuring antioxidant capacity and its application to monitoring the antioxidant status in premature neonates, *Clin. Sci.* **84**:407-412.

Milligan, N. G., Newcombe R., and Compston D. A. S., 1986, A double-blind controlled trial of high-dose methylprednisolone in patients with multiple sclerosis. 1: clinical effects, *J. Neurol. Neurosurg. Psychiatry* **50**:511-516.

Minetti, M., Mallozzi C., Di Stasi A. M. M., and Pietraforte D., 1998, Bilirubin is an effective antioxidant of peroxynitrite-mediated protein oxidation in human blood plasma, *Arch. Biochem. Biophys.* **352**:165-174.

Minokoshi, Y., Kim Y. B., Peroni O. D., Fryer L. G., Muller C., Carling D., and Kahn B. B., 2002, Leptin stimulates fatty-acid oxidation by activating AMP-activated protein kinase, *Nature* **415**:339-343.

Miroshin, S. J., and Kontorshikova C. N., 1995, The use of ozonetherapy technology in the treatment of modern war surgical trauma, in *The ozone in biology and medicine. 2nd all Russian scientific-practical conference, Septmber 6-8, 1995. Russian association of ozonetherapy, Reshetnikovskaya street 2, Nizhni Novgorod, 603006 Russia*, p.16.

Miura, T., Suzuki S., Sakurai S., Matsumoto A., and Shinriki N., 2001, Structure elucidation of ozonated olive oil, in *Proceedings of the 15th Ozone World Congress, London, UK, 11th-15th September 2001, Medical Therapy Conference* (IOA 2001, Ed.), Speedprint MacMedia Ltd, Ealing, London, UK, pp.72-76.

Miyazono, M, Garat C., Morris K.G. Jr, and Carter E.P., 2002, Decreased renal heme oxygenase-1 expression contributes to decreased renal function during cirrhosis, *Am J Physiol Renal Physiol.* **283**: F1123-F1131.

Moldofsky, H., Scarisbrick P., England R., and Smythe H. A., 1975, Musculoskeletal symptoms and non-REM sleep disturbance in patients with "Fibrositis Syndrome" and healthy subjects, *Psychosomatic Medicine* **37**:341-351.

Molina, M. J., and Rowland F. S., 1974, Stratospheric sink for chlorofluoromethanes: chlorine atom catalyzed destruction of ozone, *Nature* **249**:810-814.

Moncada, S., 1992, Nitric oxide gas: mediator, modulator, and pathophysiologic entity., *J. Lab. Clin. Med.* **120**:187-191.

Morena, M., Cristol J. P., Bosc J. Y., Tetta C., Forret G., descomps B., and Canaud B., 1998, Convective and diffusive losses of vitamin C during hemodiafiltration session: a contributive factor to oxidative stress in hemodialysis patients., *Nephrol. Dial. Transplant.* **13**:A200.

Morena, M., Cristol J.P., and Canaud B., 2000, Why hemodialysis patients are in a prooxidant state? What could be done to correct the pro/antioxidant imbalance, *Blood Purification* **18**: 191-199.

Mori, T. A., Woodman R. J., Burke V., Puddey I. B., Croft K. D., and Beilin L. J., 2003, Effect of eicosapentaenoic acid and docosahexaenoic acid on oxidative stress and inflammatory markers in treated-hypertensive type 2 diabetic subjects, *Free Radic. Biol. Med.* **35**:772-781.

Morisco, F., Verde V., Fogliano V., Ritieni A., Marmo R., De Luise G., Tuccillo C., and Caporaso N., 2004, Oxidative status in chronic hepatitis C: The influence of antiviral therapy and prognostic value of serum hydroperoxide assay, *Free Radic. Res.* **38**:573-580.

Morita, T., and Kourembanas S., 1995, Endothelial cell expression of vasoconstrictors and growth factors is regulated by smooth muscle cell-derived carbon monoxide, *J. Clin. Invest* **96**:2676-2682.

Morley, J. E., and Perry H. M., III, 2000, Androgen deficiency in aging men: role of testosterone replacement therapy, *J. Lab Clin. Med.* **135**:370-378.

Morris, C. R., Kuypers F. A., Larkin S., Sweeters N., Simon J., Vichinsky E. P., and Styles L. A., 2000, Arginine therapy: a novel strategy to induce nitric oxide production in sickle cell disease, *Br. J. Haematol.* **111**:498-500.

Morrow, J. D., and Jackson Roberts L., 1997, The isoprostanes: unique bioactive products of lipid peroxidation, *Prog. Lipid Res.* **36**:1-21.

Morrow, J. D., Frei B., Longmire A. W., Gaziano J. M., Lynch S. M., Shyr Y., Strauss W. E., Oates J. A., and Roberts L. J., 1995, Increase in circulating products of lipid peroxidation (F2-isoprostanes) in smokers. Smoking as a cause of oxidative damage, *N. Engl. J. Med.* **332**:1198-1203.

Mosmann, T. R., and Sad S., 1996, The expanding universe of T-cell subsets: Th1, Th2 and more, *Immunol. Today* **17**:138-146.

Motzer, R. J., Rakhit A., Thompson J. A., Nemunaitis J., Murphy B. A., Ellerhorst J., Schwartz L. H., Berg W. J., and Bukowski R. M., 2001, Randomized multicenter phase II trial of subcutaneous recombinant human interleukin-12 versus interferon-alpha 2a for patients with advanced renal cell carcinoma, *J. Interferon Cytokine Res.* **21**:257-263.

Mudd, J. B., Dawson P. J., and Santrock J., 1997, Ozone does not react with human erythrocyte membrane lipids, *Arch. Biochem. Biophys.* **341**:251-258.

Muller, W. A., 2002, Leukocyte-endothelial cell interactions in the inflammatory response, *Lab Invest* **82**:521-533.

Murphy, W. J., and Longo D. L., 2000, Growth hormone as an immunomodulating therapeutic agent, *Immunol. Today* **21**:211-213.

Murry, C. E., Jennings R. B., and Reimer K. A., 1986, Preconditioning with ischemia: a delay of lethal cell injury in ischemic myocardium, *Circulation* **74**:1124-1136.

Murry, C. E., Soonpaa M. H., Reinecke H., Nakajima H., Nakajima H. O., Rubart M., Pasumarthi K. B., Virag J. I., Bartelmez S. H., Poppa V., Bradford G., Dowell J. D., Williams D. A., and Field L. J., 2004, Haematopoietic stem cells do not transdifferentiate into cardiac myocytes in myocardial infarcts, *Nature* **428**:664-668.

Musselman, D. L., Lawson D. H., Gumnick J. F., Manatunga A. K., Penna S., Goodkin R. S., Greiner K., Nemeroff C. B., and Miller A. H., 2001, Paroxetine for the prevention of depression induced by high-dose interferon alfa, *N. Engl. J. Med.* **344**:961-966.

Nakao, N., Frodl E. M., Widner H., Carlson E., Eggerding F. A., Epstein C. J., and Brundin P., 1995, Overexpressing Cu/Zn superoxide dismutase enhances survival of transplanted neurons in a rat model of Parkinson's disease, *Nature Medicine* **1**:226-231.

Natelson, B. H., 2001, Chronic fatigue syndrome, *JAMA* **285**:2557-2559.

Nath, K. A., Haggard J. J., Croatt A. J., Grande J. P., Poss K. D., and Alam J., 2000, The indispensability of heme oxygenase-1 in protecting against acute heme protein-induced toxicity in vivo, *Am. J. Pathol.* **156**:1527-1535.

Nathan, C. F., and Cohn Z. A., 1981, Antitumor effects of hydrogen peroxide in vivo, *J. Exp. Med.* **154**:1539-1553.

Nathan, C. F., Brukner L. H., Silverstein S. C., and Cohn Z. A., 1979a, Extracellular cytolysis by activated macrophages and granulocytes. I. Pharmacologic triggering of effector cells and the release of hydrogen peroxide, *J. Exp. Med.* **149**:84-99.

Nathan, C. F., Silverstein S. C., Brukner L. H., and Cohn Z. A., 1979b, Extracellular cytolysis by activated macrophages and granulocytes. II. Hydrogen peroxide as a mediator of cytotoxicity, *J. Exp. Med.* **149**:100-113.

Neuhaus, O., Farina C., Yassouridis A., Wiendl H., Then B. F., Dose T., Wekerle H., and Hohlfeld R., 2000, Multiple sclerosis: comparison of copolymer-1- reactive T cell lines from treated and untreated subjects reveals cytokine shift from T helper 1 to T helper 2 cells, *Proc. Natl. Acad. Sci. U. S. A* **97**:7452-7457.

Neumann, A. U., Lam N. P., Dahari H., Gretch D. R., Wiley T. E., Layden T. J., and Perelson A. S., 1998, Hepatitis C viral dynamics in vivo and the antiviral efficacy of interferon-a therapy, *Science* **282**:103-107.

Newsome, D. A., Swartz M., Leone N. C., Elston R. C., and Miller E., 1988, Oral zinc in macular degeneration, *Arch. Ophthalmol.* **106**:192-198.

Nicolaides, N., 1974, Skin lipids:their biochemical uniqueness, *Science* **186**:19-26.

Nieva, J., and Wentworth P., Jr., 2004, The antibody-catalyzed water oxidation pathway--a new chemical arm to immune defense?, *Trends Biochem. Sci.* **29**:274-278.

Noguchi, P., 2003, Risks and benefits of gene therapy, *N. Engl. J. Med.* 348: 193-194.

Nortvedt, M. W., Riise T., Myhr K. M., Nyland H. I., and Hanestad B. R., 1999, Type I interferons and the quality of life of multiple sclerosis patients. Results from a clinical trial on interferon alfa-2a, *Mult. Scler.* **5**:317-322.

Noyer, C. M., and Brandt L. J., 1999, Hyperbaric oxygen therapy for perineal Crohn's disease, *Am. J. Gastroenterol.* **94**:318-321.

Null, 1996, Ozone: a wide-spectrum realer, *Penthouse Magazine* **January**.

O'Byrne, P. M., Inman M. D., and Adelroth E., 2004, Reassessing the Th2 cytokine basis of asthma, *Trends Pharmacol. Sci.* **25**:244-248.

Oehler, M. K., and Bicknell R., 2000, The promise of anti-angiogenic cancer therapy, *Br. J. Cancer* **82**:749-752.

O'Farrelly, C., and Crispe I. N., 1999, Prometheus through the looking glass: reflections on the hepatic immune system, *Immunol. Today* **20**:394-398.

Okabe, N., 2001, The pathogenesis of Crohn's disease, *Digestion* **63 Suppl 1**:52-59.

Olivieri, G., Bodycote J., and Wolff S., 1984, Adaptive response of human lymphocytes to low concentrations of radioactive thymidine, *Science* **223**:594-597.

Olivieri, N. F., and Brittenham G. M., 1997, Iron-chelating therapy and the treatment of thalassemia, *Blood* **89**:739-761.

Olsen, S. J., DeBess E. E., McGivern T. E., Marano N., Eby T., Mauvais S., Balan V. K., Zirnstein G., Cieslak P. R., and Angulo F. J., 2001, A nosocomial outbreak of fluoroquinolone-resistant salmonella infection, *N. Engl. J. Med.* **344**:1572-1579.

Olwin, J. H., Ratajczak H. V., and House R. V., 1997, Successful treatment of herpetic infections by autohemotherapy, *J. Altern. Complement Med.* **3**:155-158.

Onik, G., Maroon J., Helms C., Schweigel J., Mooney V., Kahanovitz N., Day A., Morris J., McCulloch J. A., and Reicher M., 1987, Automated percutaneous discectomy: initial patient experience. Work in progress, *Radiology* **162**:129-132.

O'Reilly, M. S., Boehm T., Shing Y., Fukai N., Vasios G., Lane W. S., Flynn E., Birkhead J. R., Olsen B. R., and Folkman J., 1997, Endostatin: an endogenous inhibitor of angiogenesis and tumor growth, *Cell* **88**:277-285.

Orlic, D., Kajstura J., Chimenti S., Limana F., Jakoniuk I., Quaini F., Nadal-Ginard B., Bodine D.M., Leri A., and Anversa P., 2001, Mobilized bone marrow cells repair the infarcted heart, improving function and survival, *Proc Natl Acad Sci U S A.* **98**: 10344-10349.

Orringer, E. P., Casella J. F., Ataga K. I., Koshy M., Adams-Graves P., Luchtman-Jones L., Wun T., Watanabe M., Shafer F., Kutlar A., Abboud M., Steinberg M., Adler B., Swerdlow P., Terregino C., Saccente S., Files B., Ballas S., Brown R., Wojtowicz-Praga S., and Grindel J. M., 2001, Purified poloxamer 188 for treatment of acute vaso-occlusive crisis of sickle cell disease: A randomized controlled trial, *JAMA* **286**:2099-2106.

Orta de Velasquez, Ma. T., Rojas Ma. N., Martìnez J. L., and Monje I., 2001, Destruction of helminth eggs (*Ascaris suum*) by ozone, in *Proceedings of the 15th Ozone World Congress, London, UK, 11th-15th September 2001, Medical Therapy Conference* (IOA 2001, Ed.), Speedprint MacMedia Ltd, Ealing, London, UK, pp.63-71.

Otterbein, L. E., Kolls J. K., Mantell L. L., Cook J. L., Alam J., and Choi A. M. K., 1999, Exogenous administration of heme oxygenase-1 by gene transfer provides protection against hyperoxia -induced lung injury, *J. Clin. Invest.* **103**:1047-1054.

Overgaard, J., Gonzalez G. D., Hulshof M. C., Arcangeli G., Dahl O., Mella O., and Bentzen S. M., 1995, Randomised trial of hyperthermia as adjuvant to radiotherapy for recurrent or metastatic malignant melanoma. European Society for Hyperthermic Oncology, *Lancet* **345**:540-543.

Owen, C. G., Fletcher A. E., Donoghue M., and Rudnicka A. R., 2003, How big is the burden of visual loss caused by age related macular degeneration in the United Kingdom?, *Br. J. Ophthalmol.* **87**:312-317.

Packer, L., Roy S., and Sen C. K., 1997, Alpha-lipoic acid: a metabolic antioxidant and potential redox modulator of transcription, *Adv. Pharmacol.* **38**:79-101.

Pamphilon, D., 2000, Viral inactivation of fresh frozen plasma, *Br. J. Haematol.* **109**:680-693.

Pannen, B. H. J., Köhler N., Hole B., Bauer M., Clemens M. G., and Geiger K. K., 1998, Protective role of endogenous carbon monoxide in hepatic microcirculatory dysfunction after hemorrhagic shock in rats, *J. Clin. Invest.* **102**:1220-1228.

Pantel, K., Cote R. J., and Fodstad Ø., 1999, Detection and clinical importance of micrometastatic disease, *J. Nat. Cancer Inst.* **91**:1113-1124.

Pardo, C. A., Xu Z., Borchelt D. R., Price D. L., Sisodia S. S., and Cleveland D. W., 1995, Superoxide dismutase is an abundant component in cell bodies, dendrites, and axons of motor neurons and in a subset of other neurons, *Proc. Nat. Acad. Sci. USA* **92**:954-958.

Parker, A. J., Wessely S., and Cleare A. J., 2001, The neuroendocrinology of chronic fatigue syndrome and fibromyalgia, *Psychol. Med.* **31**:1331-1345.

Parks, D. A., and Granger D. N., 1983, Ischemia-induced vascular changes: role of xanthine oxidase and hydroxyl radicals, *Am. J. Physiol* **245**:G285-G289.

Parmiani, G., Rodolfo M., and Melani C., 2000, Immunological gene therapy with ex vivo gene-modified tumor cells: a critique and a reappraisal, *Hum. Gene Ther.* **11**:1269-1275.

Parola, M., Bellomo G., Robino G., Barrera G., and Dianzani M. U., 1999, 4-Hydroxynonenal as a biological signal: molecular basis and pathophysiological implications, *Antiox. Redox Signal.* **1**:255-284.

Patterson, C., and Runge M. S., 2000, Therapeutic myocardial angiogenesis via vascular endothelial growth factor gene therapy: moving on down the road, *Circulation* **102**:940-942.

Pauleikhoff, D., and Koch J. M., 1995, Prevalence of age-related macular degeneration, *Curr. Opin. Ophthalmol.* **6**:51-56.

Pauleikhoff, D., Barondes M. J., Minassian D., Chisholm I. H., and Bird A. C., 1990, Drusen as risk factors in age-related macular disease, *Am. J. Ophthalmol.* **109**:38-43.

Paulesu, L., Luzzi E., and Bocci V., 1991, Studies on the biological effects of ozone: 2. Induction of tumor necrosis factor (TNF-a) on human leucocytes, *Lymphokine Cytokine Res.* **10**:409-412.

Pawliuk, R., Westerman K. A., Fabry M. E., Payen E., Tighe R., Bouhassira E. E., Acharya S. A., Ellis J., London I. M., Eaves C. J., Humphries R. K., Beuzard Y., Nagel R. L., and Leboulch P., 2001, Correction of sickle cell disease in transgenic mouse models by gene therapy, *Science* **294**:2368-2371.

Payne, L. C., and Krueger J. M., 1992, Interactions of cytokines with the hypothalamus-pituitary axis, *J. Immunother.* **12**:171-173.

Payr, E., 1935, Über Ozonbehandlung in der Chirurgie, *Münch. Med. Wochenschr.* **82**:220-291.

Peng, J., Jones G. L., and Watson K., 2000, Stress proteins as biomarkers of oxidative stress: effects of antioxidant supplements, *Free Rad. Biol. Med.* **28**:1598-1606.

Peralta, C., Leon O. S., Xaus C., Prats N., Jalil E. C., Planell E. S., Puig-Parellada P., Gelpi E., and Rosello-Catafau J., 1999, Protective effect of ozone treatment on the injury associated with hepatic ischemia-reperfusion: antioxidant-prooxidant balance, *Free Rad. Res.* **31**:191-196.

Peralta, C., Xaus C., Bartrons R., Leon O. S., Gelpi E., and Rosello-Catafau J., 2000, Effect of ozone treatment on reactive oxygen species and adenosine production during hepatic ischemia-reperfusion, *Free Rad. Res.* **33**:595-605.

Perdue, M. H., 1999, Mucosal immunity and inflammation III. The mucosal antigen barrier: cross talk with mucosal cytokines, *Amer. J. Physiol.* **277**:G1-G5.

Perletti, G., Concari P., Giardini R., Marras E., Piccinini F., Folkman J., and Chen L., 2000, Antitumor activity of endostatin against carcinogen-induced rat primary mammary tumors, *Cancer Res.* **60**:1793-1796.

Perry, G., Nunomura A., Hirai K., Zhu X., Perez M., Avila J., Castellani R. J., Atwood C. S., Aliev G., Sayre L. M., Takeda A., and Smith M. A., 2002, Is oxidative damage the fundamental pathogenic mechanism of Alzheimer's and other neurodegenerative diseases?, *Free Radic. Biol. Med.* **33**:1475-1479.

Petersen, K. F., Oral E. A., Dufour S., Befroy D., Ariyan C., Yu C., Cline G. W., DePaoli A. M., Taylor S. I., Gorden P., and Shulman G. I., 2002, Leptin reverses insulin resistance and hepatic steatosis in patients with severe lipodystrophy, *J. Clin. Invest* **109**:1345-1350.

Peterson, L. R., 1998, Estrogen replacement therapy and coronary artery disease, *Curr. Opin. Cardiol.* **13**:223-231.

Petralia, B., Tommasini G., Lavaroni A., and Fabris G., 2001, A tutto gas! Il "mal di schiena" curato con l'ozonoterapia, *Riv. Neuroradiol.* **14**:71-73.

Pianko, S., and McHutchison J., 1999, Chronic hepatitis B: new therapies on the horizon?, *Lancet* **354**:1662-1663.

Pickup, J., Mattock M., and Kerry S., 2002, Glycaemic control with continuous subcutaneous insulin infusion compared with intensive insulin injections in patients with type 1 diabetes: meta-analysis of randomised controlled trials, *BMJ* **324**:705.

Pierce, G. F., Tarpley J. E., Tseng J., Bready J., Chang D., Kenney W. C., Rudolph R., Robson M. C., Vande Berg J., Reid P., Kaufman S., and Farrell C. L., 1995, Detection of platelet-derived growth factor (PDGF)-AA in actively healing human wounds treated with recombinant PDGF-BB and absence of PDGF in chronic nonhealing wounds, *J. Clin. Invest.* **96**:1336-1350.

Piguet, B., Palmvang I. B., Chisholm I. H., Minassian D., and Bird A. C., 1992, Evolution of age-related macular degeneration with choroidal perfusion abnormality, *Am. J. Ophthalmol.* **113**:657-663.

Pippard, M. J., and Weatherall D. J., 2000, Oral iron chelation therapy for thalassaemia: an uncertain scene, *Br. J. Haematol.* **111**:2-5.

Pizarro, T. T., Michie M. H., Bentz M., Woraratanadharm J., Smith M. F., Jr., Foley E., Moskaluk C. A., Bickston S. J., and Cominelli F., 1999, IL-18, a novel immunoregulatory cytokine, is up-regulated in Crohn's disease: expression and localization in intestinal mucosal cells, *J. Immunol.* **162**:6829-6835.

Podda, M., Traber M. G., Weber C., Yan L.-J., and Packer L., 1998, UV-irradiation depletes antioxidants and causes oxidative damage in a model of human skin, *Free Rad. Biol. Med.* **24**:55-65.

Polidori, M. C., Mecocci P., Levine M., and Frei B., 2004, Short-term and long-term vitamin C supplementation in humans dose-dependently increases the resistance of plasma to ex vivo lipid peroxidation, *Arch. Biochem. Biophys.* **423**:109-115.

Polidori, M. C., Stahl W., Eichler O., Niestroj I., and Sies H., 2001, Profiles of antioxidants in human plasma, *Free Rad. Biol. Med.* **30**:456-462.

Polman, C. H., and Uitdehaag B. M., 2000, Drug treatment of multiple sclerosis, *Brit. Med. J.* **321**:490-494.

Polman, C., Barkhof F., Kappos L., Pozzilli C., Sandbrink R., Dahlke F., Jakobs P., and Lorenz A., 2003, Oral interferon beta-1a in relapsing-remitting multiple sclerosis: a double-blind randomized study, *Mult. Scler.* **9**:342-348.

Powell, P., Bentall R. P., Nye F. J., and Edwards R. H., 2001, Randomised controlled trial of patient education to encourage graded exercise in chronic fatigue syndrome, *BMJ* **322**:387-390.

Prengler, M., Pavlakis S. G., Prohovnik I., and Adams R. J., 2002, Sickle cell disease: the neurological complications, *Ann. Neurol.* **51**:543-552.

Present, D. H., Rutgeerts P., Targan S., Hanauer S. B., Mayer L., van Hogezand R. A., Podolsky D. K., Sands B. E., Braakman T., DeWoody K. L., Schaible T. F., and van Deventer S. J. H., 1999, Infliximab for the treatment of fistulas in patients with Crohn's disease, *N. Engl. J. Med.* **340**:1398-1405.

Prins, J. B., Bleijenberg G., Bazelmans E., Elving L. D., de Boo T. M., Severens J. L., van der Wilt G. J., Spinhoven P., and van der Meer J. W., 2001, Cognitive behaviour therapy for chronic fatigue syndrome: a multicentre randomised controlled trial, *Lancet* **357** :841-847.

Prows, D. R., Shertzer H. G., Daly M. J., Sidman C. L., and Leikauf G. D., 1997, Genetic analysis of ozone-induced acute lung injury in sensitive and resistant strains of mice, *Nat. Genet.* **17**:471-474.

Pryor, W. A., 1992, How far does ozone penetrate into the pulmonary air/tissue boundary before it reacts?, *Free Rad. Biol. Med.* **12**:83-88.

Pryor, W. A., 2000, Vitamin E and heart disease: basic science to clinical intervention trials, *Free Rad. Biol.. Med.* **28**:141-164.

Pryor, W. A., Squadrito G. L., and Friedman M., 1995, The cascade mechanism to explain ozone toxicity: the role of lipid ozonation products, *Free Rad. Biol.. Med.* **19**:935-941.

Pullar, J. M., Vissers M. C., and Winterbourn C. C., 2000, Living with a killer: the effects of hypochlorous acid on mammalian cells, *IUBMB. Life* **50**:259-266.

Purasiri, P., Mckechnie A., Heys S. D., and Eremin O., 1997, Modulation in vitro of human natural cytotoxicity, lymphocyte proliferative response to mitogens and cytokine production by essential fatty acids, *Immunology* **92**:166-172.

Puskas, F., Gergely P., Jr., Banki K., and Perl A., 2000, Stimulation of the pentose phosphate pathway and glutathione levels by dehydroascorbate, the oxidized form of vitamin C, *FASEB J.* **14**:1352-1361.

Qi, W.-N., and Scully S. P., 1997, Extracellular collagen modulates the regulation of chondrocytes by transforming growth factor-b1, *J. Orthopaed. Res.* **15**:483-490.

Radu, R. A., Mata N. L., Nusinowitz S., Liu X., Sieving P. A., and Travis G. H., 2003, Treatment with isotretinoin inhibits lipofuscin accumulation in a mouse model of recessive Stargardt's macular degeneration, *Proc. Natl. Acad. Sci. U. S. A* **100**:4742-4747.

Rafikova, O., Rafikov R., and Nudler E., 2002, Catalysis of S-nitrosothiols formation by serum albumin: the mechanism and implication in vascular control, *Proc. Natl. Acad. Sci. U. S. A* **99**:5913-5918.

Raghu, G., Brown K. K., Bradford W. Z., Starko K., Noble P. W., Schwartz D. A., and King T. E., Jr., 2004, A placebo-controlled trial of interferon gamma-1b in patients with idiopathic pulmonary fibrosis, *N. Engl. J. Med.* **350**:125-133.

Rahman, I., Clerch L. B., and Massaro D., 1991, Rat lung antioxidant enzyme induction by ozone, *Amer. J. Physiol.* **260**:L412-L418.

Ranjbar, S., and Holmes H., 1996, Influence of hydrogen peroxide on the in vitro infectivity of human immunodeficiency virus, *Free Rad. Biol. Med.* **20**:573-577.

Rasmussen, H., Chu K. W., Campochiaro P., Gehlbach P. L., Haller J. A., Handa J. T., Nguyen Q. D., and Sung J. U., 2001, Clinical protocol. An open-label, phase I, single administration, dose-escalation study of ADGVPEDF.11D (ADPEDF) in neovascular age-related macular degeneration (AMD), *Hum. Gene Ther.* **12**:2029-2032.

Rassaf, T., Preik M., Kleinbongard P., Lauer T., Heiss C., Strauer B. E., Feelisch M., and Kelm M., 2002, Evidence for in vivo transport of bioactive nitric oxide in human plasma, *J. Clin. Invest* **109**:1241-1248.

Rattan, V., Shen Y., Sultana C., Kumar D., and Kalra V. K., 1997, Diabetic RBC-induced oxidant stress leads to transendothelial migration of monocyte-like HL-60 cells, *Am. J. Physiol* **273**:E369-E375.

Re, R., Pellegrini N., Proteggente A., Pannala A., Yang M., and Rice-Evans C., 1999, Antioxidant activity applying an improved ABTS radical cation decolorization assay, *Free Rad. Biol.. Med.* **26**:1231-1237.

Reddy, S. P., Harwood R. M., Moore D. F., Grimm E. A., Murray J. L., and Vadhan-Raj S., 1997, Recombinant Interleukin-2 in combination with recombinant interferon-g in patients with advanced malignancy: a phase 1 study, *J. Immunother.* **20**:79-87.

Reeve, V. E., and Tyrrell R. M., 1999, Heme oxygenase induction mediates the photoimmunoprotective activity of UVA radiation in the mouse, *Proc. Natl. Acad. Sci. U. S. A* **96**:9317-9321.

Reichlin, S., 1993, Neuroendocrine-immune interactions, *N. Engl. J. Med.* **329**:1246-1253.

Reid, S., Chalder T., Cleare A., Hotopf M., and Wessely S., 2000, Chronic fatigue syndrome, *Brit. Med. J.* **320**:292-296.

Reimold, A. M., 2003, New indications for treatment of chronic inflammation by TNF-alpha blockade, *Am. J. Med. Sci.* **325**:75-92.

Reisberg, B., Doody R., Stoffler A., Schmitt F., Ferris S., and Mobius H. J., 2003, Memantine in moderate-to-severe Alzheimer's disease, *N. Engl. J. Med.* **348**:1333-1341.

Reiter, R. J., 1991, Pineal melatonin: cell biology of its synthesis and of its physiological interactions, *Endocr. Rev.* **12**:151-180.

Renaud, B., and Brun-Buisson C., 2001, Outcomes of primary and catheter-related bacteremia. A cohort and case- control study in critically ill patients, *Am. J. Respir. Crit Care Med.* **163**:1584-1590.

Resnick, H. E., and Howard B. V., 2002, Diabetes and cardiovascular disease, *Annu. Rev. Med.* **53**:245-267.

Reth, M., 2002, Hydrogen peroxide as second messenger in lymphocyte activation, *Nat. Immunol.* **3**: 1129-1134.

Revel, M., 2003, Interferon-beta in the treatment of relapsing-remitting multiple sclerosis, *Pharmacol. Ther.* **100**:49-62.

Rhee, S.G., Bae, Y.S., Lee, S.R., and Kwon J., 2000, Hydrogen peroxide: a key messenger that modulates protein phosphorylation through cysteine oxidation. *Sci STKE*. Oct 10. **53**: PE1

Rice, R. G., 2001, Century 21 - Pregnant with ozone, in *Proceedings of the 15th Ozone World Congress, London, UK, 11th-15th September 2001, Volume I* (IOA 2001, Ed.), Speedprint MacMedia Ltd, Ealing, London, UK, pp.1-19.

Rice-Evans, C., and Miller N. J., 1994, Total antioxidant status in plasma and body fluids, *Meth. Enzimol.* **234**:279-293.

Richards, S. C., and Scott D. L., 2002, Prescribed exercise in people with fibromyalgia: parallel group randomised controlled trial, *BMJ* **325**:185-188.

Richter, C., Gogvadze V., Laffranchi R., Schlapbach R., Schweizer M., Suter M., Walter P., and Yaffee M., 1995, Oxidants in mitochondria: from physiology to diseases, *Biochim. Biophys. Acta* **1271**:67-74.

Richter, C., Park J. W., and Ames B. N., 1988, Normal oxidative damage to mitochondrial and nuclear DNA is extensive, *Proc. Nat. Acad. Sci. USA* **85**:6465-6467.

Riedemann, N. C., Guo R. F., and Ward P. A., 2003, Novel strategies for the treatment of sepsis, *Nat. Med.* **9**: 517-524.

Riethmüller, G., Klein C. A., and Pantel K., 1999, Hunting down the seminal cells of clinical metastases, *Immunol. Today* **20**:294-296.

Riksen, N. P., Rongen G. A., Blom H. J., Russel F. G., Boers G. H., and Smits P., 2003, Potential role for adenosine in the pathogenesis of the vascular complications of hyperhomocysteinemia, *Cardiovasc. Res.* **59**:271-276.

Riva Sanseverino, E., 1989, Knee-joint disorders treated by oxygen-ozone therapy, *Eur. Medicophysica* **25** :163-170.

Roberts, A. B., Sporn M. B., Assoian R. K., Smith J. M., Roche N. S., Wakefield L. M., Heine U. I., Liotta L. A., Falanga V., Kehrl J. H., and ., 1986, Transforming growth factor type beta: rapid induction of fibrosis and angiogenesis in vivo and stimulation of collagen formation in vitro, *Proc. Natl. Acad. Sci. U. S. A* **83**:4167-4171.

Roberts, W. C., 1996, The underused miracle drugs: the statin drugs are to atherosclerosis what penicillin was to infectious disease, *Am. J. Cardiol.* **78**:377-378.

Robinson, D., Hamid Q., Bentley A., Ying S., Kay A. B., and Durham S. R., 1993, Activation of CD4+ T cells, increased TH2-type cytokine mRNA expression, and eosinophil recruitment in bronchoalveolar lavage after allergen inhalation challenge in patients with atopic asthma, *J. Allergy Clin. Immunol.* **92**:313-324.

Rocchini, A. P., 2002, Childhood obesity and a diabetes epidemic, *N. Engl. J. Med.* **346**:854-855.

Rodriguez, M. M., Menéndez S., Devesa E., Gomez M., Garcia J., and Carrasco M., 1993, Ozone therapy for senile dementia, in *Proceedings of the XI Ozone World Congress, San Francisco*, pp.M4-19-M4-25.

Roederer, M., Staal F. J. T., Raju P. A., Ela S. W., Herzenberg Le. A., and Herzenberg L. A., 1990, Cytokine-stimulated human immunodeficiency virus replication is inhibited by N-acetyl-L-cysteine, *Proc. Nat. Acad. Sci. USA* **87**:4884-4888.

Rokitansky, O., 1982, Klinik und Biochemie der Ozontherapie., *Hospitalis* **52**:643-647.
Rokitansky, O., Rokitansky A., Steiner J., Trubel W., Viebahn R., and Washüttl J., 1981, Die Ozontherapie bei peripheren, arteriellen Durchblutungs-störungen; klinik, biochemische und blutgasanalytische Untersuchungen, in *Wasser* IOA, Ozon-Weltkongress, Berlin, pp.53-75.
Romero Valdes, A., Menendez Cepero S., Gomez Moraleda M., and Ley Pozo J., 1993, Ozone therapy in the advanced stages of arteriosclerosis obliterans, *Angiologia* **45**:146-148.
Romero, A., et al., 1988, La ozonoterapia en la aterosclerosis obliterante, *CENIC Ciencias Biologicas* **20**:70-76.
Romero, A., et al., 1993, Arteriosclerosis obliterans and ozone therapy: its administration by different routes, *Angiologia* :177-179.
Romero, M. J., Bosch-Morell F., Romero B., Rodrigo J. M., Serra M. A., and Romero F. J., 1998, Serum malondialdehyde: possible use for the clinical management of chronic hepatitis C patients, *Free Rad. Biol.. Med.* **25**:993-997.
Rosa, L., Rosa E., Sarner L., and Barrett S., 1998, A close look at therapeutic touch, *JAMA* **279**:1005-1010.
Rosen, P., Nawroth P. P., King G., Moller W., Tritschler H. J., and Packer L., 2001, The role of oxidative stress in the onset and progression of diabetes and its complications: a summary of a Congress Series sponsored by UNESCO-MCBN, the American Diabetes Association and the German Diabetes Society, *Diabetes Metab Res. Rev.* **17**:189-212.
Rosenberg, G. A., 1999, Ischemic brain edema, *Prog. Cardiovasc. Dis.* **42**:209-216.
Rosenberg, S. A., 2001, Progress in human tumour immunology and immunotherapy, *Nature* **411**:380-384.
Rosenberg, S. A., Lotze M. T., Muul L. M., Chang A. E., Avis F. P., Leitman S., Linehan W. M., Robertson C. N., Lee R. E., Rubin J. T., Seipp C. A., Simpson C. G., and White D. E., 1987, A progress report on the treatment of 157 patients with advanced cancer using lymphokine-activated killer cells and interleukin-2 or high-dose interleukin-2 alone., *N. Engl. J. Med.* **316**:889-897.
Roth, J. A., and Cristiano R. J., 1997, Gene therapy for cancer: what have we done and where are we going?, *J. Nat. Cancer Inst.* **89**:21-39.
Rotilio, G., 2001, Risk from exposure to metals: deficits and excesses (Cu, Fe, Mn, Al, Cr, B), in *Nutrition and Brain* (J. D. Fernstrom, R. Uauy, and P. Arroyo, Eds.), Karger AG, Basel, pp.247-262.
Rotilio, G., Carri M. T., Rossi L., and Ciriolo M. R., 2000, Copper-dependent oxidative stress and neurodegeneration, *IUBMB. Life* **50**:309-314.
Rousseau, Y., Haeffner-Cavaillon N., Poignet J. L., Meyrier A., and Carreno M. P., 2000, In vivo intracellular cytokine production by leukocytes during haemodialysis, *Cytokine* **12**:506-517.
Rowland, L. P., and Shneider N. A., 2001, Amyotrophic lateral sclerosis, *N. Engl. J. Med.* **344**:1688-1700.
Rowland, M., 2000, Transmission of Helicobacter pylori: is it all child's play?, *Lancet* **355**:332-333.
Rubartelli, A., Poggi A., Sitia R., and Zocchi M. R., 1999, HIV-1 Tat: a polypeptide for all seasons, *Immunol. Today* **19**:543-545.
Rubin, P., Hanley J., Keys H. M., Marcial V., and Brady L., 1979, Carbogen breathing during radiation therapy-the Radiation Therapy Oncology Group Study, *Int. J. Radiat. Oncol. Biol. Phys.* **5**:1963-1970.
Rudick, R. A., Cohen J. A., Weinstock-Guttman B., Kinkel R. P., and Ransohoff R. M., 1997, Management of multiple sclerosis, *N. Engl. J. Med.* **337**:1604-1611.
Rudikoff, D., and Lebwohl M., 1998, Atopic dermatitis, *Lancet* **351**:1715-1721.

Rudman, D., Feller A. G., Nagraj H. S., Gergans G. A., Lalitha P. Y., Goldberg A. F., Schlenker R. A., Cohn L., Rudman I. W., and Mattson D. E., 1990, Effects of human growth hormone in men over 60 years old, *N. Engl. J. Med.* **323**:1-6.

Ruggenenti, P., Schieppati A., and Remuzzi G., 2001, Progression, remission, regression of chronic renal diseases, *Lancet* **357**:1601-1608.

Ruiz, L., Carcelain G., Martinez-Picado J., Frost S., Marfil S., Paredes R., Romeu J., Ferrer E., Morales-Lopetegi K., Autran B., and Clotet B., 2001, HIV dynamics and T-cell immunity after three structured treatment interruptions in chronic HIV-1 infection, *AIDS* **15**:F19-F27.

Ryan, H. E., Lo J., and Johnson R. S., 1998, HIF-1 alpha is required for solid tumor formation and embryonic vascularization, *EMBO J.* **17**:3005-3015.

Ryter, S. W., and Tyrrell R. M., 2000, The heme synthesis and degradation pathways: role in oxidant sensitivity. Heme oxygenase has both pro- and antioxidant properties, *Free Rad. Biol. Med.* **28**:289-309.

Sagara, Y., Dargusch R., Chambers D., Davis J., Schubert D., and Maher P., 1998, Cellular mechanisms of resistance to chronic oxidative stress, *Free Rad. Biol. Med.* **24**:1375-1389.

Saliou, C., Kitazawa M., McLaughlin L., Yang J. P., Lodge J. K., Tetsuka T., Iwasaki K., Cillard J., Okamoto T., and Packer L., 1999, Antioxidants modulate acute solar ultraviolet radiation-induced NF- kappa-B activation in a human keratinocyte cell line, *Free Rad. Biol. Med.* **26**:174-183.

Samanta, A., and Beardsley J., 1999, Low back pain: which is the best way forward?, *Brit. Med. J.* **318**:1122-1123.

Sands, B. E., Anderson F. H., Bernstein C. N., Chey W. Y., Feagan B. G., Fedorak R. N., Kamm M. A., Korzenik J. R., Lashner B. A., Onken J. E., Rachmilewitz D., Rutgeerts P., Wild G., Wolf D. C., Marsters P. A., Travers S. B., Blank M. A., and van Deventer S. J., 2004, Infliximab maintenance therapy for fistulizing Crohn's disease, *N. Engl. J. Med.* **350**:876-885.

Saran, M., Beck-Speier I., Fellerhoff B., and Bauer G., 1999, Phagocytic killing of microorganisms by radical processes: consequences of the reaction of hydroxyl radicals with chloride yielding chlorine atoms, *Free Rad. Biol. Med.* **26**:482-490.

Sardina, J. O., et al, 1991, Tratamiento de la giardiasis recidivante con ozono, *CENIC Ciencias Biologicas* **20**:61-64.

Sarks, J. P., Sarks S. H., and Killingsworth M. C., 1988, Evolution of geographic atrophy of the retinal pigment epithelium, *Eye* **2 (Pt 5)**:552-577.

Sarnesto, A., Linder N., and Raivio K. O., 1996, Organ distribution and molecular forms of human xanthine dehydrogenase/xanthine oxidase protein, *Lab Invest* **74**:48-56.

Sartor, R. B., 2000, New therapeutic approaches to Crohn's disease, *N. Engl. J. Med.* **342**:1664-1666.

Sasaki, H., Wakutani T., Oda S., and Yamasaki Y., 1967, Application of hydrogen peroxide infusion to maxillary cancer, *Yonago Acta Med.* **11**:141-149.

Sastre, J., Pallardo F. V., and Vina J., 2003, The role of mitochondrial oxidative stress in aging, *Free Radic. Biol. Med.* **35**:1-8.

Sato, K., Balla J., Otterbein L., Smith R. N., Brouard S., Lin Y., Csizmadia E., Sevigny J., Robson S. C., Vercellotti G., Choi A. M., Bach F. H., and Soares M. P., 2001, Carbon monoxide generated by heme oxygenase-1 suppresses the rejection of mouse-to-rat cardiac transplants, *J. Immunol.* **166**:4185-4194.

Sato, Y., Sato K., and Suzuki Y., 1999, Mechanisms of free radical-induced hemolysis of human erythrocytes: comparison of calculated rate constants for hemolysis with experimental rate constants, *Arch. Biochem. Biophys.* **366**:61-69.

Schmid, P., Cox D., Bilbe G., McMaster G., Morrison C., Stähelin H., Lüscher N., and Seiler W., 1993, TGF-bs and TGF-b type II receptor in human epidermis: differential expression in acute and chronic skin wounds, *J. Pathol.* **171**:191-197.

Schreiber, S., Heinig T., Thiele H. G., and Raedler A., 1995, Immunoregulatory role of interleukin 10 in patients with inflammatory bowel disease, *Gastroenterology* **108**:1434-1444.

Schrope, M., 2000, Successes in fight to save ozone layer could close holes by 2050, *Nature* **408**:627.

Schulz, S., 1986, The role of ozone/oxygen in clindamycin-associated enterocolitis in the Djungarian hamster (Phodopus sungorus sungorus), *Lab. Anim.* **20**:41-48.

Schwartz, R. S., Curfman G. D., 2002, Can the heart repair itself? *N. Engl. J. Med.* **346**: 2-4.

Schwarz, K. B., 1996, Oxidative stress during viral infection: a review, *Free Rad. Biol. Med.* **21**:641-649.

Scott, M. D., van den Berg J. J., Repka T., Rouyer-Fessard P., Hebbel R. P., Beuzard Y., and Lubin B. H., 1993, Effect of excess alpha-hemoglobin chains on cellular and membrane oxidation in model beta-thalassemic erythrocytes, *J. Clin. Invest* **91**:1706-1712.

Sechi, L. A., Lezcano I., Nunez N., Espim M., Dupre I., Pinna A., Molicotti P., Fadda G., and Zanetti S., 2001, Antibacterial activity of ozonized sunflower oil (Oleozon), *J. Appl. Microbiol.* **90**:279-284.

Seddon, J. M., Ajani U. A., Sperduto R. D., Hiller R., Blair N., Burton T. C., Farber M. D., Gragoudas E. S., Haller J., and Miller D. T. 1994, Dietary carotenoids, vitamins A, C, and E, and advanced age-related macular degeneration. Eye Disease Case-Control Study Group, *JAMA* **272**:1413-1420.

Seddon, J. M., Gensler G., Milton R. C., Klein M. L., and Rifai N., 2004, Association between C-reactive protein and age-related macular degeneration, *JAMA* **291**:704-710.

Seeman, T. E., and Robbins R. J., 1994, Aging and hypothalamic-pituitary-adrenal response to challenge in humans, *Endocr. Rev.* **15**:233-260.

Seifried, H.E., McDonald S.S., Anderson D.E., Greenwald P., and Milner J.A., 2003, The antioxidant conundrum in cancer, *Cancer Res.* **63**: 4295-4298.

Semenza, G. L., 2001, Hypoxia-inducible factor 1: oxygen homeostasis and disease pathophysiology, *Trends Mol. Med.* **7**:345-350.

Semenza, G. L., 2003, Targeting HIF-1 for cancer therapy, *Nat. Rev. Cancer* **3**:721-732.

Servaes, P., Verhagen C., and Bleijenberg G., 2002, Fatigue in cancer patients during and after treatment: prevalence, correlates and interventions, *Eur. J. Cancer* **38**:27-43.

Shanahan, F., 2002, Crohn's disease, *Lancet* **359**: 62-69.

Sharma, Y. K., and Davis K. R., 1997, The effects of ozone on antioxidant responses in plants, *Free Rad. Biol. Med.* **23**:480-488.

Sharpe, M. C., Archard L. C., Banatvala J. E., Borysiewicz L. K., Clare A. W., David A., Edwards R. H., Hawton K. E., Lambert H. P., Lane R. J., and ., 1991, A report--chronic fatigue syndrome: guidelines for research, *J. R. Soc. Med.* **84**:118-121.

Shaschova, N. M., Kachalina T. S., and Nevmjatullin A. L., 1995, Application of ozonotherapy in complex treatment of inner female genital inflammatory diseases, in *Proceedings Ozone in Medicine, 12th World Congress of the International Ozone Association, 15th to 18th May 1995, Lille France* (International Ozone Association, Ed.), Instaprint S.A., Tours, pp.145-155.

Sheldon, T., 2004, Netherlands to crack down on complementary medicine, *BMJ* **328**:485.

Shiba, M., Tadokoro K., Sawanobori M., Nakajima K., Suzuki K., and Juji T., 1997, Activation of the contact system by filtration of platelet concentrates with a negatively charged white cell-removal filter and measurement of venous blood bradykinin level in patients who received filtered platelets, *Transfusion* **37**:457-462.

Shinriki, N., Ishizaki K., Yoshizaki T., Miura K., and Ueda T., 1988, Mechanism of inactivation of tobacco mosaic virus with ozone, *Wat. Res.* **22**:933-938.

Shinriki, N., Suzuki T., Takama K., Fukunaga K., Ohgiya S., Kubota K., and Miura T., 1998, Susceptibilities of plasma antioxidants and erythrocyte constituents to low levels of ozone, *Haematologia* **29**:229-239.

Shiomori, T., Miyamoto H., and Makishima K., 2001, Significance of airborne transmission of methicillin-resistant Staphylococcus aureus in an otolaryngology-head and neck surgery unit, *Arch. Otolaryngol. Head Neck Surg.* **127**:644-648.

Shull, S., Heintz N. H., Periasamy M., Manohar M., Janssen Y. M. W., Marsh J. P., and Mossman B. T., 1991, Differential regulation of antioxidant enzymes in response to oxidants, *J. Biol. Chem.* **266**:24398-24403.

Siemann, D. W., Hill R. P., and Bush R. S., 1977, The importance of the pre-irradiation breathing times of oxygen and carbogen (5% CO2: 95% O2) on the in vivo radiation response of a murine sarcoma, *Int. J. Radiat. Oncol. Biol. Phys.* **2**:903-911.

Siemann, D. W., Horsman M. R., and Chaplin D. J., 1994, The radiation response of KHT sarcomas following nicotinamide treatment and carbogen breathing, *Radiother. Oncol.* **31**:117-122.

Siemsen, C.-H., 1995, Ozon-Anwendung bei akuten und chronischen Gelenkerkrankungen, in *Ozon-Handbuch. Grundlagen. Prävention. Therapie* (E. G. Beck, and R. Viebahn-Hänsler, Eds.), Ecomed, Landsberg, p.V-9.2 1-V-9.2 14.

Silver, F. H., and Glasgold A. I., 1995, Cartilage wound healing. An overview, *Otolaryngol. Clin. N. Amer.* **28**:847-864.

Simonian, N. A., and Coyle J. T., 1996, Oxidative stress in neurodegenerative diseases, *Annu. Rev. Pharmacol. Toxicol.* **36**:83-106.

Slavin, J., 1996, The role of cytokines in wound healing, *J. Pathol.* **178**:5-10.

Slonim, A. D., and Singh N., 2001, Nosocomial bloodstream infection and cost, *Crit Care Med.* **29**:1849.

Slonim, A. E., Bulone L., Damore M. B., Goldberg T., Wingertzahn M. A., and McKinley M. J., 2000, A preliminary study of growth hormone therapy for Crohn's disease, *N. Engl. J. Med.* **342**:1633-1637.

Small, D. L., Morley P., and Buchan A. M., 1999, Biology of ischemic cerebral cell death, *Prog. Cardiovasc. Dis.* **42**:185-207.

Smith, L. J., Shamsuddin M., Sporn P. H., Denenberg M., and Anderson J., 1997, Reduced superoxide dismutase in lung cells of patients with asthma, *Free Rad. Biol. Med.* **22**:1301-1307.

Smith, L., 1969, Chemonucleolysis, *Clin. Orthop.* **67**:72.

Snyder, S. H., and Baranano D. E., 2001, Heme oxygenase: a font of multiple messengers, *Neuropsychopharmacology* **25**:294-298.

Soares, C., 2004, Body building, *Sci. Am.* **290**:20, 22.

Sohal, R. S., Mockett R. J., and Orr W. C., 2002, Mechanisms of aging: an appraisal of the oxidative stress hypothesis, *Free Radic. Biol. Med.* **33**:575-586.

Soholm, B., 1998, Clinical improvement of memory and other cognitive functions by Ginkgo biloba: review of relevant literature, *Adv. Ther.* **15**:54-65.

Song, C. W., Hasegawa T., Kwon H. C., Lyons J. C., and Levitt S. H., 1992, Increase in tumor oxygenation and radiosensitivity caused by pentoxifylline, *Radiat. Res.* **130**:205-210.

Song, C. W., Lee I., Hasegawa T., Rhee J. G., and Levitt S. H., 1987, Increase in pO2 and radiosensitivity of tumors by Fluosol-DA (20%) and carbogen, *Cancer Res.* **47**:442-446.

Song, C. W., Shakil A., Griffin R. J., and Okajima K., 1997, Improvement of tumor oxygenation status by mild temperature hyperthermia alone or in combination with carbogen, *Semin. Oncol.* **24**:626-632.

Song, C. W., Shakil A., Osborn J. L., and Iwata K., 1996, Tumour oxygenation is increased by hyperthermia at mild temperatures, *Int. J. Hyperthermia* **12**:367-373.

Sorensen, P. S., Ross C., Clemmesen K. M., Bendtzen K., Frederiksen J. L., Jensen K., Kristensen O., Petersen T., Rasmussen S., Ravnborg M., Stenager E., and Koch-Henriksen N., 2003, Clinical importance of neutralising antibodies against interferon beta in patients with relapsing-remitting multiple sclerosis, *Lancet* **362**:1184-1191.

Spencer, F. A., Allegrone J., Goldberg R. J., Gore J. M., Fox K. A., Granger C. B., Mehta R. H., and Brieger D., 2004, Association of statin therapy with outcomes of acute coronary syndromes: the GRACE study, *Ann. Intern. Med.* **140**:857-866.

Sperduto, R. D., Ferris F. L. I., and Kurinij N., 1990, Do we have a nutritional treatment for age-related macular degeneration?, *Arch. Ophthalmol.* **108**:1403-1405.

Sperduto, R. D., Ferris F. L., III, and Kurinij N., 1990, Do we have a nutritional treatment for age-related cataract or macular degeneration?, *Arch. Ophthalmol.* **108**:1403-1405.

Sporn, M. B., and Roberts A. B., 1993, A major advance in the use of growth factors to enhance wound healing, *J. Clin. Invest* **92**:2565-2566.

Stadtman, E. R., and Oliver C. N., 1991, Metal-catalyzed oxidation of proteins. Physiological consequences, *J. Biol. Chem.* **266**:2005-2008.

Stamler, J. S., 2004, S-nitrosothiols in the blood: roles, amounts, and methods of analysis, *Circ. Res.* **94**:414-417.

Stamler, J. S., Singel D. J., and Loscalzo J., 1992, Biochemistry of nitric oxide and its redox-activated forms, *Science* **258**:1898-1902.

Stamm, C., Westphal B., Kleine H. D., Petzsch M., Kittner C., Klinge H., Schumichen C., Nienaber C. A., Freund M., and Steinhoff G., 2003, Autologous bone-marrow stem-cell transplantation for myocardial regeneration, *Lancet* **361**:45-46.

Stasi, R., Abriani L., Beccaglia P., Terzoli E., and Amadori S.,2003, Cancer-related fatigue: evolving concepts in evaluation and treatment, *Cancer*, **98**: 1786-1801.

Steece-Collier, K., Maries E., and Kordower J. H., 2002, Etiology of Parkinson's disease: Genetics and environment revisited, *Proc. Natl. Acad. Sci. U. S. A* **99**:13972-13974.

Steidler, L., Hans W., Schotte L., Neirynck S., Obermeier F., Falk W., Fiers W., and Remaut E., 2000, Treatment of murine colitis by Lactococcus lactis secreting interleukin- 10, *Science* **289**:1352-1355.

Stein, J. L., and Schwartzbrod J. K., 1990, Experimental contamination of vegetables with helminth eggs, *Wat. Sci. Tech.* **22**:51-57.

Steinberg, M. H., 1999, Management of sickle cell disease, *N. Engl. J. Med.* **340**:1021-1030.

Steinhart, H., Schulz S., and Mutters R., 1999, Evaluation of ozonated oxygen in an experimental animal model of osteomyelitis as a further treatment option for skull-base osteomyelitis, *Eur. Arch. Otorhinolaryngol.* **256**:153-157.

Stephan, F., Cheffi A., and Bonnet F., 2001, Nosocomial infections and outcome of critically ill elderly patients after surgery, *Anesthesiology* **94**:407-414.

Steuer-Vogt, M. K., Bonkowsky V., Ambrosch P., Scholz M., Neiss A., Strutz J., Hennig M., Lenarz T., and Arnold W., 2001, The effect of an adjuvant mistletoe treatment programme in resected head and neck cancer patients: a randomised controlled clinical trial, *Eur. J. Cancer* **37**:23-31.

Stiller, B., Sonntag J., Dahnert I., Alexi-Meskishvili V., Hetzer R., Fischer T., and Lange P.E., 2001, Capillary leak syndrome in children who undergo cardiopulmonary bypass: clinical outcome in comparison with complement activation and C1 inhibitor, *Intensive Care Med.* **27**: 193-200.

Stone, J. R., and Collins T., 2002, The role of hydrogen peroxide in endothelial proliferative responses, *Endothelium* **9**: 231-238.

Stover, B. H., Shulman S. T., Bratcher D. F., Brady M. T., Levine G. L., and Jarvis W. R., 2001, Nosocomial infection rates in US children's hospitals' neonatal and pediatric intensive care units, *Am. J. Infect. Control* **29**:152-157.

Strauer, B. E., and Kornowski R., 2003, Stem cell therapy in perspective, *Circulation* **107**: 929-934.

Strauer, B. E., Brehm M., Zeus T., Gattermann N., Hernandez A., Sorg R. V., Kogler G., and Wernet P., 2001, [Intracoronary, human autologous stem cell transplantation for myocardial regeneration following myocardial infarction], *Dtsch Med Wochenschr.* **126**: 932-938.

Su, W. Y., and Gordon T., 1997, In vivo exposure to ozone produces an increase in a 72-kDa heat shock protein in guinea pigs, *J. Appl. Physiol.* **83**:707-711.

Subarsky, P., and Hill R. P., 2003, The hypoxic tumour microenvironment and metastatic progression, *Clin. Exp. Metastasis* **20**:237-250.

Suckfull, M., 2002, Fibrinogen and LDL apheresis in treatment of sudden hearing loss: a randomised multicentre trial, *Lancet* **360**:1811-1817.

Suhadolnik, R. J., Peterson D. L., O'Brien K., Cheney P. R., Herst C. V., Reichenbach N. L., Kon N., Horvath S. E., Iacono K. T., Adelson M. E., De Meirleir K., De Becker P., Charubala R., and Pfleiderer W., 1997, Biochemical evidence for a novel low molecular weight 2-5A-dependent RNase L in chronic fatigue syndrome, *J. Interferon Cytokine Res.* **17**:377-385.

Sun, J. S., Lu F. J., Huang W. C., Hou S. M., Tsuang Y. H., and Hang Y. S., 1999, Antioxidant status following acute ischemic limb injury: a rabbit model, *Free Rad. Res.* **31**:9-21.

Swartz, M. N., 1988, The chronic fatigue syndrome--one entity or many?, *N. Engl. J. Med.* **319**:1726-1728.

Sweet, F., Kao M.-S., Lee S.-C. D., Hagar W. L., and Sweet W. E., 1980, Ozone selectively inhibits growth of human cancer cells, *Science* **209**:931-933.

Symons, M. C., Rusakiewicz S., Rees R. C., and Ahmad S. I., 2001, Hydrogen peroxide: a potent cytotoxic agent effective in causing cellular damage and used in the possible treatment for certain tumours, *Med. Hypotheses* **57**:56-58.

Szatrowski, T. P., and Nathan C. F., 1991, Production of large amounts of hydrogen peroxide by human tumor cells, *Cancer Res.* **51**:794-798.

Tabaracci, G., 2001, L'ozonoterapia con tecnica "classica" intramuscolo paravertebrale, *Riv. Neuroradiol.* **14**:67-70.

Tacchini, L., Pogliaghi G., Radice L., Bernelli-Zazzera A., and Cairo G., 1996, Post-transcriptional control of increased hepatic catalase gene expression in response to oxidative stress, *Redox Report* **2**:273-278.

Taga, K., Mostowski H., and Tosato G., 1993, Human interleukin-10 can directly inhibit T-cell growth., *Blood* **81**:2964-2971.

Tamura, Y., Peng P., Liu K., Daou M., and Srivastava P. K., 1997, Immunotherapy of tumors with autologous tumor-derived heat shock protein preparations, *Science* **278**:117-120.

Tan, S., Yokoyama Y., Dickens E., Cash T. G., Freeman B. A., and Parks D. A., 1993, Xanthine oxidase activity in the circulation of rats following hemorrhagic shock, *Free Radic. Biol. Med.* **15**:407-414.

Tarkington, B. K., Duvall T. R., and Last J. A., 1994, Ozone exposure of cultured cells and tissues, *Meth. Enzymol.* **234**:257-265.

Tateishi-Yuyama, E., Matsubara H., Murohara T., Ikeda U., Shintani S., Masaki H., Amano K., Kishimoto Y., Yoshimoto K., Akashi H., Shimada K., Iwasaka T., and Imaizumi T., 2002,

Therapeutic angiogenesis for patients with limb ischaemia by autologous transplantation of bone-marrow cells: a pilot study and a randomised controlled trial, *Lancet* **360**:427-435.

Taylor, R. S., Belli A. M., and Jacob S., 1999, Distal venous arterialisation for salvage of critically ischaemic inoperable limbs, *Lancet* **354**:1962-1965.

Tepel, M., van der G. M., Statz M., Jankowski J., and Zidek W., 2003, The antioxidant acetylcysteine reduces cardiovascular events in patients with end-stage renal failure: a randomized, controlled trial, *Circulation* **107**:992-995.

Thiele, J. J., Traber M. G., Tsang K., Cross C. E., and Packer L., 1997a, In vivo exposure to ozone depletes vitamins C and E and induces lipid peroxidation in epidermal layers of murine skin, *Free Rad. Biol. Med.* **23**:385-391.

Thiele, J. J., Traber M. T., Podda M., Tsang K., Cross C. E., and Packer L., 1997b, Ozone depletes tocopherols and tocotrienols topically applied to murine skin, *FEBS Lett.* **401**:167-170.

Thomas, J. A., Darby T. D., Wallin R. F., Garvin P. J., and Martis L., 1978, A review of the biological effects of di-(2-ethylhexyl) phthalate, *Toxicol. Appl. Pharmacol.* **45**:1-?

Thomas, T., Thomas G., McLendon C., Sutton T., and Mullan M., 1996, Beta-Amyloid-mediated vasoactivity and vascular endothelial damage, *Nature* **380**:168-171.

Thomson, A. J., Webb D. J., Maxwell S. R., and Grant I. S., 2002, Oxygen therapy in acute medical care, *BMJ* **324**:1406-1407.

Tibbles, P. M., and Edelsberg J. S., 1996, Hyperbaric-oxygen therapy, *N. Engl. J. Med.* **334**:1642-1648.

Ting, H. H., Timimi F. K., Boles K. S., Creager S. J., Ganz P., and Creager M. A., 1996, Vitamin C improves endothelium-dependent vasodilation in patients with non-insulin-dependent diabetes mellitus, *J. Clin. Invest* **97**:22-28.

Tisdale, M. J., 2002, Cachexia in cancer patients, *Nat. Rev. Cancer* **2**:862-871.

Titheradge, M. A., 1999, Nitric oxide in septic shock, *Biochim. Biophys. Acta* **1411**:437-455.

Topol, E. J., 2004, Intensive statin therapy--a sea change in cardiovascular prevention, *N. Engl. J. Med.* **350**:1562-1564.

Torri, G., Della Grazia A., and Casadei C., 1999, Clinical experience in the treatment of lumbar disk disease, with a cycle of lumbar muscle injections of an oxygen + ozone mixture, *Int. J. Med. Biol. Environ.* **27**:177-183.

Tosetti, F., Ferrari N., De Flora S., and Albini A., 2002, Angioprevention': angiogenesis is a common and key target for cancer chemopreventive agents, *FASEB J.* **16**:2-14.

Toze, S., 1999, PCR and the detection of microbial pathogens in water and wastewater, *Wat. Res.* **33**:3545-3556.

Tracey, K. J., 2002, The inflammatory reflex, *Nature* **420**:853-859.

Traverso, N., Menini S., Odetti P., Pronzato M. A., Cottalasso D., and Marinari U. M., 2002, Diabetes impairs the enzymatic disposal of 4-hydroxynonenal in rat liver, *Free Radic. Biol. Med.* **32**:350-359.

Trippel, S. B., 1995, Growth factor actions on articular cartilage, *J. Rheumatol.* **43**:129-132.

Tse, H. F., Kwong Y. L., Chan J. K., Lo G., Ho C. L., and Lau C. P., 2003, Angiogenesis in ischaemic myocardium by intramyocardial autologous bone marrow mononuclear cell implantation, *Lancet* **361**:47-49.

Tylicki, L., Biedunkiewicz B., Nieweglowski T., Chamienia A., Slizien A. D., Luty J., Lysiak-Szydlowska W., and Rutkowski B., 2004, Ozonated autohemotherapy in patients on maintenance hemodialysis: influence on lipid profile and endothelium, *Artif. Organs* **28** :234-237.

Tylicki, L., Niew g. T., Biedunkiewicz B., Burakowski S., and Rutkowski B., 2001, Beneficial clinical effects of ozonated autohemotherapy in chronically dialysed patients with atherosclerotic ischemia of the lower limbs--pilot study, *Int. J. Artif. Organs* **24** :79-82.

Tylicki, L., Nieweglowski T., Biedunkiewicz B., Chamienia A., Debska-Slizien A., Aleksandrowicz E., Lysiak-Szydlowska W., and Rutkowski B., 2003, The influence of ozonated autohemotherapy on oxidative stress in hemodialyzed patients with atherosclerotic ischemia of lower limbs, *Int. J. Artif. Organs* **26**:297-303.

Ueno, I., Hoshino M., Miura T., and Shinriki N., 1998, Ozone exposure generates free radicals in the blood samples *in vitro*. Detection by the ESR spin-trapping technique, *Free Rad. Res.* **29**:127-135.

Unger, R. H., 2002, Lipotoxic diseases, *Annu. Rev. Med.* **53**:319-336.

Urschel, H. C., 1967, Cardiovascular effects of hydrogen peroxide: current status, *Dis. Chest* **51**:180-192.

Valacchi, G., and Bocci V., 1999, Studies on the biological effects of ozone: 10. Release of factors from ozonated human platelets, *Mediat. Inflamm.* **8**:205-209.

Valacchi, G., and Bocci V., 2000, Studies on the biological effects of ozone: 11. Release of factors from human endothelial cells, *Mediat. Inflamm.* **9**:271-276.

Valacchi, G., Pagnin E., Okamoto T., Corbacho A. M., Olano E., Davis P.A., van der Vliet A., Packer L., and Cross C. E., 2003, Induction of stress proteins and MMP-9 by 0.8 ppm of ozone in murine skin, Biochem. *Biophys. Res. Commun.* **305**: 741-746.

Valacchi, G., van der Vliet A., Schock B. C., Okamoto T., Obermuller-Jevic U., Cross C.E., and Packer L., 2002, Ozone exposure activates oxidative stress responses in murine skin, *Toxicology* **179**: 163-170.

Valacchi, G., Weber S. U., Luu C., Cross C.E., and Packer L., 2000, Ozone potentiates vitamin E depletion by ultraviolet radiation in the murine stratum corneum, *FEBS Lett.* **466**: 165-168.

Valdagni, R., and Amichetti M., 1994, Report of long-term follow-up in a randomized trial comparing radiation therapy and radiation therapy plus hyperthermia to metastatic lymph nodes in stage IV head and neck patients, *Int. J. Radiat. Oncol. Biol. Phys.* **28**:163-169.

Valeri, C. R., Contreas T. J., Feingold H., Shebley R. H., and Jaeger R. J., 1973, Accumulation of di-2-ethylhexyl phthalate (DEHP) in whole blood, platelet concentrates and platelet-poor plasma. I: Effect of DEHP on platelet survival and function, *Environ. Health Perspect.* **3**:103-118

Van der Zee, J., van Beek E., Dubbelman T. M. A. R., and Van Steveninck J., 1987, Toxic effects of ozone on murine L929 fibroblasts, *Biochem. J.* **247**:69-72.

van Leeuwen, R., Vingerling J. R., Hofman A., de Jong P. T., and Stricker B. H., 2003, Cholesterol lowering drugs and risk of age related maculopathy: prospective cohort study with cumulative exposure measurement, *BMJ* **326**:255-256.

van Parijs, L., and Abbas A. K., 1998, Homeostasis and self-tolerance in the immune system: turning lymphocytes off, *Science* **280**:243-248.

Varro, J., 1966, Über das Krebsproblem und seine Therapie, *Ztschr. Int. Med. Ges. F. Blut U. Geschwulstkrankheiten* **4**:5-6.

Varro, J., 1974, Die krebsbehandlung mit ozon, *Erfahrungsheilkunde* **23**:178-181.

Varro, J., 1983, Ozone applications in cancer cases, in *Medical Applications of Ozone* (J. LaRaus, Ed.), International Ozone Association, Pan American Committee, Norwalk,Conn., pp.94-95.

Vasiliou, V., Pappa A., and Petersen D. R., 2000, Role of aldehyde dehydrogenases in endogenous and xenobiotic metabolism, *Chem. Biol. Interact.* **129**:1-19.

Vaughn, J. M., Chen Y. S., Novotny J. F., and Strout D., 1990, Effects of ozone treatment on the infectivity of hepatitis A virus, *Can. J. Microbiol.* **36**:557-560.

Vaupel, P., and Hockel M., 2000, Blood supply, oxygenation status and metabolic micromilieu of breast cancers: characterization and therapeutic relevance, *Int. J. Oncol.* **17**:869-879.

Verga, C., 1989, Nuovo approccio terapeutico alle ernie e protrusioni discali lombari, *Riv. Neuroradiol.* **2**:148.

Verma, A., Hirsch D. J., Glatt C. E., Ronnett G. V., and Snyder S. H., 1993, Carbon monoxide: a putative neural messenger, *Science* **259**:381-384.

Verrazzo, G., Coppola L., Luongo C., Sammartino A., Giunta R., Grassia A., Ragone R., and Tirelli A., 1995, Hyperbaric oxygen, oxygen-ozone therapy, and rheologic parameters of blood in patients with peripheral occlusive arterial disease, *Undersea Hyperbar. Med.* **22**:17-22.

Verteporfin in Photodynamic Therapy (VIP) Study Group, 2003, Verteporfin therapy of subfoveal choroidal neovascularization in pathologic myopia: 2-year results of a randomized clinical trial--VIP report no. 3, *Ophthalmology*. **110**:667-73.

Victorin, K., 1992, Review of the genotoxicity of ozone, *Mutat. Res.* **277**:221-238.

Videm, V., Mollnes T. E., Bergh K., Fosse E., Mohr B., Hagve T. A., Aasen A. O., Svennevig J. L., 1999, Heparin-coated cardiopulmonary bypass equipment. II. Mechanisms for reduced complement activation in vivo, *J. Thorac. Cardiovasc. Surg.* **117**: 803-809.

Viebahn, R., 1999a, *The use of ozone in medicine* , ODREI Publishers, Iffezheim, pp.1-148.

Viebahn-Hänsler, R., 1999b, Einfluss auf den erythrozytenstoffwechsel, in *Ozon-Handbuch. Grundlagen. Prävention. Therapie* (E. G. Beck, and R. Viebahn-Hänsler, Eds.), Ecomed, Landsberg, p.1- 15

Viebahn-Hänsler, R., Lell B., and Kremsner P. G., 2001, The effect of ozone on plasmodium falciparum-infected red blood cells, in *Proceedings of the 15th Ozone World Congress, London, UK, 11th-15th September 2001, Medical Therapy Conference* (IOA 2001, Ed.), Speedprint MacMedia Ltd, Ealing, London, UK, pp.26-39.

Vingerling, J. R., Hofman A., Grobbee D. E., and de Jong P. T., 1996, Age-related macular degeneration and smoking. The Rotterdam Study, *Arch. Ophthalmol.* **114**:1193-1196.

Viru, A., and Tendzegolskis Z., 1995, Plasma endorphin species during dynamic exercise in humans, *Clin. Physiol* **15**:73-79.

Vivekananthan, D.P., Penn M.S., Sapp S.K., Hsu A., and Topol E.J., 2003, Use of antioxidant vitamins for the prevention of cardiovascular disease: meta-analysis of randomised trials, *Lancet* **361**: 2017-2023.

Vollmer, T., Key L., Durkalski V., Tyor W., Corboy J., Markovic-Plese S., Preiningerova J., Rizzo M., and Singh I., 2004, Oral simvastatin treatment in relapsing-remitting multiple sclerosis, *Lancet* **363**:1607-1608.

von Harsdorf, R., Poole-Wilson P. A., and Dietz R., 2004, Regenerative capacity of the myocardium: implications for treatment of heart failure, *Lancet* **363**:1306-1313.

Wadhwa, P.D., Zielske S.P., Roth J.C., Ballas C.B., Bowman J.E., Gerson S.L., 2002, Cancer gene therapy: scientific basis. *Annu Rev Med* **53**: 437-452.

Wagner, M., Cadetg P., Ruf R., Mazzucchelli L., Ferrari P., and Redaelli C.A., 2003, Heme oxygenase-1 attenuates ischemia/reperfusion-induced apoptosis and improves survival in rat renal allografts, *Kidney Int.* **63**: 1564-1573.

Wahl, C., Liptay S., Adler G., and Schmid R. M., 1998, Sulfasalazine: a potent and specific inhibitor of nuclear factor kappa B, *J. Clin. Invest.* **101**:1163-1174.

Wang, P., Chen H., Qin H., Sankarapandi S., Becher M. W., Wong P. C., and Zweier J. L., 1998, Overexpression of human copper, zinc-superoxide dismutase (SOD1) prevents postischemic injury, *Proc. Natl. Acad. Sci. U. S. A* **95**:4556-4560.

Warkentin, T. E., 2003, Heparin-induced thrombocytopenia: pathogenesis and management, *Br J Haematol.* **121**: 535-555.

Warlow, C., Sudlow C., Dennis M., Wardlaw J., and Sandercock P., 2003, Stroke, *Lancet* **362**:1211-1224.

Warren, H. S., Suffredini A. F., Eichacker P. Q., and Munford R. S., 2002, Risks and benefits of activated protein C treatment for severe sepsis, *N. Eng. J. Med.* **347**: 1027-1030.

Warren, J. B., and Higenbottam T., 1996, Caution with use of inhaled nitric oxide, *Lancet* **348**:629-630.

Wasser, G. H., 1995a, Behandlung von Verletzungen mit ozoniertem Wasser, in *Ozon-Handbuch. Grundlagen. Prävention. Therapie* (E. G. Beck, and R. Viebahn-Hänsler, Eds.), Ecomed, Landsberg, p.V-7.4 1-V-7.4 8.

Wasser, G. H., 1995b, Zerebrale Durchblutungsstörungen, in *Ozon-Handbuch. Grundlagen. Prävention. Therapie* (E. G. Beck, and R. Viebahn-Hänsler, Eds.), Ecomed, Landsberg, p.V-6.3 1-V-6.3 12.

Weber, W., and Butcher J., 2001, Doubts over cell therapy for Parkinson's disease, *Lancet* **357**:859.

Webster, G. J., Hallett R., Whalley S. A., Meltzer M., Balogun K., Brown D., Farrington C. P., Sharma S., Hamilton G., Farrow S. C., Ramsay M. E., Teo C. G., and Dusheiko G. M., 2000, Molecular epidemiology of a large outbreak of hepatitis B linked to autohaemotherapy, *Lancet* **356**:379-384.

Weck, P. K., Buddin D. A., and Whisnant J. K., 1988, Interferons in the treatment of genital human papillomavirus infections, *Amer. J. Med.* **85**:159-164.

Wehrli, F., and Steinbart H., 1954, Erfahrungen mit der Haematogenen Oxydations - Therapie (HOT), *Ars Medici* **10**:44-51.

Weleber, R. G., 1996, The Cuban experience. False hope for a cure for retinitis pigmentosa, *Arch. Ophthalmol.* **114**:606-607.

Wells, K. H., Latino J., Gavalchin J., and Poiesz B. J., 1991, Inactivation of human immunodeficiency virus type 1 by ozone in vitro, *Blood* **78**:1882-1890.

Wenzel, R. P., and Edmond M. B., 1999, The evolving technology of venous access, *N. Engl. J. Med.* **340**:48-50.

Wenzel, R. P., and Edmond M. B., 2001, The impact of hospital-acquired bloodstream infections, *Emerg. Infect. Dis.* **7**:174-177.

Werkmeister, H., 1995, Dekubitalgeschwüre und die Behandlung mit der Ozon-Unterdruckbegasung, in *Ozon-Handbuch. Grundlagen. Prävention. Therapie* (E. G. Beck, and R. Viebahn-Hänsler, Eds.), Ecomed, Landsberg, p.V-7.1 1-V-7.1 22.

Wessely, S., 2001, Chronic fatigue: symptom and syndrome, *Ann. Intern. Med.* **134**:838-843.

West, I. C., 2000, Radicals and oxidative stress in diabetes, *Diabet. Med.* **17**:171-180.

West, S., Vitale S., Hallfrisch J., Munoz B., Muller D., Bressler S., and Bressler N. M., 1994, Are antioxidants or supplements protective for age related macular degeneration?, *Arch. Ophthalmol.* **112**:222-227.

Westendorp, M. O., Shatrov V. A., Schulze-Osthoff K., Frank R., Kraft M., Los M., Krammer P. H., Droge W., and Lehmann V., 1995, HIV-1 Tat potentiates TNF-induced NF-kappa B activation and cytotoxicity by altering the cellular redox state, *EMBO J.* **14**:546-554.

Whysner, J., Conaway C. C., Verna L., and Williams G. M., 1996, Vinyl chloride mechanistic data and risk assessment: DNA reactivity and cross-species quantitative risk extrapolation, *Pharmacol. Ther.* **71**:7-28.

Wiernsperger, N. F., 2003, Oxidative stress as a therapeutic target in diabetes: revisiting the controversy, *Diabetes Metab* **29**:579-585.

Wigley, F. M., Wise R. A., Seibold J. R., McCloskey D. A., Kujala G., Medsger T. A., Jr., Steen V. D., Varga J., Jimenez S., Mayes M., and ., 1994, Intravenous iloprost infusion in patients with Raynaud phenomenon secondary to systemic sclerosis. A multicenter, placebo-controlled, double-blind study, *Ann. Intern. Med.* **120**:199-206.

Williams, H., 2002, New treatments for atopic dermatitis, *BMJ* **324**:1533-1534.

Williamson, L. M., 2000, Leucocyte depletion of the blood supply - how will patients benefit?, *Br. J. Haematol.* **110**:256-272.

Willis, W. D. J., 1995, Il sistema somatosensoriale, in *Fisiologia* (R. M. Berne, and M. N. Levy, Eds.), Casa Editrice Ambrosiana, Milano, pp.130-151.

Wilson, P. W., and Grundy S. M., 2003a, The metabolic syndrome: practical guide to origins and treatment: Part I, *Circulation* **108**:1422-1424.

Wilson, P. W., and Grundy S. M., 2003b, The metabolic syndrome: a practical guide to origins and treatment: Part II, *Circulation* **108**:1537-1540.

Wiseman, H., and Halliwell B., 1996, Damage to DNA by reactive oxygen and nitrogen species: role in inflammatory disease and progression to cancer, *Biochem. J.* **313**:17-29.

Witko-Sarsat, V., Friedlander M., Nguyen K. T., Capeillere-Blandin C., Nguyen A. T., Canteloup S., Dayer J. M., Jungers P., Drueke T., and Descamps-Latscha B., 1998, Advanced oxidation protein products as novel mediators of inflammation and monocyte activation in chronic renal failure, *J. Immunol.* **161**:2524-2532.

Witko-Sarsat, V., Rieu P., Descamps-Latscha B., Lesavre P., and Halbwachs-Mecarelli L., 2000, Neutrophils: molecules, functions and pathophysiological aspects, *Lab Invest* **80**:617-653.

Witschi, H., Espiritu I., Pinkerton K. E., Murphy K., and Maronpot R. R., 1999, Ozone carcinogenesis revisited, *Toxicol. Sci.* **52**:162-167.

Wolff, H. H., 1974, Die Behandlung peripherer Durchblutungsstörungen mit Ozon, *Erfahr. Hk.* **23**:181-184.

Wolff, H. H., 1979, *Das medizinische Ozon. Theoretische Grundlagen, Therapeutische Anwendungen*, Verlag für Medizin, Heidelberg.

Wolff, S., 1996, Aspects of the adaptive response to very low doses of radiation and other agents, *Mutat. Res.* **358**:135-142.

Wood, M. J., Johnson R. W., McKendrick M. W., Taylor J., Mandal B. K., and Crooks J., 1994, A randomized trial of acyclovir for 7 days or 21 days with and without prednisolone for treatment of acute herpes zoster, *N. Engl. J. Med.* **330**:896-900.

Xu, R. X., 2004, *Burns regenerative medicine and therapy*, Karger Publ., Basel.

Yakes, F. M., and Van Houten B., 1997, Mitochondrial DNA damage is more extensive and persists longer than nuclear DNA damage in human cells following oxidative stress, *Proc. Natl. Acad. Sci. U. S. A* **94**:514-519.

Yamamoto, H., Yamamoto Y., Yamagami K., Kume M., Kimoto S., Toyokuni S., Uchida K., Fukumoto M., and Yamaoka Y., 2000, Heat-shock preconditioning reduces oxidative protein denaturation and ameliorates liver injury by carbon tetrachloride in rats, *Res. Exp. Med. (Berl)* **199**:309-318.

Yamamoto, Y., 2000, Fate of lipid hydroperoxides in blood plasma, *Free Radic. Res.* **33**:795-800.

Yang, J. C., Haworth L., Sherry R. M., Hwu P., Schwartzentruber D. J., Topalian S. L., Steinberg S. M., Chen H. X., and Rosenberg S. A., 2003, A randomized trial of bevacizumab, an anti-vascular endothelial growth factor antibody, for metastatic renal cancer, *N. Engl. J. Med.* **349**:427-434.

Yoritaka, A., Hattori N., Uchida K., Tanaka M., Stadtman E. R., and Mizuno Y., 1996, Immunohistochemical detection of 4-hydroxynonenal protein adducts in Parkinson disease, *Proc. Nat. Acad. Sci. USA* **93**:2696-2701.

Young, S. D., Marshall R. S., and Hill R. P., 1988, Hypoxia induces DNA overreplication and enhances metastatic potential of murine tumor cells, *Proc. Natl. Acad. Sci. U. S. A* **85**:9533-9537.

Youngman, L. D., Park J. Y., and Ames B. N., 1992, Protein oxidation associated with aging is reduced by dietary restriction of protein or calories, *Proc. Natl. Acad. Sci. U. S. A* **89**:9112-9116.

Yu, B. P., 1994, Cellular defenses against damage from reactive oxygen species, *Physiol. Rev.* **74**:139-162.

Yu, B. P., 1996, Aging and oxidative stress: modulation by dietary restriction, *Free Rad. Biol. Med.* **21** :651-668.

Zabel, W., 1960, Ganzheitsbehandlung der Geschwulsterkrankungen, *Hippokrates* **31**:751-760.

Zagury, D., Lachgar A., Chams V., Fall L. S., Bernard J., Zagury J. F., Bizzini B., Gringeri A., Santagostino E., Rappaport J., Feldman M., Burny A., and Gallo R. C., 1998, Interferon alpha and Tat involvement in the immunosuppression of uninfected T cells and C-C chemokine decline in AIDS, *Proc. Natl. Acad. Sci. U. S. A* **95**:3851-3856.

Zajicek, G., 1995, The placebo effect is the healing force of nature, *Cancer J.* **8**:44-45.

Zeuzem, S., 2004, Heterogeneous virologic response rates to interferon-based therapy in patients with chronic hepatitis C: who responds less well?, *Ann. Intern. Med.* **140**:370-381.

Zhang, Y., and Hogg N., 2004, S-nitrosohemoglobin: a biochemical perspective, *Free Radic. Biol. Med.* **36**:947-958.

Zhong, H., De Marzo A. M., Laughner E., Lim M., Hilton D. A., Zagzag D., Buechler P., Isaacs W. B., Semenza G. L., and Simons J. W., 1999, Overexpression of hypoxia-inducible factor 1alpha in common human cancers and their metastases, *Cancer Res.* **59**:5830-5835.

Zhulina, N. I., et al., 1993, Ozonetherapy efficiency in the treatment of patients with atherosclerosis of coronary and cerebral vessels, in *Ozone in Medicine: Proceedings of the 11th Ozone World Congress* (P. A. C. International Ozone Association, Ed.), Stamford, Conn., pp.M-2-9-11.

Zino, S., Skeaff M., Williams S., and Mann J., 1997, Randomised controlled trial of effect of fruit and vegetable consumption on plasma concentrations of lipids and antioxidants, *Brit. Med. J.* **314**:1787-1791.

Zoukourian, C., Wautier M. P., Chappey O., Dosquet C., Rohban T., Schmidt A. M., Stern D., and Wautier J. L., 1996, Endothelial cell dysfunction secondary to the adhesion of diabetic erythrocytes. Modulation by iloprost, *Int. Angiol.* **15**:195-200.

Zuckerbraun, B. S., and Billiar T. R., 2003, Heme oxygenase-1: a cellular Hercules, *Hepatology* **37**:742-744.

References

INDEX

MIX
Papier aus verantwortungsvollen Quellen
Paper from responsible sources
FSC® C105338
www.fsc.org

If you have any concerns about our products,
you can contact us on
ProductSafety@springernature.com

In case Publisher is established outside the EU,
the EU authorized representative is:
Springer Nature Customer Service Center GmbH
Europaplatz 3, 69115 Heidelberg, Germany

Printed by Libri Plureos GmbH
in Hamburg, Germany